Structural Steelwork

Design to Limit State Theory

Fourth Edition

Structural Steelwork

Design to Limit State Theory

Fourth Edition

Dennis Lam
Thien-Cheong Ang
Sing-Ping Chiew

CRC Press
Taylor & Francis Group
Boca Raton   London   New York

CRC Press is an imprint of the
Taylor & Francis Group, an **informa** business

A SPON PRESS BOOK

CRC Press
Taylor & Francis Group
6000 Broken Sound Parkway NW, Suite 300
Boca Raton, FL 33487-2742

© 2014 by Taylor & Francis Group, LLC
CRC Press is an imprint of Taylor & Francis Group, an Informa business

No claim to original U.S. Government works

Printed on acid-free paper
Version Date: 20130709

International Standard Book Number-13: 978-0-415-53191-7 (Paperback)

Library of Congress Cataloging-in-Publication Data

Lam, Dennis.
   Structural steelwork : design to limit state theory / authors, Dennis Lam, Thien-Cheong Ang, Sing-Ping Chiew. -- Fourth edition.
      pages cm
   Includes bibliographical references and index.
   ISBN 978-0-415-53191-7 (pbk.)
   1. Steel, Structural. 2. Building, Iron and steel. I. Title.

TA684.L25 2013
624.1'821--dc23
                                                   2013019088

Visit the Taylor & Francis Web site at
http://www.taylorandfrancis.com

and the CRC Press Web site at
http://www.crcpress.com

Printed and bound by CPI Group (UK) Ltd, Croydon, CR0 4YY

Contents

Preface

This is the fourth edition of the *Structural Steelwork: Design to Limit State Theory*, which proved to be very popular with both students and practising engineers. All the chapters have been updated and rearranged to comply with the Eurocode 3, Design of Steel Structures. In addition, it is also compliant with the other Eurocodes such as Eurocode 0, Basis of Structural Design, and Eurocode 1, Action of Structures. The book contains a detailed explanation of the principles underlying steel design and is intended for students reading for civil and/or structural engineering degrees in universities. It will also be useful for final year students involved in design projects and for practising engineers and architects who require an introduction to the Eurocodes. The topics are illustrated with fully worked examples, and problems are also provided for practice.

Dennis Lam
Thien-Cheong Ang
Sing-Ping Chiew

Authors

Dennis Lam is the chair professor of structural engineering at the University of Bradford, United Kingdom, and visiting professor at Hong Kong Polytechnic University. He is the president of the Association for International Cooperation and Research in Steel–Concrete Composite Structures and the chair of the Research Panel for the Institution of Structural Engineers. He is also a member of the British Standard Institution B525 and European Committee for Standardization CEN/T250/SC4 responsible for BS5950 and Eurocode 4.

Thien-Cheong Ang was formerly at Nanyang Technological University, Singapore.

Sing-Ping Chiew is head of the Division of Structural Engineering and Mechanics at Nanyang Technological University, Singapore.

Chapter 1

Introduction

1.1 STEEL STRUCTURES

Steel-frame buildings consist of a skeletal framework that carries all the loads to which the building is subjected. The sections through three common types of buildings are shown in Figure 1.1. These are

1. Single-storey lattice roof building
2. Single-storey portal frame building
3. Medium-rise braced multistorey building

These three types cover many of the uses of steel-frame buildings such as factories, warehouses, offices, flats and schools. A design for the lattice roof building (Figure 1.1a) is given, and the design of the elements for the braced multistorey building (Figure 1.1c) is also included. The design of portal frame is described separately in Chapter 9.

The building frame is made up of separate elements – the beams, columns, trusses and bracing – listed beside each section in Figure 1.1. These must be joined together, and the building attached to the foundations. The elements are discussed more fully in Section 1.2.

Buildings are 3D and only the sectional frame has been shown in Figure 1.1. These frames must be propped and braced laterally so that they remain in position and carry the loads without buckling out of the plane of the section. Structural framing plans are shown in Figures 1.2 and 1.3 for the building types illustrated in Figure 1.1a and c.

Various methods for analysis and design have been developed over the years. In Figure 1.1, the single-storey structure in (a) and the multistorey building in (c) are designed by the simple design method, whilst the portal frame in (b) is designed by the continuous design method. All design is based on Eurocode 3 (EN1993). Design theories are discussed briefly in Section 1.4, and design methods are set out in detail in Chapter 2.

1.2 STRUCTURAL ELEMENTS

As mentioned earlier, steel buildings are composed of distinct elements:

1. *Beams and girders*: members carrying lateral loads in bending and shear.
2. *Ties*: members carrying axial loads in tension.
3. *Struts, columns or stanchions*: members carrying axial loads in compression. These members are often subjected to bending as well as compression.
4. *Trusses and lattice girders*: framed members carrying lateral loads. These are composed of struts and ties.

1

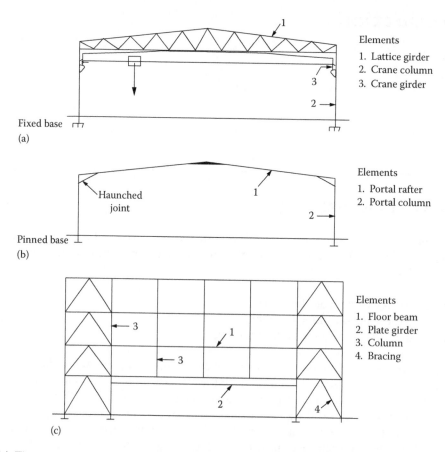

Figure 1.1 Three common types of steel buildings: (a) single-storey lattice roof building with crane; (b) single-storey rigid pinned base portal; (c) multistorey building.

5. *Purlins*: beam members carrying roof sheeting.
6. *Sheeting rails*: beam members supporting wall cladding.
7. *Bracing*: diagonal struts and ties that, with columns and roof trusses, form vertical and horizontal trusses to resist wind loads and hence provided the stability of the building.

Joints connect members together such as the joints in trusses, joints between floor beams and columns or other floor beams. Bases transmit the loads from the columns to the foundations.

The structural elements are listed in Figures 1.1 through 1.3, and the types of members making up the various elements are discussed in Chapter 3. Some details for a factory and a multistorey building are shown in Figure 1.4.

1.3 STRUCTURAL DESIGN

Nowadays, building design is usually carried out by a multidiscipline design team. An architect draws up plans for a building to meet the client's requirements. The structural engineer examines various alternative framing arrangements and may carry out

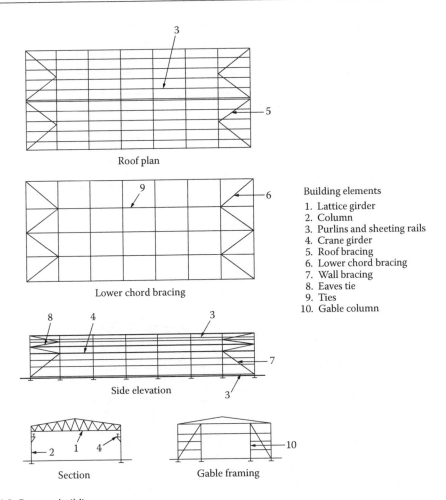

Roof plan

Lower chord bracing

Side elevation

Section

Gable framing

Building elements
1. Lattice girder
2. Column
3. Purlins and sheeting rails
4. Crane girder
5. Roof bracing
6. Lower chord bracing
7. Wall bracing
8. Eaves tie
9. Ties
10. Gable column

Figure 1.2 **Factory building.**

preliminary designs to determine which is the most economical. This is termed the 'conceptual design stage'. For a given framing arrangement, the problem in structural design consists of

1. Estimation of loading
2. Analysis of main frames, trusses or lattice girders, floor systems, bracing and connections to determine axial loads, shears and moments at critical points in all members
3. Design of the elements and connections using design data from step 2
4. Production of arrangement and detail drawings from the designer's sketches

This book covers the design of elements first. Then, to show various elements in their true context in a building, the design for the basic single-storey structure with lattice roof shown in Figure 1.2 is given.

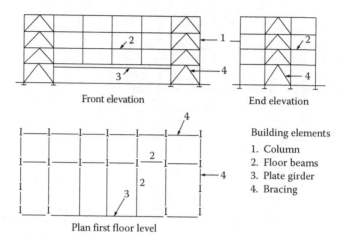

Front elevation   End elevation

Plan first floor level

Building elements
1. Column
2. Floor beams
3. Plate girder
4. Bracing

Figure 1.3 Multistorey office building.

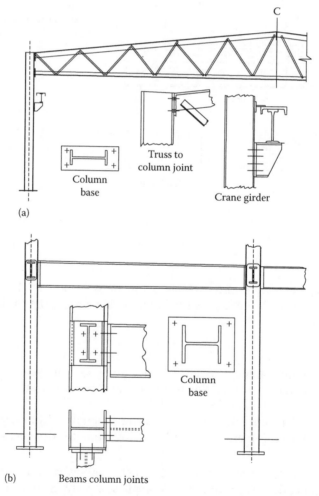

(a)

(b)   Beams column joints

Figure 1.4 (a) Factory building and (b) multistorey building.

1.4 DESIGN METHODS

Steel design may be based on three design theories:

1. Elastic design
2. Plastic design
3. Limit-state design

Elastic design is the traditional method and is still commonly used in the United States. Steel is almost perfectly elastic up to the yield point, and elastic theory is a very good method on which the method is based. Structures are analysed by elastic theory, and sections are sized so that the permissible stresses are not exceeded. This method was used in the United Kingdom in accordance with BS 449-2: 1967: *The Use of Structural Steel in Building.*

Plastic theory developed to take account of behaviour past the yield point is based on finding the load that causes the structure to collapse. Then the working load is the collapse load divided by a load factor. This too was permitted under BS 449.

Finally, limit-state design has been developed to take account of all conditions that can make the structure become unfit for use. The design is based on the actual behaviour of materials and structures in use and is in accordance with EC 3 (EN1993).

The code requirements relevant to the worked problems are noted and discussed. The complete code should be obtained and read in conjunction with this book.

The aim of structural design is to produce a safe and economical structure that fulfils its required purpose. Theoretical knowledge of structural analysis must be combined with knowledge of design principles and theory and the constraints given in the standard to give a safe design. A thorough knowledge of properties of materials, methods of fabrication and erection is essential for the experienced designer. The learner must start with the basics and gradually build up experience through doing coursework exercises in conjunction with a study of design principles and theory.

The ECs are drawn up by panels of experts from the professional institutions and include engineers from educational and research institutions, consulting engineers, government authorities and fabrication and construction industries. The standards give the design methods, factors of safety, design loads, design strengths, deflection limits and safe construction practices.

As well as the main design standard for structural steelwork, EN1993, reference must be made to other relevant standards, including

1. BS EN 10020: 2000: This gives definition and classification of grades of steel.
2. BS EN 10029: 1991 (plates); BS EN 10025: 1993 (sections); BS EN 10210-1: 1994 (hot-finished hollow sections); and BS EN 10219-1: 1997 (cold-formed hollow sections). This gives the mechanical properties for the various types of steel sections.
3. EN1991-1-1: Actions on structures (general actions), densities, self-weight, and imposed loads for buildings.
4. EN1991-1-4: Actions on structures (general actions), wind actions.
5. EN1991-1-3: Actions on structures (general actions), snow loads.

Representative loading may be taken for element design. Wind loading depends on the complete building and must be estimated using the wind code.

1.5 DESIGN CALCULATIONS AND COMPUTING

Calculations are needed in the design process to determine the loading on the structure, carry out the analysis and design the elements and joints and must be set out clearly in a standard form. Design sketches to illustrate and amplify the calculations are an integral part of the procedure and are used to produce the detail drawings.

Computing now forms an increasingly larger part of design work, and all routine calculations can be readily carried out on a PC. The use of the computer speeds up calculation and enables alternative sections to be checked, giving the designer a wider choice than would be possible with manual working. However, it is most important that students understand the design principles involved before using computer programs.

It is through doing exercises that the student consolidates the design theory given in lectures. Problems are given at the end of most chapters.

1.6 DETAILING

Chapter 12 deals with the detailing of structural steelwork. In the earlier chapters, sketches are made in design problems to show building arrangements, loading on frames, trusses, members, connections and other features pertinent to the design. It is often necessary to make a sketch showing the arrangement of a joint before the design can be carried out. At the end of the problem, sketches are made to show basic design information such as section size, span, plate sizes, drilling and welding. These sketches are used to produce the working drawings.

The general arrangement drawing and marking plans give the information for erection. The detailed drawings show all the particulars for fabrication of the elements. The designer must know the conventions for making steelwork drawings, such as the scales to be used, the methods for specifying members, plates, bolts and welding. He or she must be able to draw standard joint details and must also have the knowledge of methods of fabrication and erection. AutoCAD is becoming generally available, and the student should be given an appreciation of their use.

Chapter 2

Limit-state design

2.1 LIMIT-STATE DESIGN PRINCIPLES

The central concepts of limit-state design are as follows:

1. All separate conditions that make the structure unfit for use are taken into account. These are the separate limit states.
2. The design is based on the actual behaviour of materials and performance of structures and members in service.
3. Ideally, design should be based on statistical methods with a small probability of the structure reaching a limit state.

The three concepts are examined in more detail as follows.

Requirement (1) means that the structure should not overturn under applied loads and its members and joints should be strong enough to carry the forces to which they are subjected. In addition, other conditions such as excessive deflection of beams or unacceptable vibration, though not in fact causing collapse, should not make the structure unfit for use.

In concept (2), the strengths are calculated using plastic theory, and post-buckling behaviour is taken into account. The effect of imperfections on design strength is also included. It is recognized that calculations cannot be made in all cases to ensure that limit states are not reached. In cases such as brittle fracture, good practice must be followed to ensure that damage or failure does not occur.

Concept (3) implies recognition of the fact that loads and material strengths vary, approximations are used in design and imperfections in fabrication and erection affect the strength in service. All these factors can only be realistically assessed in statistical terms. However, it is not yet possible to adopt a complete probability basis for design, and the method adopted is to ensure safety by using suitable factors. Partial factors of safety are introduced to take account of all the uncertainties in loads, materials strengths, etc. mentioned earlier. These are discussed more fully later.

2.2 LIMIT STATES FOR STEEL DESIGN

The limit states for which steelwork is to be designed are covered in Section 2 of EN1993-1-1 and Section 2 of EN1990. These are as follows.

7

2.2.1 Ultimate limit states

The ultimate limit states include the following:

1. Strength (including general yielding, rupture, buckling and transformation into a mechanism)
2. Stability against overturning and sway
3. Fracture due to fatigue
4. Brittle fracture

When the ultimate limit states are exceeded, the whole structure or part of it collapses.

2.2.2 Serviceability limit states

The serviceability limit states consist of the following:

5. Deflection
6. Vibration (e.g. wind-induced oscillation)
7. Repairable damage due to fatigue
8. Corrosion and durability

The serviceability limit states, when exceeded, make the structure or part of it unfit for normal use but do not indicate that collapse has occurred.

All relevant limit states should be considered, but usually it will be appropriate to design on the basis of strength and stability at ultimate loading and then check that deflection is not excessive under serviceability loading. Some recommendations regarding the other limit states will be noted when appropriate, but detailed treatment of these topics is outside the scope of this book.

2.3 WORKING AND FACTORED LOADS

2.3.1 Working loads

The working loads (also known as the specified, characteristic or nominal loads) are the actual loads the structure is designed to carry. These are normally thought of as the maximum loads that will not be exceeded during the life of the structure. In statistical terms, characteristic loads have a 95% probability of not being exceeded. The main loads on buildings may be classified as

1. *Dead loads*: These are due to the weights of floor slabs, roofs, walls, ceilings, partitions, finishes, services and self-weight of steel. When sizes are known, dead loads can be calculated from weights of materials or from the manufacturer's literature. However, at the start of a design, sizes are not known accurately, and dead loads must often be estimated from experience. The values used should be checked when the final design is complete. For examples on element design, representative loading has been chosen, but for the building design examples, actual loads from EN1991-1-1 are used.
2. *Imposed loads*: These take account of the loads caused by people, furniture, equipment, stock, etc. on the floors of buildings and snow on roofs. The values of the floor loads used depend on the use of the building. Imposed loads are given in EN1993-1-1, and snow load is given in EN1993-1-3.

3. *Wind loads*: These loads depend on the location and building size. Wind loads are given in EN1991-1-4.
4. *Dynamic loads*: These are caused mainly by cranes. An allowance is made for impact by increasing the static vertical loads, and the inertia effects are taken into account by applying a proportion of the vertical loads as horizontal loads. Dynamic loads from cranes are given in EN1991-3.

Other loads on the structures are caused by waves, ice, seismic effects, etc. and these are outside the scope of this book.

2.3.2 Factored loads for the ultimate limit states

In accordance with EN1990, factored loads are used in design calculations for strength and equilibrium.

Factored load = working or nominal load × relevant partial load factor, γ_f

The partial load factor takes account of

1. The unfavourable deviation of loads from their nominal values
2. The reduced probability that various loads will all be at their nominal value simultaneously

It also allows for the uncertainties in the behaviour of materials by using material partial factors, γ_M, and of the structure as opposed to those assumed in design.

The partial load factors, γ_f, are given in Annex A1 of EN1990. The factored loads should be applied in the most unfavourable manner, and members and connections should not fail under these load conditions. Brief comments are given on some of the load combinations:

1. The main load for design of most members and structures is dead plus imposed load.
2. In light roof structures, uplift and load reversal occurs, and tall structures must be checked for overturning. The load combination of dead plus wind load is used in these cases with a load factor of 1.0 for dead and 1.5 for wind load.
3. It is improbable that wind and imposed loads will simultaneously reach their maximum values and load factors are reduced accordingly.
4. It is also unlikely that the impact and surge load from cranes will reach maximum values together, and so the load factors are reduced. Again, when wind is considered with crane loads, the factors are further reduced.

2.4 STABILITY LIMIT STATES

To ensure stability, EN1990 states that structures must be checked using factored loads for the following two conditions:

1. *Overturning*: The structure must not overturn or lift off its seat.
2. *Sway*: To ensure adequate resistance, design checks are required:
   a. Design to resist the applied horizontal loads in addition with.
   b. The design for notional horizontal loads: These are to be taken as 0.5% of the factored dead plus imposed load and are to be applied at the roof and each floor level. They are to act with 1.35 times the dead and 1.5 times the imposed load.

Sway resistance may be provided by bracing rigid-construction shear walls, stair wells or lift shafts. The designer should clearly indicate the system he or she is using. In examples in this book, stability against sway will be ensured by bracing and rigid portal action.

2.5 STRUCTURAL INTEGRITY

The provisions of Annex A of EN1991-1-7 ensure that the structure complies with the building regulations and has the ability to resist progressive collapse following accidental damage. The main parts of the clause are summarized as follows:

1. All structures must be effectively tied at all floors and roofs. Columns must be anchored in two directions approximately at right angles. The ties may be steel beams or reinforcement in slabs. End connections must be able to resist a factored tensile load of 75 kN for floors and for roofs.
2. Additional requirements are set out for certain multistorey buildings where the extent of accidental damage must be limited. In general, tied buildings will be satisfactory if the following five conditions are met:
   a. Sway resistance is distributed throughout the building.
   b. Extra tying is to be provided as specified.
   c. Column splices are designed to resist a specified tensile force.
   d. Any beam carrying a column is checked as set out in (3) later.
   e. Precast floor units are tied and anchored.
3. Where required in (2), the aforementioned damage must be localized by checking to see if at any storey, any single column or beam carrying a column may be removed without causing more than a limited amount of damage. If the removal of a member causes more than the permissible limit, it must be designed as a key element. These critical members are designed for accidental loads set out in the building regulations. The recommended value for building structures is 34 kN/m^2.

The complete section in the code and the building regulations should be consulted.

2.6 SERVICEABILITY LIMIT-STATE DEFLECTION

Deflection is the main serviceability limit state that must be considered in design. The limit state of vibration is outside the scope of this book, and fatigue was briefly discussed in Section 2.2.1 and, again, is not covered in detail. The protection for steel to prevent the limit state of corrosion being reached was mentioned in Section 2.2.4.

NA to BS EN1993-1-1 states in NA2.23 that deflection under serviceability loads of a building or part should not impair the strength or efficiency of the structure or its components or cause damage to the finishings. The serviceability loads used are the unfactored imposed loads except in the following cases: Table 2.1 gives suggested limits for calculated vertical deflections of certain members under the characteristic load combination due to variable loads and should not include permanent loads.

The structure is considered to be elastic and the most adverse combination of loads is assumed. Deflection limitations are given in NA 2.23 and NA 2.24. These are given here in Table 2.1. These limitations cover beams and structures other than pitched-roof portal frames.

Table 2.1 Deflection limits

Deflection of beams due to unfactored imposed loads	
Cantilevers	Length/180
Beams carrying plaster or other brittle finish	Span/360
All other beams (except purlins and sheeting rails)	Span/200
Purlins and sheeting rails	To suit the characteristics of particular cladding
Horizontal deflection of columns due to unfactored imposed and wind loads	
Tops of columns in single-storey buildings except portal frames	Height/300
In each storey of a building with more than one storey	Storey height/300

It should be noted that calculated deflections are seldom realized in the finished structure. The deflection is based on the beam or frame steel section only, and composite action with slabs or sheeting is ignored. Again, the full value of the imposed load used in the calculations is rarely achieved in practice.

2.7 DESIGN STRENGTH OF MATERIALS

The design strengths for steel are given in Section 3.2 of EN1993-1-1. Note that the material partial factor, γ_{M0}, part of the overall safety factor in limit-state design, is taken as 1.0 in the code. The design strength may be taken as

- The ratio f_u/f_y of the specified minimum ultimate tensile strength f_u to the specified minimum yield strength, f_y
- The elongation at failure on a gauge length of $5.65\sqrt{A_0}$ (where A_0 is the original cross-sectional area)
- The ultimate strain ε_u, where ε_u corresponds to the ultimate strength f_u

Note: The limiting values of the ratio f_u/f_y, the elongation at failure and the ultimate strain ε_u may be defined in the National Annex (NA). The following values are recommended:

- $f_u/f_y \geq 1,10$
- Elongation at failure not less than 15%
- $\varepsilon_u \geq 15\varepsilon_y$, where ε_y is the yield strain ($\varepsilon_y = f_y/E$)

The values of f_y and f_u are given in Table 3.1 of the EN1993-1-1.

The code states that the following values for the elastic properties are to be used:

Modulus of elasticity, $E = 210,000$ N/mm^2

Shear modulus, $G = \dfrac{E}{2(1+v)} \approx 81,000$ N/mm^2

Poisson's ratio, $v = 0.30$

Coefficient of linear thermal expansion $\alpha = 12 \times 10^{-6}$/K   (for $T \leq 100°C$)

2.8 DESIGN METHODS FOR BUILDINGS

The design of buildings must be carried out in accordance with one of the methods given in Clause 1.5.3 of EN1993-1-1. The design methods are as follows:

1. *Simple design*: In this method, the connections between members are assumed not to develop moments adversely affecting either the members or the structure as a whole. The structure is assumed to be pin jointed for analysis. Bracing or shear walls are necessary to provide resistance to horizontal loading.
2. *Continuous design*: The connections are assumed to be capable of developing the strength and/or stiffness required by an analysis assuming full continuity. The analysis may be made using either elastic or plastic methods.
3. *Semi-continuous design*: This method may be used where the joints have some degree of strength and stiffness but insufficient to develop full continuity. Either elastic or plastic analysis may be used. The moment capacity, rotational stiffness and rotation capacity of the joints should be based on experimental evidence. This may permit some limited plasticity, provided that the capacity of the bolts or welds is not the failure criterion. On this basis, the design should satisfy the strength, stiffness and in-plane stability requirements of all parts of the structure when partial continuity at the joints is taken into account in determining the moments and forces in the members.
4. *Experimental verification*: The code states that where the design of a structure or element by calculation in accordance with any of the aforementioned methods is not practicable, the strength and stiffness may be confirmed by loading tests. The test procedure is set out in Clause 2.5 of the code.

In practice, structures are designed to either the simple or the continuous methods of design. Semi-continuous design has never found general favour with designers. Examples in this book are generally of the simple method of design.

Chapter 3

Materials

3.1 STRUCTURAL STEEL PROPERTIES

The nominal values of the yield strength f_y and the ultimate strength f_u for structural steel may be taken from Table 3.1 in EN 1993-1-1 or direct from the product standard, that is EN 10025 for hot-rolled sections. The UK National Annex states that the nominal values for structural steel should be obtained from the product standard. The product standards give more steps in the reduction of strength with an increasing thickness of the product.

Steel is composed of about 98% of iron with the main alloying elements carbon, silicon and manganese. Copper and chromium are added to produce the weather-resistant steels that do not require corrosion protection. The rules in EN 1993-1-1 relate to structural steel grades S235, S275, S355 and S450 and thus cover all the structural steels likely to be used in buildings. In exceptional circumstances, components might use higher-strength grades; EN 1993-1-12 gives guidance on the use of EN 1993-1-1 design rules for higher-strength steels.

The stress–strain curves for the four grades of steel are shown in Figure 3.1a, and these are the basis for the design methods used for steel. Initially, the steel has a linear stress–strain curve whose slope is the Young's modulus, E. The limit of the linear elastic behaviour is yield stress f_y and the corresponding yield strain $\varepsilon_y = f_y/E$. Elastic design is kept within the elastic region, and because steel is almost perfectly elastic, design based on elastic theory is a very good method to use.

Beyond the elastic limit, the stress–strain curves show a small plateau without any increase in stress until the strain-hardening strain ε_{st} is reached and then an increase in strength due to strain hardening. The plastic range is usually accounted for the ductility of the steel. The stress increases above the yield stress f_y when the strain-hardening strain ε_{st} is exceeded, and this continues until the ultimate tensile stress f_u is reached. Plastic design is based on the horizontal part of the stress–strain shown in Figure 3.1b.

The mechanical properties for steels are set out in the respective specifications mentioned earlier. The yield strengths and ultimate strengths for the most common grades from Table 3.1 of EN 1993-1-1 and from product standard EN 10025-2 are given in Table 3.1 for comparison; and other important design properties are given in Section 2.7.

3.2 DESIGN CONSIDERATIONS

Special problems occur with steelwork, and good practice must be followed to ensure satisfactory performance in service. These factors are discussed briefly later in order to bring them to the attention of students and designers, although they are not generally of great importance in the design problems covered in this book. However, it is worth noting that the material safety factor

Table 3.1 Nominal values of yield strength f_y and ultimate tensile strength f_u for hot rolled structural steel

Steel grade	EN 1993-1-1			EN 10025-2		
	Thickness (mm)	f_y [N/mm²]	f_u [N/mm²]	Thickness (mm)	f_y [N/mm²]	f_u [N/mm²]
S235	$t \leq 40$	235	360	$t \leq 16$	235	360
				$16 < t \leq 40$	225	360
	$40 < t \leq 80$	215	360	$40 < t \leq 63$	215	360
				$63 < t \leq 80$	215	360
S275	$t \leq 40$	275	430	$t \leq 16$	275	410
				$16 < t \leq 40$	265	410
	$40 < t \leq 80$	255	410	$40 < t \leq 63$	255	410
				$63 < t \leq 80$	245	410
S355	$t \leq 40$	355	490	$t \leq 16$	355	470
				$16 < t \leq 40$	345	470
	$40 < t \leq 80$	335	470	$40 < t \leq 63$	335	470
				$63 < t \leq 80$	325	470
S450	$t \leq 40$	440	550	$t \leq 16$	450	550
				$16 < t \leq 40$	430	550
	$40 < t \leq 80$	410	550	$40 < t \leq 63$	410	550
				$63 < t \leq 80$	390	550

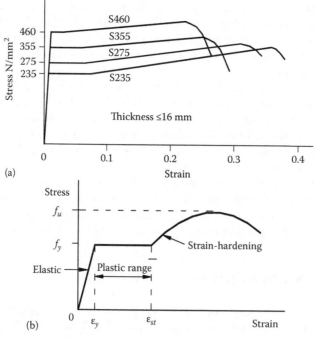

Figure 3.1 Stress–strain diagrams for structural steels: (a) stress–strain diagrams for typical structural steels; (b) stress–strain diagram for plastic design.

γ_m is set to unity in EN 1993 that implies a certain level of quality and testing in steel usage. Weld procedures are qualified by maximum carbon equivalent values. Attention to weldability should be given when dealing with special, thick and higher grade steel.

3.2.1 Fatigue

Fatigue failure can occur in members or structures subjected to fluctuating loads such as crane girders, bridges and structures that support machinery, wind and wave loading. Failure occurs through initiation and propagation of a crack that starts at a fault or structural discontinuity, and the failure load may be well below its static value.

Welded connections have the greatest effect on the fatigue strength of steel structures. Tests show that butt welds give the best performance in service, whilst continuous fillet welds are much superior to intermittent fillet welds. Bolted connections do not reduce the strength under fatigue loading. To help avoid fatigue failure, detail should be such that stress concentrations and abrupt changes of section are avoided in regions of tensile stress. Cases where fatigue could occur are noted in this book, and for further information, the reader should consult Ref. [1].

3.2.2 Ductility requirements

Ductility is the ability of a material to undergo large deformation without breaking. A measure of ductility is the percentage elongation of the gage length of the specimen during a tension test. It is calculated as 100 times the change in gage length divided by the original gage length. Thus,

$$\delta_e = \frac{L_f - L_0}{L_0} \times 100$$

where
  L_f is the final distance between the gage marks after the specimen breaks
  L_0 is the original gage length

In order to ensure structures are designed with steels that possess adequate ductility, NA to BS EN 1993-1-1 sets the following requirements:

1. Elastic global analysis
   The limiting values for the ratio f_u/f_y, the elongation at failure and the ultimate strain ε_u for elastic global analysis are given as follows:
   a. $f_u/f_y \geq 1.10$
   b. Elongation at failure not less than 15% (on a gauge length of $5.65\sqrt{A_0}$, where A_0 is the original cross-sectional area)
   c. $\varepsilon_u \geq 15\,\varepsilon_y$, where ε_u is the ultimate strain and ε_y is the yield strain
2. Plastic global analysis
   Plastic global analysis should not be used for bridges. For building the limiting values for the ratio f_u/f_y, the elongation at failure and the ultimate strain ε_u for plastic global analysis are given as follows:
   a. $f_u/f_y \geq 1.15$
   b. Elongation at failure not less than 15% (on a gauge length of $5.65\sqrt{A_0}$, where A_0 is the original cross-sectional area)
   c. $\varepsilon_u \geq 20\,\varepsilon_y$, where ε_u is the ultimate strain and ε_y is the yield strain

3.2.3 Brittle fracture

Structural steel is ductile at temperatures above 10°C, but it becomes more brittle as the temperature falls, and fracture can occur at low stresses below 0°C. The material should have sufficient fracture toughness to avoid brittle fracture of tension member at the lowest service temperature expected to occur within the intended design life of the structure. NA to BS EN 1993-1-1 sets the following requirements:

- For building and other quasi-statically loaded structures, the lowest service temperature in the steel should be taken as the lowest air temperature that may be taken as −5°C for internal steelwork and −15°C for external steelwork.
- For bridges, the lowest service temperature in the steel should be determined according to the NA to BS EN 1991-1-5 for bridge location. For structures susceptible to fatigue, it is recommended that the requirements for bridges should be applied.
- In other cases, the lowest service temperature in the steel should be taken as the lowest air temperature expected to occur within the intended design life of the structure.

Brittle fracture is initiated by the existence or formation of a small crack in a region of high local stress. Once initiated, the crack may propagate in a ductile fashion for which the external forces must supply the energy required to tear the steel. The ductility of a structural steel depends on its composition, heat treatment and thickness and varies with temperature. Figure 3.2 shows the increase with temperature of the capacity of the steel to absorb energy during impact. At low temperatures, the energy absorption is low, and initiation and propagation of brittle fractures are comparatively easy, whilst at high temperatures, the energy absorption is high because of ductile yielding, and the propagation of the cracks can be arrested.

In design, brittle fracture should be avoided by using steel quality grade with adequate impact toughness. Quality steels are designated JR, J0, J2, K2 and so forth in order of increasing resistance to brittle fracture. The Charpy impact fracture toughness is specified for the various steel quality grades: for example Grade S275 J0 steel is to have a minimum fracture toughness of 27 J at a test temperature of 0°C.

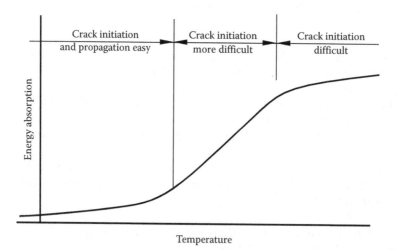

Figure 3.2 Effect of temperature on resistance to brittle fracture.

Steel with large grain size tends to be more brittle, and this is significantly influenced by heat treatment of the steel and by its thickness. The procedure for determination of the steel subgrade is given in EN 1993-1-10, which provides values of the maximum thickness t_1 for different steel grades and minimum service temperatures. For common structures, the maximum stress in the structure for accidental combination of loading σ_{Ed} and the reference temperature T_{ed} in a location of potential crack is usually determined. The required steel subgrade for a given element thickness can then be determined from Table 3.2.

Cases where brittle fracture may occur in design of structural elements are noted in this book. For further information, the reader should consult Ref. [2].

3.2.4 Fire protection

Structural steelwork performs badly in fires, with the strength decreasing with increase in temperature. At 550°C, the yield stress has fallen to approximately 0.7 of its value at normal temperatures; that is it has reached its working stress and failure occurs under working loads.

The statutory requirements for fire protection are usually set out clearly in the approved documents from the local building regulations [3] or fire safety authority. These lay down the fire-resistance period that any load-bearing element in a given building must have and also give the fire-resistance periods for different types of fire protection. Fire protection can be provided by encasing the member in concrete, fire board or cementitious fibre materials. The main types of fire protection for columns and beams are shown in Figure 3.3. More recently, intumescent paint is being used especially for exposed steelwork.

All multistorey steel buildings require fire protection. Single-storey factory buildings normally do not require fire protection for the steel frame. Further information is given in Ref. [4].

3.2.5 Corrosion protection

Exposed steelwork can be severely affected by corrosion in the atmosphere, particularly if pollutants are present, and it is necessary to provide surface protection in all cases. The type of protection depends on the surface conditions and length of life required.

The main types of protective coatings are

1. Metallic coatings: A sprayed-on in-line coating of either aluminium or zinc is used, or the member is coated by hot dipping it in a bath of molten zinc in the galvanizing process.
2. Painting, where various systems are used: One common system consists of using a primer of zinc chromate followed by finishing coats of micaceous iron oxide. Plastic and bituminous paints are used in special cases.

The single most important factor in achieving a sound corrosion-protection coating is surface preparation. Steel is covered with mill scale when it cools after rolling, and this must be removed before the protection is applied; otherwise, the scale can subsequently loosen and break the film. Blast cleaning makes the best preparation prior to painting. Acid pickling is used in the galvanizing process. Other methods of corrosion protection that can also be considered are sacrificial allowance, sherardizing, concrete encasement and cathodic protection.

Careful attention to design detail is also required (e.g. upturned channels that form a cavity where water can collect should be avoided), and access for future maintenance should also be

Table 3.2 Maximum permissible values of element thickness t in mm

Reference temperature T_{Ed} (°C)

		AC2 KV AC2 at T [°C]	J_{min}	$\sigma_{Ed} = 0.75\, f_y(t)$							$\sigma_{Ed} = 0.50\, f_y(t)$							$\sigma_{Ed} = 0.25\, f_y(t)$						
Steel grade	Subgrade	at T [°C]	J_{min}	10	0	−10	−20	−30	−40	−50	10	0	−10	−20	−30	−40	−50	−50	−40	−30	−20	−10	0	10
S235	JR	20	27	60	50	40	35	30	25	20	90	75	65	55	45	40	35	60	65	75	85	100	115	135
	J0	0	27	90	75	60	50	40	35	30	125	105	90	75	65	55	45	75	85	100	115	135	155	175
	J2	−20	27	125	105	90	75	60	50	40	170	145	125	105	90	75	65	100	115	135	155	175	200	200
S275	JR	20	27	55	45	35	30	25	20	15	80	70	55	50	40	35	30	55	60	70	80	95	110	125
	J0	0	27	75	65	55	45	35	30	25	115	95	80	70	55	50	40	70	80	95	110	125	145	165
	J2	−20	27	110	95	75	65	55	45	35	155	130	115	95	80	70	55	95	110	125	145	165	190	200
	M,N	−20	40	135	110	95	75	65	55	45	180	155	130	115	95	80	70	110	125	145	165	190	200	200
	ML,NL	−50	27	185	135	110	90	75	60	50	200	180	155	130	115	95	80	145	165	190	200	200	200	230
S355	JR	20	27	40	35	25	20	15	15	10	65	55	45	40	30	25	20	40	45	55	65	80	95	110
	J0	0	27	60	50	40	35	25	20	15	95	80	65	55	45	40	30	60	65	80	95	110	130	150
	J2	−20	27	90	75	60	50	40	35	25	135	110	95	80	65	55	45	80	95	110	130	150	175	200
	K2,M,N	−20	40	110	95	75	60	50	40	35	155	130	110	95	80	65	55	95	110	130	150	175	200	200
	ML,NL	−50	27	155	130	110	90	75	60	50	200	175	150	130	110	95	80	130	150	175	200	200	200	210
S420	M,N	−20	40	95	80	70	60	50	40	35	140	120	100	85	70	60	50	85	100	120	140	160	185	200
	ML,NL	−50	27	135	115	95	80	70	55	45	190	165	140	120	100	85	70	120	140	160	175	185	200	200
S460	Q	−20	30	70	60	50	40	35	25	20	110	95	75	65	55	45	35	70	80	95	115	130	155	175
	M,N	−20	40	90	75	65	55	45	35	25	130	110	95	75	65	55	45	80	95	115	130	155	175	200
	QL	−40	30	105	90	70	60	50	40	30	155	130	110	95	75	65	55	95	115	130	155	175	200	200
	ML,NL	−50	27	125	105	90	75	65	55	40	180	155	130	110	95	75	65	115	130	155	175	200	200	200
	QL1	−60	30	150	130	110	95	75	65	50	200	180	155	130	110	95	75	130	155	175	200	200	200	215
S690	Q	0	40	40	30	25	20	15	10	10	65	55	45	35	30	25	20	45	50	60	75	85	100	120
	Q	−20	30	50	40	30	25	20	15	10	80	65	55	45	35	30	25	50	60	75	85	100	120	140
	QL	−20	40	60	50	40	30	25	20	15	95	80	65	55	45	35	30	60	75	85	100	120	140	165
	QL	−20	40	60	50	40	30	25	20	15	95	80	65	55	45	35	30	75	85	100	120	140	165	190
	QL	−40	30	75	60	50	40	30	25	20	115	95	80	65	55	45	35	60	75	85	100	120	140	165
	QL	−40	40	75	60	50	40	30	25	20	115	95	80	65	55	45	35	75	85	100	120	140	165	190
	QL1	−40	40	90	75	60	50	40	30	25	135	115	95	80	65	55	45	85	100	120	140	165	190	200
	QL1	−60	30	110	90	75	60	50	40	30	160	135	115	95	80	65	55	100	120	140	165	190	200	200

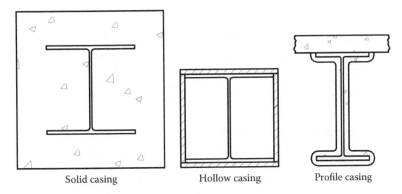

Solid casing   Hollow casing   Profile casing

Figure 3.3 Fire protections for columns and beams.

provided. For further information the reader should consult EN ISO 12944 (corrosion protection of steel structures by protective paint systems) and EN ISO 14713 (zinc coatings – guidelines and recommendations for the protection against corrosion of iron and steel in structures).

3.3 STEEL SECTIONS

3.3.1 Rolled and formed sections

Rolled and formed sections are produced in steel mills from steel blooms, beam blanks or coils by passing them through a series of rollers. The more commonly used hot-rolled sections are shown in Figure 3.4, and their principal properties and uses are discussed briefly as follows:

1. *Universal beams*: These are very efficient sections for resisting bending moment about the major axis.
2. *Universal columns*: These are sections produced primarily to resist axial load with a high radius of gyration about the minor axis to prevent buckling in that plane.
3. *Channels*: These are used for beams, bracing members, truss members and compound members.
4. *Equal and unequal angles*: These are used for bracing members, truss members and for purlins, side and sheeting rails.
5. *Structural tees*: The sections shown are produced by cutting a universal beam or column into two parts. Tees are used for truss members, ties and light beams.
6. *Circular, square and rectangular hollow sections*: These are mostly produced from hot-rolled coils and may be hot finished or cold formed. A welded mother tube is first formed, and then it is rolled to its final square or rectangular shape. In the hot process, the final shaping is done at the steel normalizing temperature, whereas in the cold process, it is done at ambient room temperature. These sections make very efficient compression members and are used in a wide range of applications as members in roof trusses, lattice girders, building frames and for purlins, sheeting rails, etc.

Note that the range in serial sizes is given for the members shown in Figure 3.4. A number of different members are produced in each serial size by varying the flange,

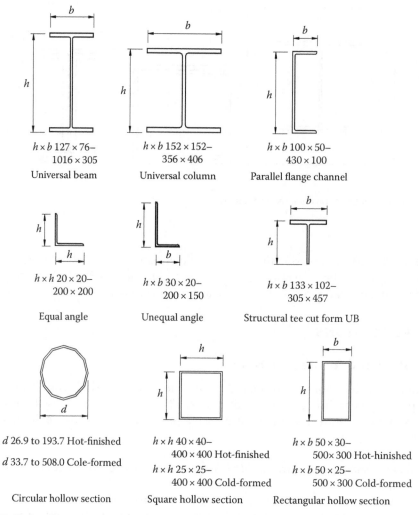

Figure 3.4 Rolled and formed sections.

web, leg or wall thicknesses. The material properties, tolerances and dimensions of the structural sections referred to in this book can be found in the following standards given in Table 3.3.

3.3.2 Compound sections

Compound sections are formed by the following means (Figure 3.5):

1. Strengthening a rolled section such as a universal beam by welding on cover plates, as shown in Figure 3.5a.
2. Combining two separate rolled sections, as in the case of the crane girder in Figure 3.5b: The two members carry loads from separate directions.
3. Connecting two members together to form a strong combined member: Examples are the laced and battened members shown in Figure 3.5c and d.

Table 3.3 Structural steel products

Product	Technical delivery requirements		Dimensions	Tolerances
	Non-alloy steels	*Fine-grain steels*		
Universal beams, universal columns and universal bearing piles	BS EN 10025-2	BS EN 10025-3 BS EN 10025-4	BS 4-1	BS EN 10034
Joists	BS EN 10025-2	BS EN 10025-3 BS EN 10025-4	BS 4-1	BS 4-1 BS EN 10024
Parallel flange channels	BS EN 10025-2	BS EN 10025-3 BS EN 10025-4	BS 4-1	BS EN 10279
Angles	BS EN 10025-2	BS EN 10025-3 BS EN 10025-4	BS EN 10056-1	BS EN 10056-2
Structural tees cut from universal beams and universal columns	BS EN 10025-2	BS EN 10025-3 BS EN 10025-4	BS 4-1	—
ASB (asymmetric beams *Slimflor®* beam	Generally BS EN 10025 but see note[b]		See note[a]	Generally BS EN 10034 but also see note[b]
Hot-finished structural hollow sections	BS EN 10210-1		BS EN 10210-2	BS EN 10210-2
Cold-formed hollow sections	BS EN 10219-2		BS EN 10219-2	BS EN 10219-2

Note that EN 1993 refers to the product standards by their CEN designation, e.g. EN 10025-2. The CEN standards are published in the United Kingdom by BSI with their prefix to the designation, e.g. BS EN 10025-2.

[a] See Ref. [5].
[b] For further details, consult Tata Steel.

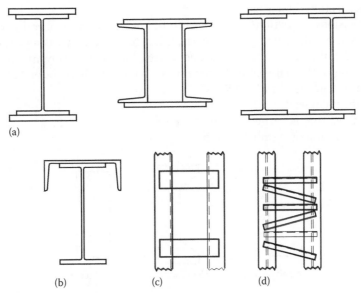

Figure 3.5 Compound sections: (a) compound beam; (b) crane girder; (c) battened member; (d) laced member.

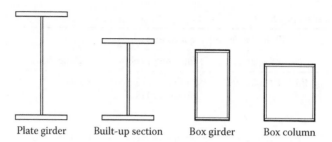

Plate girder   Built-up section   Box girder   Box column

Figure 3.6 Built-up sections.

3.3.3 Built-up sections

Built-up sections are made by welding plates together to form I, H or box members that are termed plate girders, built-up columns, box girders or columns, respectively. These members are used where heavy loads have to be carried and in the case of plate and box girders where long spans may be required. Examples of built-up sections are shown in Figure 3.6.

3.3.4 Cold-rolled open sections

Thin steel plates can be formed into a wide range of sections by cold rolling. The most important uses for cold-rolled open sections in steel structures are for purlins, side and sheeting rails. Three common sections – the zed, sigma and lipped channel – are shown in Figure 3.7. Reference should be made to manufacturer's specialized literature for the full range of sizes available and the section properties. Some members and their properties are given in Section 4.12 in design of purlins and sheeting rails.

3.4 SECTION PROPERTIES

For a given member serial size, the section properties are

1. The exact section dimensions
   The dimensions of sections are given in millimetres (mm). Generally, the centimetre (cm) is used for the calculated properties, but for surface areas and for the warping constant (I_w), the metre (m) and the decimetre (dm), respectively, are used.
2. The location of the centroid if the section is asymmetrical about one or both axes
   The axis system used in EN 1993 is
   x along the member
   y major axis, or axis perpendicular to web
   z minor axis, or axis parallel to web

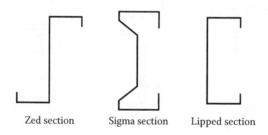

Zed section   Sigma section   Lipped section

Figure 3.7 Cold-rolled sections.

3. Area of cross section (A)
4. Second moments of area about various axes
   The second moment of area has been calculated taking into account all tapers, radii and fillets of the sections. Values are given about both the y–y and the z–z axes, named I_y and I_z, respectively.
5. Radii of gyration about various axes
   The radius of gyration is a parameter used in the calculation of buckling resistance and is derived as follows:

$$i = \left[\frac{I}{A}\right]^{1/2}$$

6. Moduli of section for various axes, both elastic and plastic
   The elastic section modulus is used to calculate the elastic design resistance for bending based on the yield strength of the section and the partial factor γ_M or to calculate the stress at the extreme fibre of the section due to a moment. It is derived as follows:

$$W_{el,y} = \frac{I_y}{z}$$

$$W_{el,z} = \frac{I_z}{y}$$

where z, y are the distance to the extreme fibres of the section from the elastic y–y and z–z axes, respectively.

The plastic section modulus about both y–y and z–z axes of the plastic cross sections is tabulated for all sections except angle sections.

For compound and built-up sections, the properties must be calculated from the first principles. The section properties for the symmetrical I section with dimensions as shown in Figure 3.8a are as follows:

1. Elastic properties

Area	$A = 2bt_f + dt_w$
Moment of inertia y–y axis	$I_y = \dfrac{bh^3}{12} - \dfrac{(b - t_w)d^3}{12}$
Moment of inertia z–z axis	$I_z = \dfrac{2t_f b^3}{12} + \dfrac{dt_w^3}{12}$
Radius of gyration y–y axis	$i_y = \left(\dfrac{I_y}{A}\right)^{0.5}$
Radius of gyration z–z axis	$i_z = \left(\dfrac{I_z}{A}\right)^{0.5}$
Modulus of section y–y axis	$W_{el,y} = \dfrac{2I_y}{h}$
Modulus of section z–z axis	$W_{el,z} = \dfrac{2I_z}{b}$

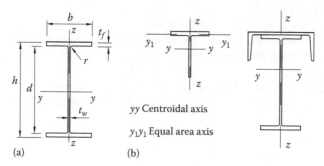

Figure 3.8 Beam section: (a) symmetrical I-section; (b) asymmetrical I-section.

2. Plastic moduli of section

The plastic modulus of section is equal to the algebraic sum of the first moments of area about the equal area axis. For the I section shown,

$$W_{pl,y} = \frac{2bt_f(h-t_f)}{2} + \frac{t_w d^2}{4}$$

$$W_{pl,z} = \frac{2t_f b^2}{4} + \frac{dt_w^2}{4}$$

For asymmetrical sections such as those shown in Figure 3.8b, the neutral axis must be located first. In elastic analysis, the neutral axis is the centroidal axis, whilst in plastic analysis, it is the equal area axis. The other properties may then be calculated using procedures from strength of materials [7]. Calculations of properties for unsymmetrical sections are given in various parts of this book.

Other properties of universal beams, columns and joists, used for determining the buckling resistance moment, are

Buckling parameter (U)
Torsional index (X)
Warping constant (I_W)
Torsional constant (I_T)

$$U = \left(\frac{W_{pl,y}g}{A}\right)^{0.5} \times \left(\frac{I_z}{I_W}\right)^{0.5}$$

$$X = \sqrt{\frac{\pi^2 EAI_W}{20GI_T I_z}}$$

$$I_W = \frac{I_z h_s^2}{4}$$

$$I_T = \frac{2}{3}bt_f^3 + \frac{1}{3}(h-2t_f)t_w^3 + 2\alpha_1 D_1^4 - 0.42t_f^4$$

where

$$g = \sqrt{1 - \frac{I_z}{I_y}}$$

$$G = \frac{E}{2(1+\nu)}$$

h_s is the distance between shear centres of flanges (i.e. $h_s = h - t_f$)

$$\alpha_1 = -0.042 + 0.2204\frac{t_w}{t_f} + 0.1355\frac{r}{t_f} - 0.0865\frac{rt_w}{t_f^2} - 0.0725\frac{t_w^2}{t_f^2}$$

$$D_1 = \frac{(t_f + r)^2 + (r + 0.25t_w)t_w}{2r + t_f}$$

More properties may be given in Ref. [5].

The most important normative references on design are provided as follows:

EN 10025 (six parts)	Hot-rolled steel products
EN 10210	Hot-finished structured hollow sections
EN 10219	Cold-formed structured hollow sections
EN 10024	Hot-rolled taper flange I sections
EN 10034	Structural steel I and H sections
EN 10279	Hot-rolled steel channels
EN 10056	Specification for equal and unequal angles
EN 14399	High-strength bolting assemblies for preloading
EN 15048	Non-preloaded structural bolting assemblies
EN 1090	Execution of steel structures

Chapter 4

Beams

4.1 TYPES AND USES

Beams span between supports to carry lateral loads that are resisted by bending and shear. However, deflections and local stresses are also important.

Beams may be cantilevered, simply supported, fixed ended or continuous, as shown in Figure 4.1a. The main uses of beams are to support floors and columns and carry roof sheeting as purlins and side cladding as sheeting rails.

Any member may serve as a beam, and common beam sections are shown in Figure 4.1b. Some comments on the different sections are given as follows:

1. The universal beam (UB) where the material is concentrated in the flanges is the most efficient section to resist uniaxial bending.
2. The universal column (UC) may be used where the depth is limited, but it is less efficient.
3. The compound beam consisting of a UB and flange plates is used where the depth is limited and the UB itself is not strong enough to carry the load.
4. The crane beam consists of a UB and channel. It must resist bending in two directions.

Beams may be of uniform or non-uniform section. Rolled beams may be strengthened in regions of maximum moment by adding cover plates or haunches. Some examples are shown in Figure 4.2.

4.2 BEAM ACTIONS

Loads are referred to as actions in the structural Eurocodes (ECs) and should be taken from EN 1991, whilst partial factors and the combination of actions are covered in EN 1990.

Types of beam actions are

1. Concentrated loads from secondary beams and columns
2. Distributed loads from self-weight and floor slabs

The actions are *further classified* by their variation in time as follows:

1. Permanent actions (G): for example, self-weight of the beams, slabs, finishes and fixed equipment and indirect actions caused by shrinkage and uneven settlements
2. Variable actions (Q): for example, imposed loads on building floors and beams, wind actions or snow loads
3. Accidental actions (A): for example, explosions or impact from vehicles

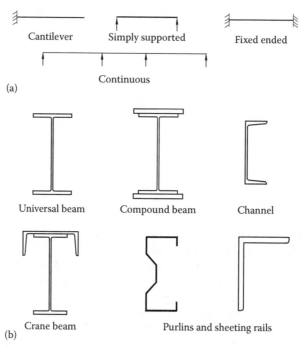

Figure 4.1 (a) Types of beams and (b) beam sections.

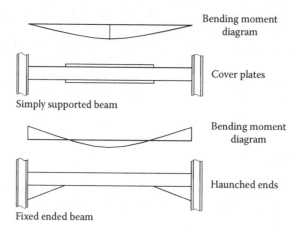

Figure 4.2 Non-uniform beam.

Actions on floor beams in a steel frame building are shown in Figure 4.3a. The figure shows actions from a two-way spanning slab that gives trapezoidal and triangular loads on the beams. One-way spanning floor slabs give uniform actions.

An actual beam with the floor slab and members it supports is shown in Figure 4.3b. The load diagram and shear force and bending moment diagrams constructed from it are also shown.

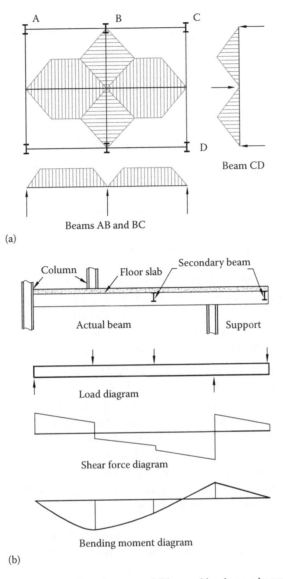

Figure 4.3 Beam loads: (a) slab loads on floor beams and (b) actual loads on a beam.

4.3 CLASSIFICATION OF BEAM CROSS SECTIONS

4.3.1 Definition of classes

The projecting flange of an I beam will buckle prematurely if it is too thin. Webs will also buckle under compressive stress from bending and from shear. This problem is discussed in more detail in Section 5.2 of Chapter 5.

To prevent local buckling from occurring, the material yield strength, loading arrangement, limiting outstands/thickness ratios for flanges and depth/thickness ratios for webs are given in EN 1993-1-1 which accounts for the effects of local buckling through cross-sectional

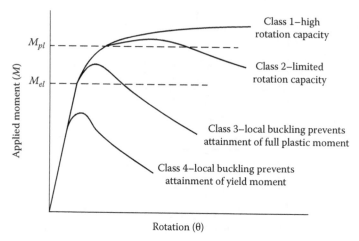

Figure 4.4 Four behavioural classes of cross section defined by EC 3.

classification, as described in Clause 5.5. Cross-sectional resistances may then be determined from Clause 6.2. The EN 1993 definitions of the four beam cross sections are classified as follows in accordance with their behaviour in bending:

Class 1 cross sections are those that can form a plastic hinge with rotation capacity required from plastic analysis without reduction of the resistance.

Class 2 cross sections are those that can develop their plastic moment resistance but have limited rotation capacity because of local buckling.

Class 3 cross sections are those in which the elastically calculated stress in the extreme compression fibre of the steel member assuming an elastic distribution of stresses can reach the yield strength, but local buckling is liable to prevent development of the plastic moment resistance.

Class 4 cross sections are those in which local buckling will occur before the attainment of yield stress in one or more parts of the cross section.

The moment–rotation characteristics of the four classes are shown in Figure 4.4.

Flat elements in a cross section are classified as

1. Internal elements supported on both longitudinal edges
2. Outside elements attached on one edge with the other free

4.3.2 Assessment of individual parts

The classification limits provided in Table 5.2 in EN 1993-1-1 are compared with c/t ratios (compressive width-to-thickness ratios), with the appropriate dimensions for c and t taken from the accompanying diagrams. The compression widths c always adopt the dimensions of the flat portions of the cross sections, that is root radii and welds are explicitly excluded from the measurement, as shown in Figure 4.5.

The limiting width-to-thickness ratios are modified by a factor ε that is a dependent upon the material yield strength. ε is defined as

$$\varepsilon = \sqrt{\frac{235}{f_y}}$$

where f_y is the nominal yield strength of the steel.

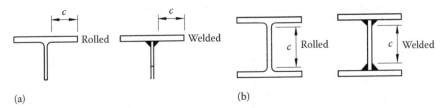

Figure 4.5 Definition of compression width c for common cases: (a) outstand flanges; (b) internal web.

The definition of the ε in EN 1993-1-1 utilizes a base value of 235 N/mm², simply because Grade S235 steel is regarded as the normal grade because it is still commonly used through-out Central Europe.

The normal yield strength depends upon the steel grade, the standard to which the steel is produced. Two thickness categories are defined in EN 1993-1-1. The first is up to and including 40 mm, and the second greater than 40 mm and less than 80 mm (for hot-rolled structural steel) or less than 65 mm (for structural hollow sections). However, the UK National Annex (NA) is likely to specify that material properties are taken as from the relevant product standard.

The cross-sectional classification of a beam member is given in Table 4.1, which is part of Table 5.1 of EN 1993-1-1.

The ratios of the flange outstand to thickness (c_w/t_w) and the web depth to thickness (c_f/t_f) are given for I, H and channel sections.

For I and H sections, $c_f = \dfrac{1}{2}\big[b - (t_w + 2r)\big]$

For channels, $c_f = \big[b - (t_w + r)\big]$

For I, H and channel sections, $c_w = d = \big[h - 2(t_f + r)\big]$

For hot-finished and cold-formed square and rectangular hollow sections, the ratios (c_w/t_w) and (c_f/t_f) are given where

$c_f = b - 3t$ and $c_w = h - 3t$

The dimension c is not precisely defined in EN 1993-1-1 and the internal profile of the corners is not specified in either EN 10210-2 or EN 10219-2. The preceding expressions give conservative values of the ratio for both hot-finished and cold-formed sections.

4.3.3 Overall cross-sectional classification

Once the classification of the individual parts of the cross section is determined, EN 1993 allows the overall section classification to be defined in one of three ways:

1. The classification of the overall cross section is taken as the least favourable of its component parts. For example, a cross section with a Class 2 flange and Class 1 web has an overall classification of Class 2.
2. Cross section with a Class 3 web and Class 1 or 2 flanges are classified as Class 2 overall cross section with an effective web (defined in Clause 6.2.2.4 in EN 1993-1-1).
3. Alternatively the classification of a cross section is defined by quoting both the flange classification and the web classification.

Table 4.1 Maximum width-to-thickness ratios for compression parts

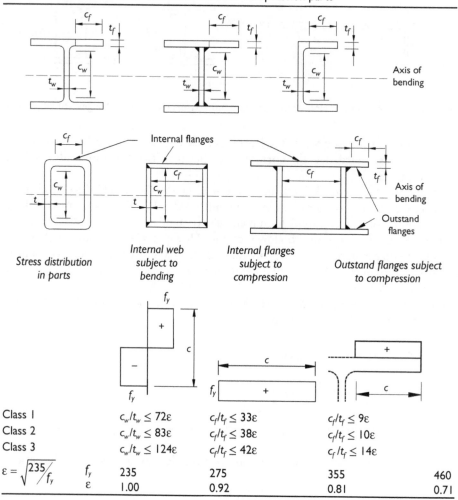

		Internal web subject to bending	Internal flanges subject to compression	Outstand flanges subject to compression
Class 1		$c_w/t_w \le 72\varepsilon$	$c_f/t_f \le 33\varepsilon$	$c_f/t_f \le 9\varepsilon$
Class 2		$c_w/t_w \le 83\varepsilon$	$c_f/t_f \le 38\varepsilon$	$c_f/t_f \le 10\varepsilon$
Class 3		$c_w/t_w \le 124\varepsilon$	$c_f/t_f \le 42\varepsilon$	$c_f/t_f \le 14\varepsilon$
$\varepsilon = \sqrt{235/f_y}$	f_y	235	275	355	460
	ε	1.00	0.92	0.81	0.71

4.4 BENDING STRESSES AND MOMENT CAPACITY

Both elastic and plastic theories are discussed here. Short or restrained beams are considered in this section. Plastic properties are used for plastic and compact sections, and elastic properties for semi-compact sections to determine moment capacities.

4.4.1 Elastic theory

1. Uniaxial bending

   The bending stress distributions for an I-section beam subjected to uniaxial moment are shown in Figure 4.6a. Define terms for the I section:

   M = applied bending moment.
   I_y = moment of inertia about y–y axis.
   $W_{el,y} = 2I_y/h$ = modulus of section for y–y axis.
   h = overall depth of beam.

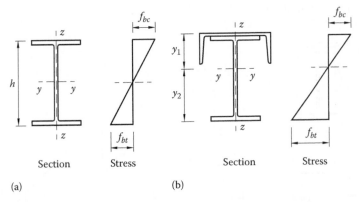

Figure 4.6 Beams in uniaxial bending: (a) T-section with two axes of symmetry; (b) crane beam with one axis of symmetry.

The maximum stress in the extreme fibres top and bottom is

$$f_{bc} = f_{bt} = \frac{M_y}{W_{el,y}}$$

The moment capacity

$$M_c = \sigma_b W_{el,y}$$

where σ_b is the allowable stress.

The design resistance for bending about one principal axis for Class 3 section is determined in Clause 5.2.5.2 of EN 1993-1-1 as follows:

$$M_{el,Rd} = \frac{W_{el} f_y}{\gamma_{M0}} \tag{4.1}$$

where

W_{el} is the elastic section modulus about one principal axis
f_y is the nominal values of yield strength
γ_{M0} is the partial factor

For the asymmetrical crane beam section shown in Figure 4.6b, the additional terms require definition as follows:

$W_{el,y1} = I_y/y_1$ = modulus of section for top flange.
$W_{el,y2} = I_y/y_2$ = modulus of section for bottom flange.
y_1, y_2 = distance from centroid to top and bottom fibres.

The bending stresses are
Top fibre in compression

$$f_{bc} = \frac{M_y}{W_{el,y2}}$$

Bottom fibre in tension

$$f_{bt} = \frac{M_y}{W_{el,y2}}$$

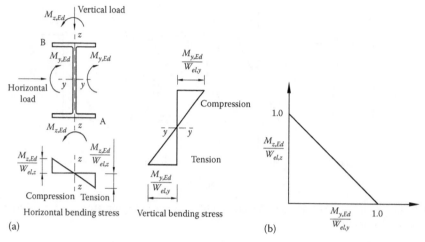

Figure 4.7 Biaxial bending: (a) bending stress; (b) interaction diagram.

The moment capacity controlled by the stress in the bottom flange is

$$M_{el,Rd} = \frac{W_{el,y2}f_y}{\gamma_{M0}}$$

2. Biaxial bending

Consider that I section in Figure 4.7a, which is subject to bending about both axes. Define the following terms:

$M_{y,Ed}$ = design bending moment about the y–y axis.
$M_{z,Ed}$ = design bending moment about the z–z axis.
$W_{el,y}$ = elastic section modulus for the y–y axis.
$W_{el,z}$ = elastic section modulus for the z–z axis.

The maximum stress at A or B is

$$f_A = f_B = \frac{M_{y,Ed}}{W_{el,y}} + \frac{M_{z,Ed}}{W_{el,z}}$$

For Class 3 cross sections, the maximum fibre stress limits by the yield stress divided by the partial factor γ_{M0} as follows:

$$f_A = f_B \le \frac{f_y}{\gamma_{M0}}$$

The design resistance for bending moment about y–y and z–z axes is given as follows:

$$M_{el,y,Rd} = \frac{W_{el,y}f_y}{\gamma_{M0}} \tag{4.2}$$

$$M_{el,z,Rd} = \frac{W_{el,z}f_y}{\gamma_{M0}} \tag{4.3}$$

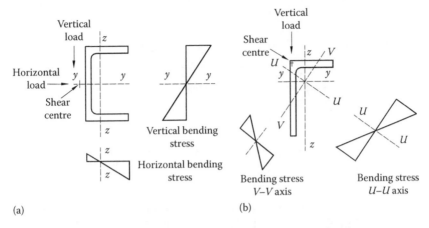

Figure 4.8 Bending of asymmetrical sections: (a) channel section; (b) unequal angle.

As a conservative approximation for all cross-sectional classes, a linear summation of the utilization ratios for each stress resultant may be used. For Class 3 cross sections subjected to the combination of $M_{y,Ed}$ and $M_{z,Ed}$, this method may be applied by using the following criteria:

$$\frac{M_{y,Ed}}{M_{el,y,Rd}} + \frac{M_{z,Ed}}{M_{el,z,Rd}} \leq 1 \tag{4.4}$$

This is shown graphically in Figure 4.7b.

3. Asymmetrical sections

Note that with the channel section shown in Figure 4.8a, the vertical load must be applied through the shear centre for bending in the free member to take place about the y–y axis; otherwise, twisting and biaxial bending occurs. However, a horizontal load applied through the centroid causes bending about the z–z axis only.

For an asymmetrical section such as the unequal angle shown in Figure 4.8b, bending takes place about the principle axes UU and VV in the free member when the load is applied through the shear centre. When the angle is used as a purlin, the cladding restrains the member so that it bends about the y–y axis.

4.4.2 Plastic theory

1. *Uniaxial bending*

The stress–strain curve for steel on which plastic theory is based is shown in Figure 4.9a. In the plastic region after yield, the strain increases without increase in stress. Consider the I section shown in Figure 4.9b. Under moment the stress first follows an elastic distribution. As the moment increases, the stress at the extreme fibre reaches the yield stress and the plastic region proceeds inwards as shown until the full plastic moment is reached and a plastic hinge is formed.

For single-axis bending, the following terms are defined:

$M_{pl,Rd}$ = the design resistance for bending moment.
W_{pl} = plastic section modulus.
f_y = yield strength.
γ_{M0} = the partial factor for resistance of cross section.

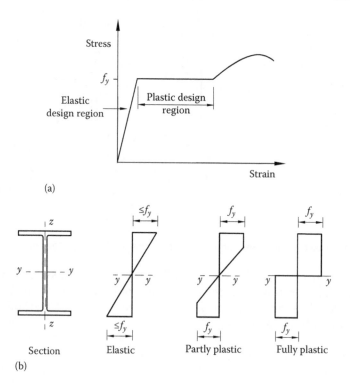

Figure 4.9 (a) Simplified stress-strain curve and (b) behaviour in bending.

The design resistance for bending moment given in Clause 6.2.5(1) of EN 1993-1-1, about one principal axis for Class 1 and 2 cross sections is determined as follows:

$$M_{pl,Rd} = \frac{W_{pl}f_y}{\gamma_{M0}} \tag{4.5}$$

For single-axis bending for a section with one axis of symmetry, consider the T-section shown in Figure 4.10. In the elastic range, bending takes place about the centroidal axis and there are two values for the elastic modulus of section.

In the plastic range, bending takes place about the equal area axis and there is one value for the plastic modulus of section:

$$W_{pl} = \frac{\gamma_{M0}M_{pl,Rd}}{f_y} = \frac{Ab}{2} \tag{4.6}$$

where
A is the area of cross section
b is the lever arm between the tension and compression forces

2. Biaxial bending
When a beam section is bent about both axes, the neutral axis will lie at an angle to the rectangular axes that depends on the section properties and values of the moments. Solutions have been obtained for various cases and a relationship established between the ratios of the applied moments and the moment capacities about each axis.

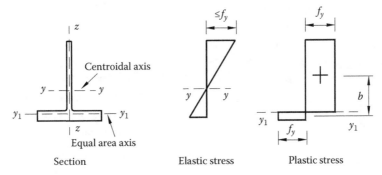

Figure 4.10 Section with one-axis symmetry.

EN 1993-1-1 treats biaxial bending as a subset of the rules for combined bending and axial force. Checks for Class 1 and 2 cross sections subjected to biaxial bending are given in Clause 6.2.9.1(6). The following equation (equation (6.41) in EN 1993-1-1) represents a more sophisticated convex interaction expression:

$$\left(\frac{M_{y,Ed}}{M_{N,y,Rd}} \right)^{\alpha} + \left(\frac{M_{z,Ed}}{M_{N,z,Rd}} \right)^{\beta} \leq 1 \tag{4.7}$$

where

$M_{y,Ed}$ is the design bending moment about the y–y axis

$M_{z,Ed}$ is the design bending moment about the z–z axis

$M_{N,y,Rd}$ is the reduced design values of the resistance to bending moment about the y–y axis

$M_{N,z,Rd}$ is the reduced design values of the resistance to bending moment about the z–z axis

α and β are constants, as defined in the following:

For I and H sections,

$\alpha = 2$ and $\beta = 5n$ but $\beta \geq 1$

For circular hollow sections,

$\alpha = 2$ and $\beta = 2$

For rectangular hollow sections,

$$\alpha = \beta = \frac{1.66}{1 - 1.13n^2}$$

but $\alpha = \beta \leq 6$

α and β can be taken as unity, thus reverting to a conservative linear interaction

$$n = \frac{N_{Ed}}{N_{pl,Rd}}$$

Figure 4.11 shows the biaxial bending interaction curves (for Class 1 and 2 cross sections) for some common cases.

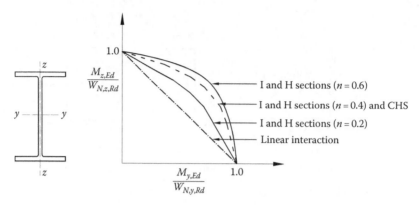

Figure 4.11 Interaction diagram for biaxial bending with axial force.

3. Unsymmetrical sections

For sections with no axis of symmetry, plastic analysis for bending is complicated, but solutions have been obtained. In many cases where such sections are used, the member is constrained to bend about the rectangular axis [see Section 4.4.1(3)]. Such cases can also be treated by elastic theory using factored loads with the maximum stress limited to the design strength.

4.5 LATERAL-TORSIONAL BUCKLING

4.5.1 General considerations

The compression flange of an I beam acts like a column and will buckle sideways if the beam is not sufficiently stiff or the flange is not restrained laterally. The load at which the beam buckles can be much less than that causing the full moment capacity to develop. Only a general description of the phenomenon and factors affecting it are set out here.

Lateral-torsional buckling (LTB) is a neutral equilibrium behaviour that the beam will initially bend in the plane of the web under the action of the applied moment. As the moment is increased, a limit state will be reached at which the in-plane bending of the beam becomes unstable and a slightly deflected and twisted form of the beam becomes possible. So the LTB involves both lateral deflection and twist rotation, as shown in Figure 4.12.

To summarize, factors influencing LTB are

1. The unrestrained length of compression flange. The longer this is, the weaker the beam. Lateral buckling is prevented by providing props at intermediate points.
2. The end conditions. Rotational restraint in plan helps to prevent buckling.
3. Section shape. Sections with greater lateral bending and torsional stiffnesses have greater resistance to buckling.
4. Note that lateral restraint to the tension flange also helps to resist buckling.
5. The application of the loads and shape of the bending moment diagram between restraints.

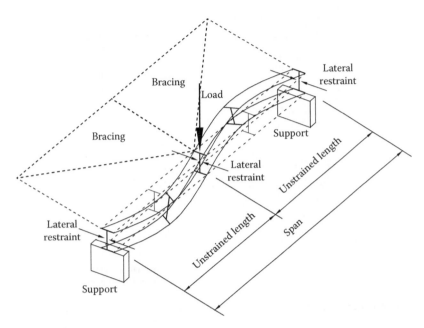

Figure 4.12 LTB of a beam with intermediate constraint.

Laterally unrestrained beams subjected to bending about their major axis have to be checked for LTB resistance. EN 1993-1-1 contains three methods for checking the lateral-torsional stability of a structural member:

1. The primary method adopts the lateral buckling curves given in Clause 6.3.2.2 (general case) and Clause 6.3.2.3 (just for rolled sections and equivalent welded sections). This method is discussed particularly in the following sections.
2. The second is a simplified assessment method for beams with restraints in buildings and is set out in Clause 6.3.2.4 of EN 1993-1-1. This method is discussed particularly in the following sections.
3. The third is a general method for lateral and LTB of structural components, such as single members with monosymmetric cross sections, built-up, non-uniform or plane frames and subframes, given in Clause 6.3.4. This method is just mentioned in this book.

4.5.2 Lateral restraints

The code states in Clause 6.3.2.1(2) that 'beams with sufficient lateral restraint to the compression flange are not susceptible to LTB', though there is little guidance on what is to be regarded as 'sufficient'. Lateral restraints need to possess adequate stiffness and strength to inhibit lateral deflection of the compression flange. In structures, full lateral restraint is provided by a floor slab if the friction or shear connection is capable of resisting a lateral force.

The following two types of restraints are defined in Sections 4.3.2 and 4.3.3 of the code:

1. Lateral restraint, which prevents sideways movement of the compression flange
2. Torsional restraint, which prevents movement of one flange relative to the other

Restraints are provided by floor slabs, end joints, secondary beams, stays, sheeting, etc., and some restraints are shown in Figure 4.13.

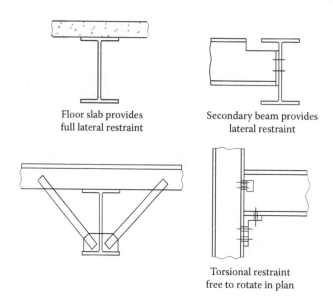

Figure 4.13 Lateral and torsional restraints.

Beams with sufficient restraint to the compression flange are not susceptible to LTB. In the following cases, the LTB can be ignored in the design:

- The fully laterally restrained beams
- Beams with certain types of cross sections, such as square or circular hollow sections, fabricated circular tubes or square box sections
- Beams with minor axis (z–z axis) bending
- The non-dimensional slenderness $\bar{\lambda}_{LT} < 0.4$ for rolled sections and hot-finished and cold-formed hollow sections or $\bar{\lambda}_{LT} < 0.2$ for welded sections

4.5.3 Elastic critical moment

Elastic theory is used to set up equilibrium equations to equate the disturbing effect to the lateral bending and torsional resistances of the beam. The solution of this equation gives the elastic critical moment M_{cr}. However, EN 1993-1-1 offers no formulations and gives no guidance on how M_{cr} should be calculated, except to say that M_{cr} should be based on gross cross-sectional properties and should take into account the loading condition, the real moment distribution and the lateral restraints, which refer to Clause 6.3.2.2(2). The formulation of elastic critical moment given in this book is provided in informative Annex F in ENV 1993-1-1 (1992).

The elastic critical moment for LTB of a beam of uniform symmetrical cross section with equal flanges, under standard conditions of restraint at each end, is loaded through the shear centre and subjected to uniform moment. Initially the beam deflects in the vertical plane due to bending, but as the moment increases, it reaches a critical value M_{cr}, less than the moment capacity, where it buckles sideways, twists and collapses.

The standard conditions of restraint at each end of the beam are as follows (as shown in Figure 4.14):

- Restrained against lateral movement
- Restrained against rotation about the longitudinal axis
- Free to rotate on plan

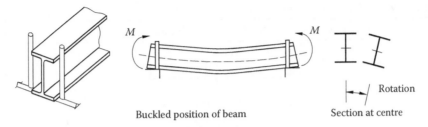

Figure 4.14 Laterally and torsionally simply supported I beam under uniform moment.

The elastic critical moment $M_{cr,0}$ is given by equation

$$M_{cr,0} = \frac{\pi^2 E I_Z}{L_{cr}^2} \sqrt{\frac{I_W}{I_Z} + \frac{L_{cr}^2 G I_t}{\pi^2 E I_z}}$$ (4.8)

where

$$G = \frac{E}{2[1+v]}$$

I_t is the torsional constant
I_w is the warping constant
I_z is the second moment of area about the minor axis
L_{cr} is the length of the beam between points that have lateral restraint

Numerical solutions have also been calculated for a number of other loading conditions. For uniform doubly symmetric cross sections, loaded through the shear centre at the level of the centroidal axis and with the standard conditions of the restrained above, M_{cr} is given as follows:

$$M_{cr} = C_1 \frac{\pi^2 E I_Z}{L_{cr}^2} \sqrt{\frac{I_W}{I_Z} + \frac{L_{cr}^2 G I_t}{\pi^2 E I_z}}$$ (4.9)

where C_1 is the factor depending on the loading and end restraint conditions.

Value of C_1 is given in Table 4.2 for end-moment loading and Table 4.3 for transverse loading. The factor C_1 is used to modify $M_{cr,0}$ to take account of the shape of the bending moment diagram over the length L_{cr} between lateral restraint.

The value of C_1 for any ratio of end-moment loading as indicated in Table 4.2 is given approximately by

$$C_1 = 1.88 - 1.40\psi + 0.52\psi^2 \quad \text{but} \quad C_1 \le 2.70$$

where ψ is the ratio of the end moment.

The theoretical solution applies to a beam subjected to a uniform moment. In other cases where the moment varies, the tendency to buckling is reduced. If the load is applied to the top flange and can move sideways, it is destabilizing, and buckling occurs at lower loads than if the load were applied at the centroid or to the bottom flange.

In the theoretical analysis, the beam was assumed to be straight. Practical beams have initial curvature and twisting, residual stresses, and the loads are applied eccentrically.

Table 4.2 C_1 Values for end-moment loading

Loading and support conditions	Bending moment diagram		Value of C_1
$M \curvearrowright$ ——— $\curvearrowleft \psi M$		$\psi = +1$	1.000
		$\psi = +0.75$	1.141
		$\psi = +0.5$	1.323
		$\psi = +0.25$	1.563
		$\psi = 0$	1.879
		$\psi = -0.25$	2.281
		$\psi = -0.5$	2.704
		$\psi = -0.75$	2.927
		$\psi = -1.0$	2.752

Table 4.3 C_1 Values for transverse loading

Loading and support conditions	Bending moment diagram	Value of C_1
w (simply supported)		1.132
w (fixed ends)		1.285
F (point load, simply supported)		1.365
F (point load, fixed ends)		1.565
F F (two point loads)		1.046

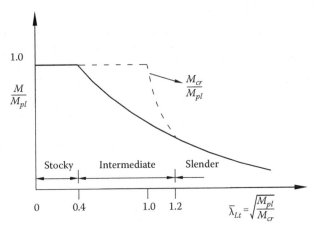

Figure 4.15 Effect of slenderness.

The theory set out previously requires modification to cover actual behaviour. Theoretical studies and tests show that slender beams fail at the elastic critical moment M_{cr} and short or restrained beams fail at the plastic moment capacity $M_{pl,Rd}$. A lower-bound curve running between the two extremes can be drawn to contain the behaviour of intermediate beams. Beam behaviour as a function of slenderness is shown in Figure 4.15.

4.5.4 Code design procedure

4.5.4.1 Lateral-torsional buckling resistance

A lateral unrestrained beam (segments of the beam that are between the points where lateral restraint exists) subjected to major axis bending should be verified against LTB as follows:

$$\frac{M_{Ed}}{M_{b,Rd}} \leq 1.0$$

where

M_{ed} is the design value of the moment

$M_{b,Rd}$ is the design buckling resistance moment

The design buckling resistance of a laterally untrained beam (or segment of the beam) should be taken as

$$M_{b,Rd} = \chi_{LT} W_y \frac{f_y}{\gamma_{M1}} \tag{4.10}$$

where W_y is the section modulus appropriate for the classification of the cross section as follows:

$W_y = W_{pl,y}$ for Class 1 or 2 cross sections.

$W_y = W_{el,y}$ for Class 3 cross sections.

$W_y = W_{eff,y}$ for Class 4 cross sections.

χ_{LT} is the reduction factor for LTB

γ_{M1} is the partial factor assigned to resistance of member to buckling, $\gamma_{M1} = 1.0$ recommend by NA to BS EN 1993-1-1

Note: In determining W_y, holes for fasteners at the beam end need not be taken into account.

4.5.4.2 Lateral-torsional buckling curves

EN 1993-1-1 provides the designer with four LTB curves from which a 'buckling resistance moment' can be derived. The buckling curves plot strength reduction against slenderness, and an important difference from BS 5950 is that the slenderness in EC buckling curves is 'non-dimensional', as shown in Figure 4.16. The advantage of the non-dimensional presentation is that separate curves are not needed for S235 and S355 nor for the stepwise reduction in yield strength at element thickness (i.e. 16 mm and 40 mm in EN 10025). The buckling curve in EN 1993-1-1 is selected on the basis of the overall height-to-width ratio of the cross section, the type of cross section and whether the cross section is rolled or welded.

EN 1993-1-1 defines LTB curves for two cases:

1. The general case
2. Rolled sections or equivalent welded sections

1. General case

The general case may be applied to all common section types, including rolled sections, and also be applied to outside the standard range of the rolled sections. For example, it may be applied to plate girders (of larger dimensions than standard rolled sections) and to castellated and cellular beams.

LTB curves for the general case can be described through the following equation:

$$\chi_{LT} = \frac{1}{\phi_{LT} + \sqrt{\phi_{LT}^2 - \overline{\lambda}_{LT}^2}} \quad \text{but } \chi_{LT} \leq 1.0 \tag{4.11}$$

where

$$\phi_{LT} = 0.5\left[1 + \alpha_{LT}\left(\overline{\lambda}_{LT} - 0.2\right) + \overline{\lambda}_{LT}^2\right]$$

$$\overline{\lambda}_{LT} = \sqrt{\frac{W_y f_y}{M_{cr}}}$$

α_{LT} is an imperfection factor from Table 4.4
M_{cr} is the elastic critical moment for LTB

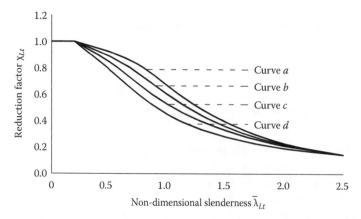

Figure 4.16 Four curves for LTB.

Table 4.4 Imperfection factors for LTB curves

Buckling curve	a	b	c	d
Imperfection factor α_{LT}	0.21	0.34	0.49	0.76

Table 4.5 LTB curves for general cases

Cross section	Limits	Buckling curve
Rolled I sections	$h/b \leq 2$	a
	$h/b > 2$	b
Welded I sections	$h/b \leq 2$	c
	$h/b > 2$	d
Other cross sections	—	d

Selection of the appropriate LTB curves for a given cross-sectional type and dimensions may be made with reference to Table 4.5

2. Rolled sections or equivalent welded sections

   For rolled sections or equivalent welded sections in bending, the LTB curves are described through the following equation:

$$\chi_{LT} = \frac{1}{\phi_{LT} + \sqrt{\phi_{LT}^2 - \beta \bar{\lambda}_{LT}^2}} \quad \text{but} \quad \chi_{LT} \leq 1.0 \text{ and } \chi_{LT} \leq \frac{1}{\bar{\lambda}_{LT}^2} \tag{4.12}$$

where

$$\phi_{LT} = 0.5 \left[1 + \alpha_{LT} \left(\bar{\lambda}_{LT} - \bar{\lambda}_{LT,0} \right) + \beta \bar{\lambda}_{LT}^2 \right]$$

$$\bar{\lambda}_{LT} = \sqrt{\frac{W_y f_y}{M_{cr}}}$$

α_{LT} is an imperfection factor from Table 4.4
M_{cr} is the elastic critical moment for LTB
β and $\bar{\lambda}_{LT,0}$ are limit parameters, which values are given in the NA

The following values are recommended from NA to BS EN 1993-1-1:

- For rolled sections and hot-finished and cold-formed hollow sections,

$$\bar{\lambda}_{LT,0} = 0.4$$

$$\beta = 0.75$$

Table 4.6 LTB curve for rolled and
equivalent welded cases (EC 3)

Cross section	Limits	Buckling curve
Rolled I sections	$h/b \leq 2$	b
	$h/b > 2$	c
Welded I sections	$h/b \leq 2$	c
	$h/b > 2$	d

Table 4.7 LTB curve for rolled and equivalent welded cases
(NA to BS EN 1993-1-1)

Cross section	Limits	Buckling curve
Rolled doubly symmetric I and H sections and hot-finished hollow sections	$h/b \leq 2.0$	b
	$2.0 < h/b \leq 3.1$	c
	$h/b > 3.1$	d
Angles		d
All other hot-rolled sections		d
Welded doubly symmetric sections and cold-formed hollow sections	$h/b \leq 2.0$	c
	$2.0 < h/b \leq 3.1$	d

- For welded sections,

$$\overline{\lambda}_{LT,0} = 0.2$$

$$\beta = 1.0$$

Selection of the appropriate LTB curves for a given cross-sectional type and dimensions may be made with reference to Table 4.6. It should be noted that NA to BS EN 1993-1-1 prescribes a maximum limit on the height-to-width ratio h/b for welded I sections, as shown in Table 4.7, beyond which the LTB rules may not be applied.

The aforementioned equation is just for the uniform moment case, taking into account the moment distribution between the lateral restraints of the beam the reduction factor χ_{LT} that may be modified by the factor f:

$$\chi_{LT,mod} = \frac{\chi_{LT}}{f} \quad \text{but} \quad \chi_{LT,mod} \leq 1.0 \text{ and } \chi_{LT,mod} \leq \frac{1}{\overline{\lambda}_{LT}^2} \tag{4.13}$$

The factor f was derived on the basis of a numerical study as

$$f = 1 - 0.5(1 - k_c)\left[1 - 2.0\left(\overline{\lambda}_{LT,0} - 0.8\right)^2\right] \tag{4.14}$$

And f is dependent upon the shape of the bending moment diagram between lateral restraints. k_c is a correction factor according to Table 4.8.

Take an I section ($h/b > 2$), for instance, the LTB curves of the general case in comparison with the buckling curves of the rolled sections or equivalent welded sections are given in

Table 4.8 Correction factor k_c

Moment distribution		k_c
	$\psi = +1$	1.0
	$-1 \leq \psi \leq 1$	$\dfrac{1}{1.33 - 0.33\psi}$
		0.94
		0.90
		0.91
		0.86
		0.77
		0.82

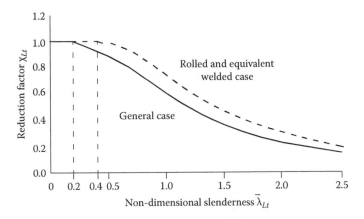

Figure 4.17 LTB curves for general case and for rolled or equivalent welded sections.

Figure 4.17. The imperfection factor α_{LT} for buckling curve b has been used for the comparison. It can be seen that the curve for the rolled and equivalent welded case is more favourable than that for the general case. The greatest difference of the two curves is the plateau length of the rolled and equivalent welded case; no LTB checks are required within this plateau length.

4.5.4.3 Simplified assessment methods for beams with restraints in buildings

EC 3 provides a quick and conservative method of determining whether the beam with discrete lateral restraint to the compression flange is susceptible to LTB or not. The LTB checks will be ignored if the lengths of a beam between points of effective lateral restraints L_c or the resulting slenderness $\bar{\lambda}_f$ of the equivalent compression flange satisfies

$$\bar{\lambda}_f = \frac{k_c L_c}{i_{f,z}\lambda_1} \le \bar{\lambda}_{c0}\frac{M_{c,Rd}}{M_{y,Ed}} \tag{4.15}$$

where

$M_{y,Ed}$ is the maximum design value of the bending moment within the restraint spacing

$$M_{c,Rd} = W_y\frac{f_y}{\gamma_{M1}}$$

W_y is the appropriate section modulus corresponding to the compression flange

k_c is a slenderness correction factor for moment distribution between restraints (see Table 4.8)

$i_{f,z}$ is the radius of gyration of the compression flange plus one-third of the compressed part of the web, about the minor axis of the section

$\bar{\lambda}_{c0}$ is a slenderness limit of the equivalent compression flange. The value of $\bar{\lambda}_{c0}$ may be given in the National Annex. A limit value $\bar{\lambda}_{c0} = \bar{\lambda}_{LT,0} + 0.1$ is recommended

For I, H, channel and box sections used in building, the limit value of $\bar{\lambda}_{c0}$ should be taken as 0.4 (NA to BS EN 1993-1-1):

$$\lambda_1 = \pi\sqrt{\frac{E}{f_y}} = 93.9\varepsilon$$

$$\varepsilon = \sqrt{\frac{235}{f_y}} \quad (f_y \text{ in N/mm}^2)$$

The method treats the compression flange of the beam and part of the web as a strut, as shown in Figure 4.18.

For Class 1, 2 and 3 cross sections, $i_{f,z}$ can be taken as

$$i_{f,z} = \sqrt{\frac{I_{z,comp.f+(1/3)comp.w}}{A_{z,comp.f+(1/3)comp.w}}} \tag{4.16}$$

where

$$I_{z,comp.f+(1/3)comp.w} = \frac{t_f b^3}{12} + \frac{(1/3)\left(h_w/2\right)t_w^3}{12}$$

$$A_{z,comp.f+(1/3)comp.w} = t_f b + \frac{1}{3}\left(\frac{h_w}{2}\right)t_w$$

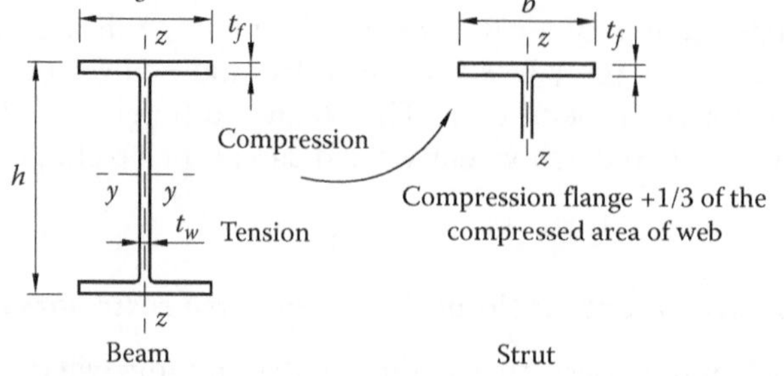

Figure 4.18 Simplified method.

For Class 4 cross sections, $i_{f,z}$ can be taken as

$$i_{f,z} = \sqrt{\frac{I_{eff,f}}{A_{eff,f} + \frac{1}{3}A_{eff,w,c}}} \tag{4.17}$$

where

$I_{eff,f}$ is the effective second moment of area of the compression flange about the minor axis of the section

$A_{eff,f}$ is the effective area of the compression flange

$A_{eff,w,c}$ is the effective area of the compressed part of the web

For the simplest case when the steel strength $f_y = 235$ N/mm^2, and such $\varepsilon = 1.0$, $M_{y,Ed}$ is equal to $M_{c,Rd}$, and uniform moment loading is assumed ($k_c = 1.0$), Equation 4.15 can be simplified as the following:

$$L_c \leq 37.6 i_{f,z}$$

When the steel strength $f_y = 275$ N/mm^2, and other cases are the same as the aforementioned, the equation can be simplified as the following:

$$L_c \leq 34.7 i_{f,z}$$

If the slenderness of the compression flange $\bar{\lambda}_f$ exceeds the limit given in Equation 4.15, the design buckling resistance moment may be taken as

$$M_{b,Rd} = k_{fl} \chi M_{c,Rd} \quad \text{but} \leq M_{c,Rd} \tag{4.18}$$

where χ is the reduction factor of the equivalent compression flange determined with $\bar{\lambda}_f$ (Figure 4.16 or Equation 4.11).

k_{fl} is the modification factor accounting for the conservation of the equivalent compression method. A value $k_{fl} = 1.0$ is recommended. However, modification factor may be given in the National Annex. From NA to BS EN 1993-1-1, the following values of k_{fl} should be taken as:

- $k_{fl} = 1.0$ for hot-rolled I sections
- $k_{fl} = 1.0$ for welded I sections with $h/b \leq 2$
- $k_{fl} = 0.9$ for other sections

The buckling curves to be used to calculate the reduction factor χ should be taken as follows:

- For welded sections provided that $h/t_f \leq 44\varepsilon$, use curve d.
- For all other sections, use curve c.

where h is the overall depth of the cross section and t_f is the thickness of the compression flange.

4.6 SHEAR IN BEAMS

4.6.1 Shear resistance check

When shear force is applied to a beam, the design value of the shear force V_{Ed} at each cross section shall satisfy

$$\frac{V_{Ed}}{V_{c,Rd}} \leq 1.0$$

where $V_{c,Rd}$ is the design shear resistance.

For plastic design, $V_{c,Rd}$ is the design plastic shear resistance $V_{pl,Rd}$. For elastic design, $V_{c,Rd}$ is the design elastic shear resistance that is calculated based on the elastic shear stress distribution.

4.6.2 Elastic design resistance

A simple beam is shown in Figure 4.19. At a distance x from the left end and at the neutral axis of the cross section, the state of the shear stress is shown in Figure 4.19d. Because this element is located at the neutral axis, it is not subjected to flexural stress. From elementary mechanics of materials, the shear stress is taken as

$$\tau_{Ed} = \frac{V_{Ed}S}{It} \tag{4.19}$$

where
  V_{Ed} is the design value of the shear force
  S is the first moment of area of the cross section about the centroidal axis between the point at which the shear is required and the boundary of the cross section
  I is the second moment of area of the whole cross section
  t is the width of the cross section at the examined point

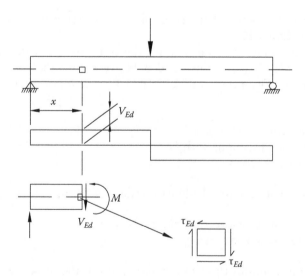

Figure 4.19 Shear stress of a simple beam.

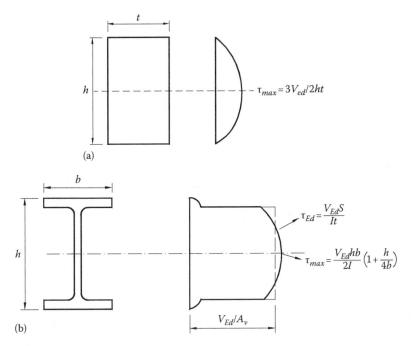

Figure 4.20 Distribution of shear stresses in beams subjected to a shear force: (a) rectangular cross-section; (b) I- or H-section.

The shear stress distribution as well as their maximum stress in a rectangular section and in an I section, based on purely elastic behaviour, is shown in Figure 4.20. The shear stress varies parabolically with depth, with the maximum value occurring at the neutral axis. However, for the I section, the difference between maximum and minimum values for the web is relatively small. Superimposed on the actual distribution is the average stress in the web, V_{Ed}/A_w, which does not differ much from the maximum web stress. Clearly, the web will completely yield long before the flanges begin to yield. Because of this, yielding of the web represents one of the shear limit states. Since the shear yield stress is approximately $1/\sqrt{3}$ of the tensile yield stress, the following equation can be written for the stress in the web at failure as

$$\tau_y = \frac{V_{Ed}}{A_w} = \frac{f_y}{\sqrt{3}}$$

As shown is Figure 4.21, the web behaves elastically in shear until first yield occurs at $\tau_y = f_y/\sqrt{3}$ and then undergoes increasing plastification until the web is fully yielded in shear. Because the shear stress distribution at first yield in nearly uniform, the nominal first yield and fully plastic loads are nearly equal, and the shear shape factor is usually very close to 1.0. Stocky unstiffened webs in steel beams reach first yield before they buckle elastically, so that their resistances are determined by the shear stress τ_y.

For verifying the design elastic shear resistance $V_{c,Rd}$, the following criterion for a critical point of the cross section may be used:

$$\frac{\tau_{Ed}}{f_y/\left(\sqrt{3}\gamma_{M0}\right)} \leq 1.0 \tag{4.20}$$

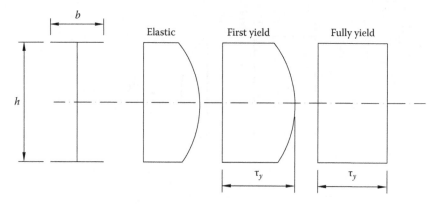

Figure 4.21 Plastification of an I section web in shear.

where

$$\tau_{Ed} = \frac{V_{Ed}S}{It}$$

For I or H sections, the shear stress in the web may be taken as uniformly distributed:

$$\tau_{Ed} = \frac{V_{Ed}}{A_w} \quad \text{if} \quad \frac{A_f}{A_w} \geq 0.6$$

where
A_f is the area of one flange
A_w is the area of the web, $A_w = h_w t_w$

4.6.3 Plastic design resistance

The ratio web depth to width h_w/t_w is an important conception to determine if the web is a stocky web or a slenderness web:

- For unstiffened webs, cross sections for which $h_w/t_w \leq 72(\varepsilon/\eta)$ are stocky.
- For webs with intermediate stiffeners, cross sections for which $h_w/t_w \leq 31(\varepsilon/\eta)\sqrt{k_\tau}$ are stocky.

where
h_w is the depth of the web (measured between the flanges)
t_w is the web thickness (taken as the minimum value if the web is not of constant thickness)
$$\varepsilon = \sqrt{\frac{235}{f_y}}$$

k_τ is a shear buckling coefficient (discussed in Chapter 5)
η is the shear area factor. It is recommended in Clause 5.1 of EN 1993-1-5 that η be taken as 1.20 (for steel grades higher than S460, $\eta = 1.0$ is recommended). However, the National Annex will define the value of η. From NA to BS EN 1993-1-5, $\eta = 1.0$ for all cases

The webs of all UBs and UCs in Grade 275 steel satisfy $h_w/t_w \leq 72(\varepsilon/\eta)$.

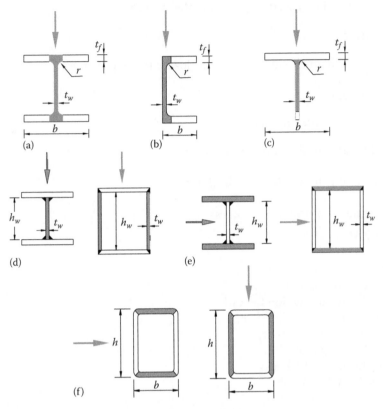

Figure 4.22 Effective shear area of the cross section: (a) rolled I- and H-section; (b) rolled channel section; (c) rolled T-section; (d) welded sections, load parallel to web; (e) welded sections, load parallel to flange; (f) rolled rectangular hollow section.

The plastic design is applied to stocky beams. The design plastic shear resistance $V_{pl,Rd}$ follows the assumption of uniform shear stress distribution along shear area A_v:

$$V_{pl,Rd} = \frac{A_v \left(f_y / \sqrt{3} \right)}{\gamma_{m0}} \tag{4.21}$$

where

$\gamma_{M0} = 1.0$

A_v is the effective shear area of the cross section that can be mobilized to resist the applied shear force (Figure 4.22). The shear area may be taken as follows:

1. Rolled I and H sections, load parallel to web

$A_v = A - 2bt_f + (t_w + 2r)t_f \quad \text{but} \geq \eta h_w t_w$

2. Rolled channel sections, load parallel to web

$A_v = A - 2bt_f + (t_w + r)t_f$

3. Rolled T-section, load parallel to web

$A_v = 0,9(A - bt_f)$

4. Welded I, H and box sections, load parallel to web

$$A_v = \eta \Sigma (h_w t_w)$$

5. Welded I, H, channel and box sections, load parallel to flanges

$$A_v = A - \Sigma (h_w t_w)$$

6. Rolled RHS of uniform thickness

Load parallel to depth: $A_v = Ah/(b+h)$

Load parallel to width: $A_v = Ab/(b+h)$

7. CHS and tubes of uniform thickness

$$A_v = 2A/\pi$$

where
  A is the cross-sectional area
  b is the overall section breadth
  h is the overall section depth
  h_w is the overall web depth
  r is the root radius
  t_f is the flange thickness
  t_w is the web thickness

The shear resistance of slender webs for which $h_w/t_w > 72(\varepsilon/\eta)$ or $h_w/t_w > 31(\varepsilon/\eta)\sqrt{k_\tau}$ will be discussed in Chapter 5 (Plate Girder).

4.7 BENDING AND SHEAR

For sections subject to coexistent bending moment and shear force, allowance is made for the effect of shear force on moment resistance. If the shear force is less than half the plastic shear resistance, its effect on the moment resistance may be ignored, unless shear buckling reduces the section resistance.

For cases where the applied shear force is greater than half the plastic shear resistance of the cross section, the moment resistance should be calculated using reduced yield strength for the shear area, given by

$$f_{yr} = (1 - \rho)f_y \tag{4.22}$$

where

$$\rho = \left(\frac{2V_{Ed}}{V_{pl,Rd}} - 1 \right)^2$$

$$V_{pl,Rd} = \frac{A_v \left(f_y / \sqrt{3} \right)}{\gamma_{M0}}$$

For I sections with equal flanges and bending about the major axis, the reduced design plastic resistance moment allowing for the shear force effect is given by

$$M_{y,V,Rd} = \frac{\left(W_{pl,y} - \rho \frac{A_w^2}{4t_w}\right)f_y}{\gamma_{M0}} \quad \text{but} \quad M_{y,V,Rd} \leq M_{y,c,Rd} \tag{4.23}$$

where

$A_w = h_w t_w$

$M_{y,c,Rd}$ is determined according to the class of the cross section

For a UB section, the plastic section modulus about the $y-y$ axis is defined by

$$W_{pl,y} = bt_f h_w + \frac{A_w^2}{4t_w} \tag{4.24}$$

Clearly, the reduced plastic resistance moment $M_{y,V,Rd} = W_{pl,y}f_{yr}/\gamma_{M0}$ calculated using Equation 4.22 is more conservative than Equation 4.23.

4.8 RESISTANCE OF THE BEAM TO TRANSVERSE LOADING

Transverse loading denotes a load that is applied perpendicular to the flange in the plane of the web. The loading is usually free and transient. However, the rules of EN 1993-1-5 are assumed that the compression flange has an adequate lateral and torsional restraint. EN 1993-1-5 covers three types of transverse load applied through a flange to the web:

1. Load application through one flange and resisted by shear forces in the web (see Figure 4.23a)
2. Load application through one flange and transferred through the web directly to the other flange (see Figure 4.23b)
3. Load application through one flange adjacent to an unstiffened end (see Figure 4.23c)

4.8.1 Length of stiff bearing

The length of stiff bearing s_s corresponds to the loaded length on top of the flange. In Figure 4.24a and b, the determination of s_s is shown for different cases that assume a load distribution with a slope of 45°(1:1) in the load-introducing element. However, s_s should not be taken as larger than h_w. For a load introduced via rollers, each roller should be verified with $s_s = 0$ and with s_s according to the layout in Figure 4.24d. If the load-introducing

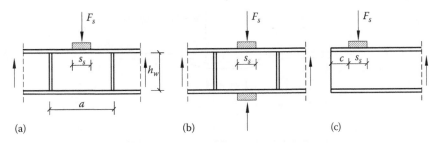

Figure 4.23 Types of transverse load application: (a) load application type a; (b) load application type b; (c) load application type c.

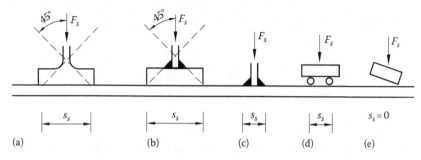

Figure 4.24 Length of stiff bearing: (a) rolled section; (b) welded section; (c) plate directly; (d) crane wheel; (e) no deformation of girder.

element is not able to follow the deformation of the beam, s_s should be set to zero (see Figure 4.24e). Additional load spreading through the flange is considered in the formula for the effective loaded length.

4.8.2 Design resistance

For unstiffened or stiffened webs, the design resistance to local buckling under transverse loading should be taken as

$$F_{Rd} = \frac{f_{yw} L_{eff} t_w}{\gamma_{M1}} = \chi_F \frac{f_{yw} l_y t_w}{\gamma_{M1}} \tag{4.25}$$

where
    t_w is the thickness of the web
    f_{yw} is the yield strength of the web
    χ_F is the reduction factor due to local buckling
    l_y is the effective loaded length appropriate to the length of the stiff bearing s_s

For type (a), the relationship of l_y and s_s is shown in Figure 4.25.

4.8.3 Reduction factor χ_F for effective length for resistance

The reduction factor χ_F should be obtained from

$$\chi_F = \frac{0.5}{\lambda_F} \leq 1.0 \tag{4.26}$$

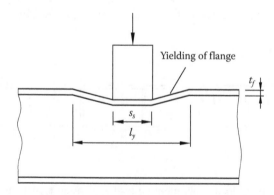

Figure 4.25 The effective loaded length.

where

$$\bar{\lambda}_F = \sqrt{\frac{l_y t_w f_{yw}}{F_{cr}}}$$ (4.27)

F_{cr} is the elastic critical load of the web. F_{cr} is determined as follows:

$$F_{cr} = 0.9 k_F E \frac{t_w^3}{h_w}$$ (4.28)

where k_F is the buckling value for transverse loading.

For beams or girders without longitudinal stiffeners, the buckling value k_F is determined depending on the type of transverse load application as follows:

For type (a),

$$k_F = 6 + 2\left(\frac{h_w}{a}\right)^2$$

For type (b),

$$k_F = 3.5 + 2\left(\frac{h_w}{a}\right)^2$$

For type (c),

$$k_F = 2 + 6\left(\frac{s_s + c}{h_w}\right) \le 6$$

The parameters a and c can be in accord with Figure 4.23.

For girders with longitudinal stiffeners and load application through one flange [type (a) and (c) in Figure 4.23], the buckling value k_F is determined according to the following equation:

$$k_F = 6 + 2\left(\frac{h_w}{a}\right)^2 + \left(5.44 \frac{b_1}{a} - 0.21\right)\sqrt{\gamma_s}$$ (4.29)

where

$$\gamma_s = 10.9 \frac{I_{sl,1}}{h_w t_w^3} \le 13\left(\frac{a}{h_w}\right)^3 + 210\left(0.3 - \frac{b_1}{a}\right)$$ (4.30)

Equation 4.29 is valid for

$$0.05 \le \frac{b_1}{a} \le 0.3 \quad \text{and} \quad \frac{b_1}{h_w} \le 0.3$$

where
b_1 is the depth of the loaded subpanel taken as the clear distance between the loaded flange and the stiffener
$I_{sl,1}$ is the second moment of the area of the stiffener closest to the loaded flange including contributing parts of the web according to Figure 4.26

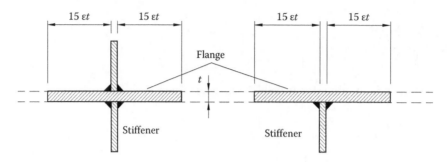

Figure 4.26 Effective cross section of stiffener.

According to Clause 9.1(2) of EN 1003-1-5, when checking the buckling resistance, the section of a stiffener may be taken as the gross area comprising the stiffener plus a width of plate equal to $15\varepsilon t$ but not more than the actual dimension available, on each side of the stiffener avoiding any overlap of contributing parts to adjacent stiffeners.

4.8.4 Effective loaded length

The effective loaded length l_y corresponds to the effective loaded length of the web taking into account the influence of the flange. For load application through one flange or both flanges, the four-hinge mechanical model is used (see Figure 4.25). The model assumes two inner and two outer plastic flange M_i and M_0, the resistance of which depends on several flange and web parameters covered by the dimensional parameters m_1 and m_2. The contributing width of the flange should be limited to $15\varepsilon t$ on each side of the web.

The effective loaded length l_y is determined depending on the type of transverse load application as follows:

- For type (a) and (b),

$$l_y = s_s + 2t_f\left(1 + \sqrt{m_1 + m_2}\right) \le a \qquad (4.31)$$

  where a is the distance between adjacent transverse stiffeners
- For type (c),

$$l_y = \min\left(l_{y1}, l_{y2}\right)$$

$$l_{y1} = l_e + t_f\sqrt{\frac{m_1}{2} + \left(\frac{l_e}{t_f}\right)^2 + m_2} \qquad (4.32)$$

$$l_{y2} = l_e + t_f\sqrt{m_1 + m_2} \qquad (4.33)$$

  where

$$l_e = \frac{k_F E t_w^2}{2 f_{yw} b_w} \le s_s + c \qquad (4.34)$$

The dimensionless parameters m_1 and m_2 are determined according to the following equations:

$$m_1 = \frac{f_{yf}b_f}{f_{yw}t_w} \qquad (4.35)$$

$$m_2 = 0 \quad \text{for } \bar{\lambda}_F \leq 0.5 \qquad (4.36)$$

$$m_2 = 0.02\left(\frac{h_w}{t_f}\right)^2 \quad \text{for } \bar{\lambda}_F \leq 0.5 \qquad (4.37)$$

4.8.5 Verification

The verification of a beam subject to transverse loading is

$$\eta_2 = \frac{F_{Ed}}{F_{Rd}} = \frac{F_{Ed}}{f_{yw}L_{eff}t_w/\gamma_{M1}} \leq 1.0$$

where F_{Ed} is the design transverse loading.

4.9 SERVICEABILITY DESIGN OF BEAMS

The design of a beam is often governed by the serviceability limit state, for which the behaviour of the beam should be so limited as to give a high probability that the beam will provide the serviceability necessary for it to carry out its intended function. The most common serviceability criteria are associated with the stiffness of the beam, which governs its deflections under loads. The deflection limits for beams are specified in Section 3.4 of EN 1990.

Serviceability design against unsightly appearance should be based on the total sustained load and so should include the effects of permanent actions and long-term imposed loads. Three categories of combinations of loads (actions) are specified in EN 1990 for serviceability checks: characteristic, frequent and quasi-permanent. Each combination contains a permanent action component, a leading variable component and other variable components. The characteristic combination of actions would generally be used when considering the function of the structure and damage to structural and nonstructural elements. The vertical deflection limits given in EN 1993-1-1 are based on the characteristic combination of actions.

However, Clause NA.2.23 of NA to BS EN 1993-1-1 proposes that permanent actions should be taken as zero in serviceability checks. The serviceability loads are just the unfactored variable actions (combination).

In general, beam deflection is a function of the span length, end restraints, modulus of elasticity of the material, moment of inertia of the cross section and loading. For the common case of a simply supported, uniformly loaded beam such as that in Figure 4.27, the maximum vertical deflection is

$$\delta = \frac{5}{384} \frac{wL^4}{EI}$$

Deflection formulae for a variety of beams and loading conditions are given in design manuals. Deflections for some common load cases for simply supported beams, both end-fixed beam and cantilever beam together with the maximum moments, are given in Table 4.9. For general-load cases, deflections can be calculated by the moment-area method.

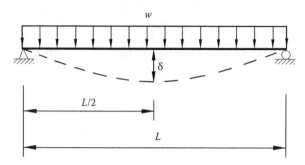

Figure 4.27 Deflection of a simply supported beam.

4.10 EXAMPLES OF BEAM DESIGN

4.10.1 Floor beams for an office building

The steel beams for part of the floor of a library with book storage are shown in Figure 4.28a. The floor is a reinforced concrete slab supported on UBs. The design loading has been estimated as

Permanent action – slab, self-weight of steel, finishes ceiling, partitions services and fire protection = 6 kN/m²
Imposed load from Table 6.2 of EN 1991-1-1 (the category of building is C4) = 5 kN/m²

Determine the section required for beams BE and DE with Grade S275 steel.
The distribution of the floor loads to the two beams assuming two-way spanning slabs is shown in Figure 4.28:

1. Beam BE
   Service permanent action = 6 × 3 = 18 kN/m
   Service imposed load = 5 × 3 = 15 kN/m
   Factored load = 1.35 × 18 + 1.5 × 15 = 46.8 kN/m
   The diagram of actions is shown in Figure 4.28b.
   The design shear force

$$V_{Ed} = \left(\frac{93.6}{2}\right) + 35.1 = 81.9 \text{ kN}$$

The design bending moment

$$M_{Ed} = 81.9 \times 2.5 - 35.1 \times 1.5 - 46.8 \times 0.5 = 128.7 \text{ kN-m}$$

Design strength, S275 steel, thickness <16 mm

$$f_y = 275 \text{ N/mm}^2$$

Plastic modulus

$$W_{pl} = \frac{M_{Ed}}{f_y} = \frac{128.7 \times 10^6}{275} = 468 \times 10^3 \text{ mm}^3$$

Table 4.9 Deflection of beams under some conditions

Beam and load	Maximum deflection and position

$\delta = \dfrac{1}{3} \dfrac{FL^3}{EI}$ (B)

$\delta = \dfrac{1}{8} \dfrac{wL^4}{EI}$ (B)

$\delta = \dfrac{5}{384} \dfrac{wL^4}{EI}$ (B)

$\delta = \dfrac{1}{48} \dfrac{Fb\left(3L^2 - 4b^2\right)}{EI}$ $(a \geq b)$ (C)

$\delta = \dfrac{1}{48} \dfrac{FL^3}{EI}$ (C)

$\delta = \dfrac{6.81}{384} \dfrac{FL^3}{EI}$ (C)

$\delta = \dfrac{1}{384} \dfrac{wL^4}{EI}$ (C)

$\delta = \dfrac{1}{192} \dfrac{FL^3}{EI}$ (C)

$\delta = \dfrac{1}{384} \dfrac{wb\left(8L^3 - 4Lb^2 + b^3\right)}{EI}$ (C)

$\delta = \dfrac{1}{120} \dfrac{wa^2\left(16a^2 + 20ab + 5b^2\right)}{EI}$ (C)

$\delta = \dfrac{1}{120} \dfrac{wL^4}{EI}$ (C)

$\delta = \dfrac{1}{146.3} \dfrac{wL^4}{EI}$ (C)

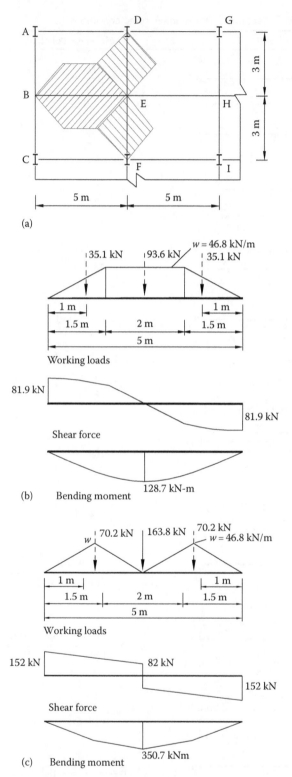

Figure 4.28 Library: (a) part floor plan and load distribution; (b) loading arrangement and internal force; (c) the loading and internal stress distribution of beam DF.

Try356 × 127UB33

The properties of the section are as follows:

$$h = 349.0 \text{ mm} \qquad b = 125.4 \text{ mm}$$

$$t_w = 6.0 \text{ mm} \qquad t_f = 8.5 \text{ mm}$$

$$h_w = h - 2t_f = 332 \text{ mm}$$

$$\frac{c_f}{t_f} = 5.82 \qquad \frac{c_w}{t_w} = 51.9$$

$$W_{el,y} = 473 \text{ cm}^3 \qquad W_{pl,y} = 543 \text{ cm}^3$$

$$I_y = 8250 \text{ cm}^4 \qquad A = 4210 \text{ mm}^2$$

$$f_y = 275 \text{ N/mm}^2$$

$$\varepsilon = \sqrt{\frac{235}{275}} = 0.924$$

Classification checks

For flanges subjected to compressive stresses,

$$\frac{c_f}{t_f} = 5.82 < 9\varepsilon = 8.32$$

For web subjected to bending moment,

$$\frac{c_w}{t_w} = 51.9 < 72\varepsilon = 66.5$$

The section is Class 1.

Shear resistance check

For rolled I and H sections, load parallel to web,

$$A_v = A - 2bt_f + (t_w + 2r)t_f \geq \eta h_w t_w$$

$$= 4210 - 2 \times 125 \cdot 4 \times 8 \cdot 5 + (6 + 2 \times 10.2) \times 8.5 \geq 1.0 \times 332 \times 6$$

$$= 2302.6 \text{ mm}^2 \geq 1992 \text{ mm}^2$$

The shear resistance of the section is given by

$$V_{pl,Rd} = \frac{A_v f_y}{\sqrt{3}\gamma_{M0}} \quad (\gamma_{M0} = 1.0)$$

$$= \frac{2302.6 \times 275}{\sqrt{3}} \times 10^{-3} = 365.6 \text{ kN}$$

$$V_{pl,Rd} > V_{Ed} = 81.9 \text{ kN}$$

Shear resistance is satisfied.

Shear buckling check
Check web-plate buckling from shear at the ultimate limit state.
Shear buckling need not be considered if

$$\frac{h_w}{t_w} \leq 72\frac{\varepsilon}{\eta}$$

$$\frac{h_w}{t_w} = \frac{332}{6} = 55.3$$

$$72\frac{\varepsilon}{\eta} = 72 \times \frac{0.924}{1.0} = 66.5 > 55.3$$

No shear buckling check is required.

Bending resistance check
The applied shear force is zero in the section with maximum bending moment (mid-span); therefore, the reduced moment resistance should not be calculated.
The design bending resistance of the cross section

$$M_{c,y,Rd} = \frac{W_{pl,y}f_y}{\gamma_{M0}} \quad (\gamma_{M0} = 1.0)$$

$$= \frac{543 \times 275}{1.0} \times 10^{-3} = 149 \text{ kN-m}$$

$$M_{c,y,Rd} > M_{Ed} = 128.7 \text{ kN-m}$$

Deflection check
The deflection due to the unfactored imposed load using formulae for Table 4.9 is

$$\delta = \frac{1}{384}\frac{wb\left(8L^3 - 4Lb^2 + b^3\right)}{EI} + \frac{1}{120}\frac{wa^2\left(16a^2 + 20ab + 5b^2\right)}{EI}$$

where

$$w = 15 \text{ kN/m}$$

$$L = 5 \text{ m} \quad a = 1.5 \text{ m} \quad b = 2 \text{ m}$$

$$\delta = \frac{1}{384}\frac{15 \times 2 \times \left(8 \times 5^3 + 4 \times 5 \times 2^2 + 2^3\right) \times 10^{12}}{210,000 \times 8,250 \times 10^4}$$

$$+ \frac{1}{120}\frac{15 \times 1.5^2 \times \left(16 \times 1.5^2 - 20 \times 1.5 \times 2 + 5 \times 2^2\right) \times 10^{12}}{210,000 \times 8,250 \times 10^4}$$

$$= 4.18 + 1.88 = 6.06 \text{ mm}$$

Assuming beam is carrying plaster or other brittle finishes,

$$\text{Limiting deflection } \delta_{\lim t} = \frac{L}{360} = \frac{5000}{360} = 13.9 \text{ mm}$$

$$\delta_{\lim t} > \delta = 6.06 \text{ mm}$$

Deflection check is satisfied.

2. Beam DF

Service permanent action = 6 × 3 = 18 kN/m
Service imposed load = 5 × 3 = 15 kN/m
Factored load w = 1.35 × 18 + 1.5 × 15 = 46.8 kN/m
  Applied shear force transferred from beam BE and EH

$$F = 2 \times 81.9 = 163.8 \text{ kN}$$

In which the imposed load = 2 × 1.5 × (15 × 1 + (1/2) × 15 × 1.5) = 78.8 kN
  The diagram of actions is shown in Figure 4.28c.
  The design shear force

$$V_{Ed} = \left(\frac{163.8}{2}\right) + \left(\frac{1}{2} \times 3 \times 46.8\right) = 152 \text{ kN}$$

The design bending moment

$$M_{Ed} = 152 \times 3 - 70.2 \times 1.5 = 350.7 \text{ kN-m}$$

Design strength, S275steel, thickness <16 mm

$$f_y = 275 \text{ N/mm}^2$$

$$\text{Plastic modulus } W_{pl} = \frac{M_{Ed}}{f_y} = \frac{350.7 \times 10^6}{275} = 1275 \times 10^3 \text{mm}^3$$

Try 457 × 152 UB 60

The properties of the section are as follows:

$h = 454.6$ mm $\qquad\qquad b = 152.9$ mm

$t_w = 8.1$ mm $\qquad\qquad t_f = 13.3$ mm

$h_w = h - 2t_f = 428$ mm

$\dfrac{c_f}{t_f} = 4.68 \qquad\qquad \dfrac{c_w}{t_w} = 50.3$

$W_{el,y} = 1120$ cm^3 $\qquad\qquad W_{pl,y} = 1290$ cm^3

$I_y = 25,500$ cm^4 $\qquad\qquad A = 7,620$ mm^2

$f_y = 275$ N/mm^2

$\varepsilon = \sqrt{\dfrac{235}{275}} = 0.924$

Classification checks

For flanges subjected to compressive stresses,

$$\frac{c_f}{t_f} = 4.68 < 9\varepsilon = 8.32$$

For web subjected to bending moment,

$$\frac{c_w}{t_w} = 50.3 < 72\varepsilon = 66.5$$

The section is Class 1.

Shear resistance check

For rolled I and H sections, load parallel to web,

$$A_v = A - 2bt_f + (t_w + 2r)t_f \geq \eta h_w t_w$$

$$= 7620 - 2 \times 152.9 \times 13.3 + (8.1 + 2 \times 10.2) \times 13.3 \geq 1.0 \times 428 \times 8.1$$

$$= 3933 \text{ mm}^2 \geq 3466.8 \text{ mm}^2$$

The shear resistance of the section is given by

$$V_{pl,Rd} = \frac{A_v f_y}{\sqrt{3}\gamma_{M0}} \quad (\gamma_{M0} = 1.0)$$

$$= \frac{3933 \times 275}{\sqrt{3}} \times 10^{-3} = 624.5 \text{ kN}$$

$$V_{pl,Rd} > V_{Ed} = 152 \text{ kN}$$

Shear resistance is satisfied.

Shear buckling check

Check web-plate buckling from shear at the ultimate limit state.
  Shear buckling need not be considered if

$$\frac{h_w}{t_w} \leq 72\frac{\varepsilon}{\eta}$$

$$\frac{h_w}{t_w} = \frac{428}{8.1} = 52.8$$

$$72\frac{\varepsilon}{\eta} = 72 \times \frac{0.924}{1.0} = 66.5 > 52.8$$

No shear buckling check is required.

Bending resistance check

$$\frac{1}{2}V_{pl,Rd} = 312.3 \text{ kN} > V_{mid} = 82 \text{ kN}$$

The reduced moment resistance should not be calculated.
   The design bending resistance of the cross section

$$M_{c,y,Rd} = \frac{W_{pl,y}f_y}{\gamma_{M0}} \quad (\gamma_{M0} = 1.0)$$

$$= \frac{1290 \times 275}{1.0} \times 10^{-3} = 355 \text{ kN-m}$$

$$M_{c,y,Rd} > M_{Ed} = 350.7 \text{ kN-m}$$

Deflection check
The deflection due to the unfactored imposed load using formulae for Table 4.9 is

$$\delta = \frac{1}{48}\frac{FL^3}{EI} + \frac{1}{146.3}\frac{wL^4}{EI}$$

where

$$w = 15 \text{ kN/m} \quad F = 78.8 \text{ kN} \quad L = 6 \text{ m}$$

$$\delta = \frac{1}{48}\frac{78.8 \times 10^3 \times 6^3 \times 10^9}{210,000 \times 25,500 \times 10^4} + \frac{1}{146.3}\frac{15 \times 6^4 \times 10^{12}}{210,000 \times 25,500 \times 10^4}$$

$$= 6.6 + 2.5 = 9.1 \text{ mm}$$

Assuming beam is carrying plaster or other brittle finishes

$$\text{Limiting deflection } \delta_{\lim t} = \frac{L}{360} = \frac{6000}{360} = 16.7 \text{ mm}$$

$$\delta_{\lim t} > \delta = 9.1 \text{ mm}$$

Deflection check is satisfied.

4.10.2 Web bearing check

A simply supported beam carrying a uniformly distributed load and a point load is shown in Figure 4.29. Assume beam is fully laterally restrained and that it sits a 130 mm bearing at each end. Select a suitable section using S275 steel to support the actions.
The applied shear force

$$V_{Ed} = \frac{1}{2} \times (100 + 10 \times 6) = 80 \text{ kN}$$

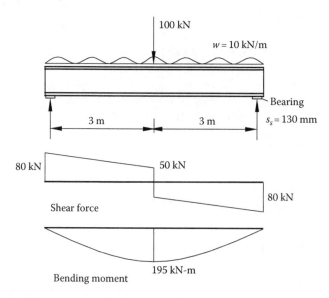

Figure 4.29 Beam subject to transverse force.

The design (maximum) bending moment

$$M_{Ed} = 80 \times 3 - 10 \times 3 \times 1.5 = 195 \text{ kN}$$

Assume suitable section belongs to Class 1. Using Grade S275 steel, $f_y = 275$ MPa, the minimum required plastic moment of resistance about the major$(y-y)$ axis, $W_{pl,y}$, is given by

$$W_{pl,y} \geq \frac{M_{Ed}}{f_y/\gamma_{M0}} = \frac{195 \times 10^3}{275/1} = 709 \text{ cm}^3$$

Try356 × l71 UB45
The properties of the section are as follows:

$h = 351.4$ mm $b = 171.1$ mm

$t_w = 7.0$ mm $t_f = 9.7$ mm

$h_w = h - 2t_f = 332$ mm

$\dfrac{c_f}{t_f} = 7.41$ $\dfrac{c_w}{t_w} = 44.5$

$W_{el,y} = 687 \text{ cm}^3$ $W_{pl,y} = 775 \text{ cm}^3$

$I_y = 12{,}100 \text{ cm}^4$ $A = 5{,}730 \text{ mm}^2$

$f_y = 275 \text{ N/mm}^2$

$\varepsilon = \sqrt{\dfrac{235}{275}} = 0.924$

$$\frac{c_f}{t_f} = 7.41 < 9\varepsilon = 8.32$$

$$\frac{c_w}{t_w} = 44.5 < 72\varepsilon = 66.5$$

The section is Class 1:

$$A_v = A - 2bt_f + (t_w + 2r)t_f \geq \eta h_w t_w$$

$$= 5730 - 2 \times 171.1 \times 9.7 + (7.0 + 2 \times 10.2) \times 9.7 \geq 1.0 \times 332 \times 7.0$$

$$= 2676.4 \text{ mm}^2 \geq 2324 \text{ mm}^2$$

The shear resistance of the section is given by

$$V_{pl,Rd} = \frac{A_v f_y}{\sqrt{3}\gamma_{M0}} \quad (\gamma_{M0} = 1.0)$$

$$= \frac{2676.4 \times 275}{\sqrt{3}} \times 10^{-3} = 736 \text{ kN}$$

$$V_{pl,Rd} > V_{Ed} = 80 \text{ kN}$$

Shear resistance is satisfied:

$$\frac{1}{2}V_{pl,Rd} = 368 \text{ kN} > V_{mid} = 80 \text{ kN}$$

The reduced moment resistance should not be calculated.
Check web-plate buckling from shear at the ultimate limit state.

$$\frac{h_w}{t_w} = \frac{332}{7} = 47.4$$

$$72\frac{\varepsilon}{\eta} = 72 \times \frac{0.924}{1.0} = 66.5 > 47.4$$

No shear buckling check is required.
The design bending resistance of the cross section

$$M_{c,y,Rd} = \frac{W_{pl,y}f_y}{\gamma_{M0}} \quad (\gamma_{M0} = 1.0)$$

$$= \frac{775 \times 275}{1.0} \times 10^{-3} = 213 \text{ kN-m}$$

$$M_{c,y,Rd} > M_{Ed} = 195 \text{ kN-m}$$

The moment resistance is satisfied.

Transverse force resistance check

From Figure 4.29, $s_s = 130$ mm, $c = 0$ and the type of load application is type (c).
The buckling coefficient for the load application is given by

$$k_F = 2 + 6\left(\frac{s_s + c}{h_w}\right) \leq 6$$

$$= 2 + 6\left(\frac{130 + 0}{332}\right) = 4.3 \leq 6$$

Effective load length

$$l_e = \frac{k_F E t_w^2}{2 f_{yw} h_w} \leq s_s + c$$

$$= \frac{4.3 \times 210,000 \times 7^2}{2 \times 275 \times 332} \leq 130 \text{ mm}$$

$$= 242 \text{ mm}$$

Hence, use $l_e = 130$ mm
The critical force F_{cr} is given by

$$F_{cr} = \frac{0.9 k_F E t_w^3}{h_w} = \frac{0.9 \times 4.3 \times 210,000 \times 7^3}{332} \times 10^{-3} = 840 \text{ kN}$$

The dimensionless parameters

$$m_1 = \frac{f_{yb} b_f}{f_{yw} t_w} = \frac{275 \times 171.1}{275 \times 7.0} = 24.4$$

Assuming $\overline{\lambda}_F > 0.5$,

$$m_2 = 0.02\left(\frac{h_w}{t_f}\right)^2 = 0.02\left(\frac{332}{9.7}\right)^2 = 23.4$$

For load type (c),

$$l_{y1} = l_e + t_f \sqrt{\frac{m_1}{2} + \left(\frac{l_e}{t_f}\right)^2} + m_2$$

$$= 130 + 9.7\sqrt{\frac{24.4}{2} + \left(\frac{130}{9.7}\right)^2} + 23.4 = 272.3 \text{ mm}$$

$$l_{y2} = l_e + t_f \sqrt{m_1 + m_2}$$

$$= 130 + 9.7\sqrt{24.4 + 23.4} = 197 \text{ mm}$$

Hence, $l_y = \min \{l_{y1}:l_{y2}\} = 197$ mm

$$\bar{\lambda}_F = \sqrt{\frac{l_y t_w f_{yw}}{F_{cr}}} = \sqrt{\frac{197 \times 7 \times 275}{840 \times 10^3}} = 0.67 > 0.5 \text{ Satisfactory}$$

Reduction factor $\chi_F = \dfrac{0.5}{\bar{\lambda}_F} = \dfrac{0.5}{0.67} = 0.75 < 1$

The effective length L_{eff} is given by

$$L_{eff} = \chi_F l_y = 0.75 \times 197 = 148 \text{ mm}$$

The resistance to transverse force

$$F_{Rd} = \frac{f_{yw} L_{eff} t_w}{\gamma_{M1}} = \frac{275 \times 148 \times 7}{1.0} \times 10^{-3} = 285 \text{ kN}$$

$$F_{Rd} > V_{Ed} = 195 \text{ kN}$$

The resistance to transverse force is satisfied.

4.10.3 Beam with unrestrained compression flange

Design the simply supported primary beam for the loading shown in Figure 4.30. The beam is required to span 12 m and to support two secondary beams. The beam ends and two second beam joints are restrained against torsion with the compression flange free to rotate in plan. The compression flange is unrestrained between supports. Use Grade S275 steel.

Factored actions

At B: $1.35 \times 100 + 1.5 \times 70 = 240$ kN

At C: $1.35 \times 60 + 1.5 \times 50 = 156$ kN

The design shear force

$$V_{Ed} = \frac{1}{12}(240 \times 9 + 156 \times 5) = 245 \text{ kN}$$

The design bending moment

$$M_{Ed} = 151 \times 5 = 755 \text{ kN-m}$$

The diagram of actions is shown in Figure 4.30.
  LTB checks will be carried out on the critical segments BC and CD.
  The check for shear resistance and deflection is dropped.

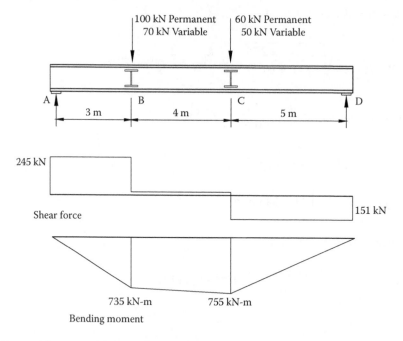

Figure 4.30 Beam with unrestrained compression flange.

Try 610 × 229 UB 140
The properties of the section are as follows:

$h = 617.2$ mm $\qquad b = 230.2$ mm

$t_w = 13.1$ mm $\qquad t_f = 22.1$ mm

$\dfrac{c_f}{t_f} = 4.34$ $\qquad \dfrac{c_w}{t_w} = 41.8$

$W_{el,y} = 3620$ cm^3 $\qquad W_{pl,y} = 4140$ cm^3

$I_y = 112{,}000$ cm^4 $\qquad I_z = 4{,}510$ cm^4

$I_w = 3.99$ dm^6 $\qquad I_T = 216$ cm^4

$t_f = 22.1$ mm >16 mm

Hence, $f_y = 265$ N/mm^2 $\quad \varepsilon = \sqrt{\dfrac{235}{265}} = 0.94$

$\dfrac{c_f}{t_f} = 4.34 < 9\varepsilon = 8.46$

$\dfrac{c_w}{t_w} = 41.8 < 72\varepsilon = 67.68$

The section is Class 1.

The design bending resistance of the cross section

$$M_{c,y,Rd} = \frac{W_{pl,y}f_y}{\gamma_{M0}} \quad (\gamma_{M0} = 1.0)$$

$$= \frac{4140 \times 265}{1.0} \times 10^{-3} = 1097 \text{ kN-m}$$

$$M_{c,y,Rd} > M_{Ed} = 755 \text{ kN-m}$$

Lateral-torsional buckling check for segment BC: General case
ψ is the ratio of end moments: $\psi = 735/755 = 0.97$

$$C_1 = 1.88 - 1.4\psi + 0.52\psi^2 = 1.01$$

$$L_{cr} = 4000 \text{ mm}$$

The critical moment for segment BC

$$M_{cr} = C_1 \frac{\pi^2 EI_z}{L_{cr}^2} \sqrt{\frac{I_W}{I_Z} + \frac{L_{cr}^2 GI_T}{\pi^2 EI_z}}$$

$$= 1.01 \frac{\pi^2 \times 210,000 \times 451 \times 10^5}{4,000^2} \sqrt{\frac{3.99 \times 10^{12}}{451 \times 10^5} + \frac{4,000^2 \times 81,000 \times 216 \times 10^4}{\pi^2 \times 210,000 \times 451 \times 10^5}} \times 10^{-6}$$

$$= 2029 \text{ kN-m}$$

Non-dimensional lateral-torsional slenderness for segment BC

$$\bar{\lambda}_{LT} = \sqrt{\frac{W_y f_y}{M_{cr}}} = \sqrt{\frac{4140 \times 265 \times 10^{-3}}{2029}} = 0.74$$

$$\bar{\lambda}_{LT} = 0.74 > \bar{\lambda}_{LT,0} = 0.4$$

So LTB must be checked.
Using Table 4.5,

$$\frac{h}{b} = \frac{617.2}{230.2} = 2.68 > 2$$

For a rolled I section, use buckling curve b.
For buckling curve b, $\alpha_{LT} = 0.34$ from Table 4.4:

$$\phi_{LT} = 0.5\left[1 + \alpha_{LT}\left(\bar{\lambda}_{LT} - 0.2\right) + \bar{\lambda}_{LT}^2\right]$$

$$= 0.5\left[1 + 0.34 \times (0.74 - 0.2) + 0.74^2\right] = 0.87$$

Reduction factor for LTB χ_{LT}

$$\chi_{LT} = \frac{1}{\phi_{LT} + \sqrt{\phi_{LT}^2 - \overline{\lambda}_{LT}^2}} \qquad \chi_{LT} \leq 1.0$$

$$= \frac{1}{0.87 + \sqrt{0.87^2 - 0.74^2}} = 0.75$$

$$M_{b,y,Rd} = \chi_{LT} \frac{W_{pl,y}f_y}{\gamma_{M1}} = 0.75 \times \frac{4140 \times 265}{1.0} \times 10^{-3} = 823 \text{ kN-m}$$

$$M_{b,y,Rd} > M_{Ed} = 755 \text{ kN-m}$$

The LTB resistance for segment BC is satisfied.

Lateral-torsional buckling check for segment BC: Rolled or equivalent welded sections case
For segment BC,

$\overline{\lambda}_{LT,0} = 0.4$ (maximum value)

$\beta = 0.75$ (maximum value)

$\overline{\lambda}_{LT} = 0.74$

Using Table 4.7,

$$3.1 > \frac{h}{b} = \frac{617.2}{230.2} = 2.68 > 2$$

For a rolled I section, use buckling curve c.
   For buckling curve c, $\alpha_{LT} = 0.49$ from Table 4.4:

$$\phi_{LT} = 0.5\left[1 + \alpha_{LT}\left(\overline{\lambda}_{LT} - \overline{\lambda}_{LT,0}\right) + \beta\overline{\lambda}_{LT}^2\right]$$

$$= 0.5\left[1 + 0.49 \times (0.75 - 0.4) + 0.75 \times 0.75^2\right] = 0.8$$

Reduction factor for LTB χ_{LT}

$$\chi_{LT} = \frac{1}{\phi_{LT} + \sqrt{\phi_{LT}^2 - \beta\overline{\lambda}_{LT}^2}} \qquad \text{but } \chi_{LT} \leq \begin{cases} 1.0 \\ \dfrac{1}{\overline{\lambda}_{LT}^2} \end{cases}$$

$$= \frac{1}{0.8 + \sqrt{0.8^2 - 0.75 \times 0.74^2}} = 0.79 \leq \begin{cases} 1.0 \\ 1.78 \end{cases}$$

$$M_{b,y,Rd} = \chi_{LT}\frac{W_{pl,y}f_y}{\gamma_{M1}} = 0.79 \times \frac{4140 \times 265}{1.0} \times 10^{-3} = 867 \text{ kN-m}$$

$$M_{b,y,Rd} > M_{Ed} = 755 \text{ kN-m}$$

The LTB resistance for segment BC is satisfied.
Rolled and equivalent welded sections method produces a less conservative result.

Lateral-torsional buckling check for segment BC: Simplified method
For Class 1 cross sections, $i_{f,z}$ can be taken as

$$i_{f,z} = \sqrt{\frac{I_{z,comp.f+(1/3)comp.w}}{A_{z,comp.f+(1/3)comp.w}}}$$

$$h_w = h - 2t_f = 573 \text{ mm}$$

$$\begin{aligned} I_{z,comp.f+(1/3)comp.w} &= \frac{t_f b^3}{12} + \frac{1/3(h_w/2)t_w^3}{12} \\ &= \frac{22.1 \times (230.2)^3}{12} + \frac{[(1/3) \times (573/2)](13.1)^3}{12} = 2248.4 \text{ cm}^4 \end{aligned}$$

$$\begin{aligned} A_{z,comp.f+(1/3)comp.w} &= t_f b + \frac{1}{3}\left(\frac{h_w}{2}\right)t_w \\ &= 22.1 \times 230.2 + \frac{1}{3}\left(\frac{573}{2}\right)13.1 = 6338 \text{ mm}^2 \end{aligned}$$

The radius of gyration of the equivalent compression parts

$$i_{f,z} = \sqrt{\frac{I_{z,comp.f+(1/3)comp.w}}{A_{z,comp.f+(1/3)comp.w}}} = \sqrt{\frac{2248.4 \times 10^4}{6338}} = 59.6 \text{ mm}$$

The appropriate section modulus of compression flange

$$W_y = A_f.y = (t_f b)(h - t_f) = (22.1 \times 230.2) \times (617.2 - 22.1) = 3027.5 \text{ cm}^3$$

$$M_{c,Rd} = \frac{W_y f_y}{\gamma_{M1}} = \frac{3027.5 \times 265}{1} \times 10^{-3} = 802 \text{ kN-m}$$

$$M_{y,Rd} = 1097 \text{ kN-m}$$

$$\lambda_1 = 93.9\varepsilon = 93.9 \times 0.94 = 88.3$$

Correction factor k_c is taken from Table 4.8:

$$k_c = \frac{1}{1.33 - 0.33\psi} = \frac{1}{1.33 - 0.33(735/755)} = 0.99$$

Slenderness limit of the equivalent compression flange

$$\bar{\lambda}_{c0} = \bar{\lambda}_{LT,0} + 0.1 = 0.5$$

The resulting slenderness

$$\bar{\lambda}_f = \frac{k_c l_c}{i_{f,z} \lambda_1} \leq \bar{\lambda}_{co} \frac{M_{c,Rd}}{M_{y,Rd}}$$

$$= \frac{0.99 \times 4000}{59.6 \times 88.3} \leq 0.5 \times \frac{802}{1097}$$

$$= 0.75 \leq 0.37$$

Hence, LTB must be checked.

For rolled sections, buckling curve c is used to calculate the reduction factor χ (from Figure 4.16 or Figure 6.4 in EN 1993-1-1):

$$\bar{\lambda}_f = 0.75$$

$$\phi = 0.5\left(1 + 0.49(0.75 - 0.2) + 0.75^2\right) = 0.92$$

$$\chi = \frac{1}{\phi + \sqrt{\phi^2 - \bar{\lambda}^2}}$$

$$= \frac{1}{0.92 + \sqrt{0.92^2 - 0.75^2}} = 0.69$$

$$K_{fl} = 1.0 \text{ From UK National Annex}$$

The design buckling resistance moment

$$M_{b,Rd} = k_{fl} \chi M_{c,Rd} = 1.0 \times 0.69 \times 802 = 553 \text{ kN}$$

$$M_{b,Rd} = 553 \text{ kN} < M_{Ed} = 755 \text{ kN}$$

LTB check for segment BC is not satisfied.

The simplified method gives a very conservative result.

Lateral-torsional buckling check for segment CD: General case
From Table 4.2,

$$C_1 = 1.88$$

$$L_{cr} = 5000 \text{ mm}$$

The critical moment for segment CD

$$M_{cr} = C_1 \frac{\pi^2 E I_z}{L_{cr}^2} \sqrt{\frac{I_w}{I_z} + \frac{L_{cr}^2 G I_T}{\pi^2 E I_z}}$$

$$= 1.88 \frac{\pi^2 \times 210,000 \times 451 \times 10^5}{5,000^2} \sqrt{\frac{3.99 \times 10^{12}}{451 \times 10^5} + \frac{5,000^2 \times 81,000 \times 216 \times 10^4}{\pi^2 \times 210,000 \times 451 \times 10^5}} \times 10^{-6}$$

$$= 2585 \text{ kN-m}$$

Non-dimensional lateral-torsional slenderness for segment CD

$$\bar{\lambda}_{LT} = \sqrt{\frac{W_y f_y}{M_{cr}}} = \sqrt{\frac{4140 \times 265 \times 10^{-3}}{2585}} = 0.65$$

$$\bar{\lambda}_{LT} = 0.65 > \bar{\lambda}_{LT,0} = 0.4$$

So LTB must be checked.
Using Table 4.5,

$$\frac{h}{b} = 2.68 > 2$$

For a rolled I section, use buckling curve b.
For buckling curve b, $\alpha_{LT} = 0.34$ from Table 4.4:

$$\phi_{LT} = 0.5\left[1 + \alpha_{LT}\left(\bar{\lambda}_{LT} - 0.2\right) + \bar{\lambda}_{LT}^2\right]$$

$$= 0.5\left[1 + 0.34 \times (0.65 - 0.2) + 0.65^2\right] = 0.79$$

Reduction factor for LTB χ_{LT}

$$\chi_{LT} = \frac{1}{\phi_{LT} + \sqrt{\phi_{LT}^2 - \bar{\lambda}_{LT}^2}} \qquad \chi_{LT} \leq 1.0$$

$$= \frac{1}{0.79 + \sqrt{0.79^2 - 0.65^2}} = 0.81$$

$$M_{b,y,Rd} = \chi_{LT} \frac{W_{pl,y} f_y}{\gamma_{M1}} = 0.81 \times \frac{4140 \times 265}{1.0} \times 10^{-3} = 889 \text{ kN-m}$$

$$M_{b,y,Rd} > M_{Ed} = 755 \text{ kN-m}$$

The LTB resistance for segment BC is satisfied.

Lateral-torsional buckling check for segment CD: Rolled or equivalent welded sections case
For segment CD,

$$\bar{\lambda}_{LT,0} = 0.4 \text{ (maximum value)}$$

$$\beta = 0.75 \text{ (maximum value)}$$

$$\bar{\lambda}_{LT} = 0.65$$

Using Table 4.7,

$$3.1 > \frac{h}{b} = 2.68 > 2$$

For a rolled I section, use buckling curve *c*.

For buckling curve *c*, $\alpha_{LT} = 0.49$ from Table 4.4:

$$\phi_{LT} = 0.5\left[1 + \alpha_{LT}(\bar{\lambda}_{LT} - \bar{\lambda}_{LT,0}) + \beta\bar{\lambda}_{LT}^2\right]$$

$$= 0.5\left[1 + 0.49 \times (0.65 - 0.4) + 0.75 \times 0.65^2\right] = 0.72$$

Reduction factor for LTB χ_{LT}

$$\chi_{LT} = \frac{1}{\phi_{LT} + \sqrt{\phi_{LT}^2 - \beta\bar{\lambda}_{LT}^2}} \quad \text{but} \quad \chi_{LT} \leq \begin{cases} 1.0 \\ \dfrac{1}{\bar{\lambda}_{LT}^2} \end{cases}$$

$$= \frac{1}{0.72 + \sqrt{0.72^2 - 0.75 \times 0.65^2}} = 0.86 \leq \begin{cases} 1.0 \\ 2.37 \end{cases}$$

$$M_{b,y,Rd} = \chi_{LT}\frac{W_{pl,y}f_y}{\gamma_{M1}} = 0.86 \times \frac{4140 \times 265}{1.0} \times 10^{-3} = 943 \text{ kN-m}$$

$$M_{b,y,Rd} > M_{Ed} = 755 \text{ kN-m}$$

The LTB resistance for segment CD is satisfied.
Again rolled and equivalent welded sections method produces a less conservative result.

4.10.4 Beam subjected to bending about two axes

A beam of span 6 m with simply supported ends restrained against torsion has its major principal axis inclined at 30° to the horizontal, as shown in Figure 4.31. The beam is supported at its ends on sloping roof girders. If the beam is 457 × 152 UB 52 with Grade S355 steel, find the maximum factored load that can be carried at the centre. The load is applied by slings to the top flange.

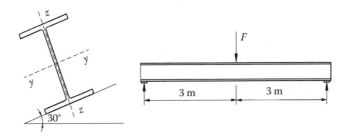

Figure 4.31 Beam in biaxial bending.

Let the centre factored load = FkN. The beam self-weight is ignored:

Moments: $M_{y,Ed} = \left(F \times \dfrac{6}{4}\right)\cos 30 = 1.3F$ kN-m

$M_{z,Ed} = \left(F \times \dfrac{6}{4}\right)\sin 30 = 0.75F$ kN-m

Properties for 457 × 152 UB 52

$h = 449.8$ mm $\qquad b = 152.4$ mm

$t_w = 7.6$ mm $\qquad t_f = 10.9$ mm

$\dfrac{c_f}{t_f} = 5.71 \qquad \dfrac{c_w}{t_w} = 53.6$

$W_{pl,y} = 1,100$ cm^3 $\quad W_{pl,z} = 133$ cm^3

$I_y = 21,400$ cm^4 $\quad I_z = 645$ cm^4

The section is Class 1. The design strength $f_y = 355$ MPa.
For cross section where there are no fastener holes,

$$M_{N,y,Rd} = M_{pl,y,Rd} \dfrac{1-n}{1-0.5a}$$

$$n = \dfrac{N_{Ed}}{N_{pl,Rd}} = 0$$

$$a = \dfrac{A-2bt_f}{A} = \dfrac{6660 - 2\times152.4\times10.9}{6660} = 0.5$$

$$\dfrac{1-n}{1-0.5a} = \dfrac{1}{1-0.5\times0.5} = 1.33 \quad \text{but} \quad \le1.0$$

Hence,

$$M_{N,y,Rd} = M_{pl,y,Rd} = \frac{W_{pl,y}f_y}{\gamma_{M0}}$$

$$= \frac{1100 \times 355}{1.0} \times 10^{-3} = 391 \text{ kN}$$

$$M_{N,z,Rd} = M_{pl,z,Rd} = \frac{W_{pl,z}f_y}{\gamma_{M0}}$$

$$= \frac{133 \times 355}{1.0} \times 10^{-3} = 47 \text{ kN}$$

For biaxial bending, the criterion is used:

$$\left(\frac{M_{y,Ed}}{M_{N,y,Rd}}\right)^{\alpha} + \left(\frac{M_{z,Ed}}{M_{N,z,Rd}}\right)^{\beta} \leq 1$$

For I section,

$\alpha = 2$

$\beta = 5n = 0$ but $\beta \geq 1$

Hence, $\beta = 1$

$$\left(\frac{M_{y,Ed}}{M_{N,y,Rd}}\right)^{\alpha} + \left(\frac{M_{z,Ed}}{M_{N,z,Rd}}\right)^{\beta} = \left(\frac{1.3F}{391}\right)^2 + \left(\frac{0.75F}{47}\right)^1 \leq 1$$

$F \leq 60 \text{ kN}$

Therefore, the maximum factored load that the beam can be carried at the centre is 60 kN.

4.11 CRANE BEAMS

4.11.1 Types and uses

Crane beams carry hand-operated or electric overhead cranes in industrial buildings such as factories, workshops and steelworks. Types of beams used are shown in Figure 4.32a and b. These beams are subjected to vertical and horizontal loads due to the weight of the crane, the hook load and the dynamic loads. Because the beams are subjected to horizontal loading, a larger flange or a horizontal beam is provided at the top on all but beams for very light cranes.

Light crane beams consist of a UB only or of a UB and channel, as shown in Figure 4.32a. Heavy cranes require a plate girder with surge girder, as shown in Figure 4.32b. Only light crane beams are considered in this book.

Some typical crane rails and the fixing of a rail to the top flange are shown in Figure 4.32c. If a crane rail is rigidly fixed to the top flange of the runway beam, it may be included as part

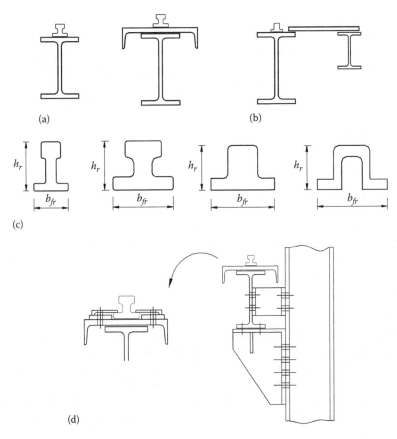

Figure 4.32 Type of crane beams and rails and connection to column: (a) light crane beam; (b) heavy crane girder; (c) crane rails; (d) fixing of crane beam.

of the cross section that is taken into account to calculate the resistance. To allow for wear, the normal height of the rail should be reduced when calculating the cross-sectional properties. This reduction should generally be taken as 25% of the minimum nominal thickness t_r below the wearing surface (see Figure 4.32c). The connection of a crane girder to the bracket and column is shown in Figure 4.32d. The size of crane rails depends on the capacity and use of the crane.

4.11.2 Crane data

Crane data can be obtained from the manufacturer's literature. The data required for crane beam design are

Span between crane rails: l_c
Weight of crane: Q_c
Hoist load: Q_h
End carriage wheel centres: χ_w
Minimum hook approach: e_{min}
Maximum static wheel load: $Q_{r,max}$
Minimum static wheel load: $Q_{r,min}$
Span of crane runway beam: l

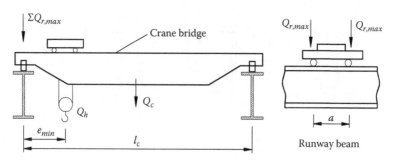

Figure 4.33 Crane data and load arrangement.

The crane data and load arrangement are shown in Figure 4.33.

1. Actions induced by cranes on runway beams

   A crane runway beam is subjected to general actions for which the relevant parts of EN 1991-1 are presumed. In addition, crane actions act on the crane beam according to EN 1991-3. The actions associated with cranes on runway beams include variable and accidental actions.

   Variable actions

   Variable actions result from variation in time and location. These actions are usually separated in vertical crane actions caused by gravity loads and hoist loads and in horizontal crane actions caused by acceleration or deceleration and by skewing and other dynamic effects. The different actions induced by cranes are taken into account by combining them in groups of loads according to Table 4.10. Each of these groups of loads defines one crane action for the combination of non-crane loads.

   Accidental actions

   Cranes can generate accidental actions due to collision with buffers or collision of lifting attachments with obstacles. These actions should be considered for the structural design where appropriate protection is not provided.

   For the plate buckling verifications of the beam, the vertical forces induced by the self-weight and hoist load are relevant. Usually load group 1 governs the design under wheel load.

Table 4.10 Recommended values of γ-factors

Action			Symbol	Situation P/T	Situation A
Permanent crane actions	Unfavourable		γ_{Gsup}	1.35	1.00
	Favourable		γ_{Ginf}	1.00	1.00
Variable crane actions	Unfavourable		γ_{Qsup}	1.35	1.00
	Favourable	Crane present	γ_{Qinf}	1.00	1.00
		Crane not present		0.00	0.00
Other variable actions	Unfavourable		γ_Q	1.00	1.00
	Favourable			0.00	0.00
Accidental actions			γ_A	—	1.00

P, persistent situation; *T*, transient situation; *A*, accidental situation.

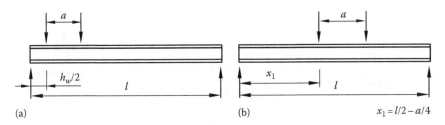

Figure 4.34 Rolling loads: (a) maximum shear; (b) maximum moment.

At ultimate limit sates, the partial factors on actions in the persistent, transient and accidental design situations should be defined. For the design runway beams, the values of γ-factors are given in Table 4.10.

2. Maximum shear and moment

The wheel loads are rolling loads and must be placed in position to give maximum shear and moment. For two equal wheel loads

a.   The maximum shear occurs when one load is nearly over a support.

b.   The maximum moment occurs when the centre of gravity of the loads and one load are placed equidistant about the centreline of the girder. The maximum moment occurs under the wheel load nearest the centre of the girder.

The load cases are shown in Figure 4.34.

Note that if the spacing between the loads is greater than 0.586 of the span of the beam, the maximum moment will be given by placing one wheel load at the centre of the beam.

4.11.3 Crane beam design

1. Local vertical compressive stresses

The local vertical compressive stresses $\sigma_{oz,Ed}$ at the underside of the top flange should be determined from

$$\sigma_{oz,Ed} = \frac{F_{z,Ed}}{l_{eff}t_w} \tag{4.38}$$

where

$F_{z,Ed}$ is the design value of the wheel load

$F_{z,Ed} = \gamma_Q Q_{r,max}$

γ_Q is the partial factor for variable actions

l_{eff} is the effective loaded length (see Figure 4.35)

t_w is the thickness of the web plate

The effective loaded length l_{eff} over which the local vertical stress $\sigma_{oz,Ed}$ due to a single wheel load may be assumed to be uniformly distributed and should be calculated according to Table 4.11.

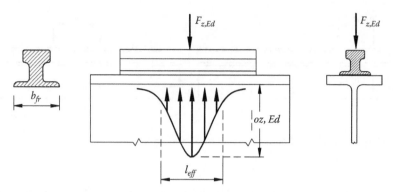

Figure 4.35 Effective loaded length l_{eff}.

Table 4.11 Effective loaded length l_{eff}

Case	Description of crane rail	Effective loaded length
(a)	Rigidly fixed to the flange	$l_{eff} = 3.25\sqrt[3]{\dfrac{I_{rf}}{t_w}}$
(b)	Not rigidly fixed to the flange	$l_{eff} = 3.25\sqrt[3]{\dfrac{I_r + I_{f,eff}}{t_w}}$
(c)	Mounted on a suitable resilient elastomeric bearing pad at least 6 mm thick	$l_{eff} = 4.25\sqrt[3]{\dfrac{I_r + I_{f,eff}}{t_w}}$

where
$I_{f,eff}$ is the second moment of area of a flange with an effective width b_{eff}

$$I_{eff} = \frac{b_{eff} t_f^3}{12}$$

I_r is the second moment of area of the rail
I_{rf} is the second moment of the combined cross section comprising
the rail and a flange with an effective width b_{eff}
t_w is the web thickness

$$b_{eff} = b_{fr} + h_f + t_f \leq b$$

where
b is the overall width of the top flange
b_{fr} is the width of the foot of the rail
h_r is the height of the rail
t_f is the flange thickness

2. Local bending stresses in the bottom flange
The bending stresses due to wheel loads applied at locations more than b from the end of the beam can be determined at three locations indicated in Figure 4.36
   - Location 0: the web-to-flange transition
   - Location 1: centreline of the wheel load
   - Location 2: outside edge of the flange

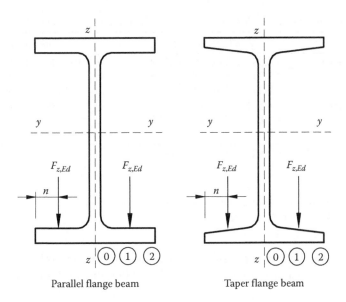

Figure 4.36 Locations for determining stresses due to wheel loads.

If the distance χ_w is not less than 1.5b, the local longitudinal bending stress $\sigma_{ox,Ed}$ and transverse bending stress $\sigma_{oy,Ed}$ in the bottom flange due to the application of a wheel load should be obtained from

$$\sigma_{ox,Ed} = c_x \frac{F_{z,Ed}}{t_1^2}$$

$$\sigma_{oy,Ed} = c_y \frac{F_{z,Ed}}{t_1^2}$$

(4.39)

where
  $F_{z,Ed}$ is the vertical crane wheel load
  t_1 is the thickness of the flange at the centreline of the wheel load
  c_x and c_y are the coefficients for determining the longitudinal and transverse bending stresses at locations 0, 1 and 2

The coefficients may be determined from Table 4.12.
3. Resistance of cross section
  The verifications about moment capacity, shear capacity and web bearing capacity are similar to the common beam as described in the previous subsection.
4. Buckling resistance of member
  The LTB resistance of a simply supported runway beam may be verified by checking the compression flange plus one-fifth of the web against flexural buckling as a compression member. It should be checked for an axial compressive force equal to the bending moment due to the vertical actions, divided by the depth between the centroid of the flanges.

Table 4.12 Coefficients c_{xi} and c_{yi} for calculating stresses at points $i = 0$, 1 and 2

Stress	Parallel flange beams	Taper flange beam
Longitudinal bending stress $\sigma_{ox,Ed}$	$c_{x0} = 0.05 - 0.58\mu + 0.148e^{3.015\mu}$ $c_{x1} = 2.23 - 1.49\mu + 1.39e^{-18.33\mu}$ $c_{x2} = 0.73 - 1.58\mu + 2.91e^{-6.0\mu}$	$c_{x0} = -0.981 - 1.479\mu + 1.12e^{1.322\mu}$ $c_{x1} = 1.81 - 1.15\mu + 1.06e^{-7.7\mu}$ $c_{x2} = 1.99 - 2.81\mu + 0.84e^{-4.69\mu}$
Transverse bending stress $\sigma_{oy,Ed}$	$c_{y0} = -2.11 + 1.977\mu + 0.0076e^{6.53\mu}$ $c_{y1} = 10.108 - 7.408\mu - 10.108e^{-1.364\mu}$ $c_{y2} = 0$	$c_{y0} = -1.096 + 1.095\mu + 0.192e^{-6.0\mu}$ $c_{y1} = 3.965 - 4.835\mu - 3.965e^{-2.675\mu}$ $c_{y2} = 0$

Note: The coefficients for taper flange beams are for a slope of 14% or 8°

$$\mu = \frac{2n}{b - t_w}$$

where

  n is the distance from the centreline of the wheel load to the flange edge
  t_w is the thickness of the web

The NA of Singapore gives an alternative assessment method.

Members that are subjected to combined bending and torsion should satisfy

$$\frac{M_{y,Ed}}{\chi_{LT}M_{y,Rk}/\gamma_{M1}} + \frac{C_{mz}M_{z,Ed}}{M_{z,Rk}/\gamma_{M1}} + \frac{k_w k_{zw} k_\alpha B_{Ed}}{B_{Rk}/\gamma_{M1}} \leq 1 \tag{4.40}$$

where

  C_{mz} is the equivalent uniform moment factor for bending about the z–z axis, according to EN 1993-1-1 Table B.3

$$k_w = 0.7 - \frac{0.2B_{Ed}}{B_{Rk}/\gamma_{M1}}$$

$$k_{zw} = 1 - \frac{M_{z,Ed}}{M_{z,Rk}/\gamma_{M1}}$$

$$k_\alpha = \frac{1}{1 - \left(M_{y,Ed}/M_{y,cr}\right)}$$

  $M_{y,Rk}$ and $M_{z,Rk}$ are the characteristic values of the resistance moment of the cross section about its y–y and z–z axis, respectively, from EN 1993-1-1 Table 6.7
  $M_{y,cr}$ is the elastic critical LTB moment about y–y axis
  B_{Ed} is the design value of the warping torsional moment
  B_{Rk} is the characteristic value of the warping torsional resistance moment
  χ_{LT} is the reduction factor for LTB, which is determined for rolled or equivalent welded sections with equal flanges

4.11.4 Limits for deformation and displacements

The vertical deflection limitations for crane beams are given in Table 4.13.

The formula for deflection at the centre of the beam is given in Figure 4.37 for crane wheel loads placed in the position to give the maximum moment. The deflection should

Table 4.13 Limiting values of vertical deflections

	Description	Deflection
(a)	Vertical deflection due to vertical loads	$\delta_z \leq L/600$ and $\delta_z \leq 25$ mm
(b)	The difference between the vertical deformations of two beams forming a crane runway	$\Delta h_c \leq s/600$ s is the span between two beams.
(c)	Vertical deflection for a monorail hoist block, relative to its supports, due to the payload only	$\delta_{pay} \leq L/500$

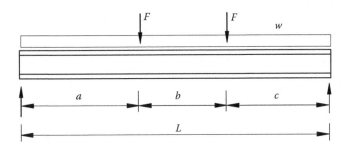

Figure 4.37 Crane beam deflection.

also be checked with the loads placed equidistant about the centre of the beam, when $a = c$ in the formula given:

$$\delta_{max} = \frac{FL^3}{48EI}\left[\frac{3(a+c)}{L} - \frac{4(a^3+c^3)}{L^3}\right] + \frac{5wL^4}{384EI} \tag{4.41}$$

4.11.5 Design of a crane beam

Design a simply supported beam to carry an electric overhead crane, as shown in Figure 4.38. The design data are as follows:

Span between crane rails = 20 m
Weight of crane = 180 kN
Hoist load = 250 kN
Minimum hook approach = 1.0 m
Wheel base = 3.0 m

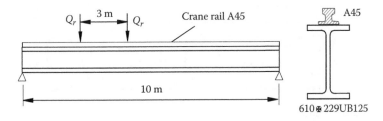

Figure 4.38 Crane runway beam and crane rail.

Use 610 × 229UB125 with Grade S275 steel and crane rail with foot flange, type A45 according to DIN 536-1. The section of the runway beam is shown in Figure 4.38. The self-weight of the crane runway beam with rail is 1.47 kN/m.

Section properties

Crane rail

$h_r = 55$ mm $\qquad\qquad b_{fr} = 125$ mm

610 × 229UB125

$h = 612.2$ mm $\qquad\qquad b = 229$ mm

$t_w = 11.9$ mm $\qquad\qquad t_f = 19.6$ mm

$h_w = h - 2t_f = 573$ mm

$W_{pl,y} = 3,680$ cm^3 $\qquad\qquad I_y = 98,600$ cm^4

Maximum wheel loads, moments and shear

The crane loads are shown in Figure 4.39. The maximum characteristic wheel loads for one wheel at A

$$Q_{r,max} = \frac{1}{2} \times \frac{1}{20}(250 \times 19 + 180 \times 10) = 163.75 \text{ kN}$$

The maximum internal force induced by the wheel loads occurs at different locations since it is a moving load. The location of the maximum bending moment $x_1 = l/2 - a/4 = 10/2 - 3/4 = 4.25$ m (see Figure 4.40).

The maximum shear force is determined for the wheel-load position at $x_2 = h_w/2 = 0.573/2 \approx 0.3$ m next to the support, as shown in Figure 4.41.

The characteristic bending moment and shear force distribution due to self-weight of the crane runway beam are shown in Figure 4.42.

Classification of the cross section

The thickness of the flanges is larger than 16 mm; hence, the yield strength of S275 steel is $f_y = 265$ MPa

$$\varepsilon = \sqrt{\frac{235}{265}} = 0.94$$

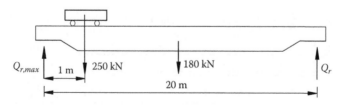

Figure 4.39 Crane beam loads.

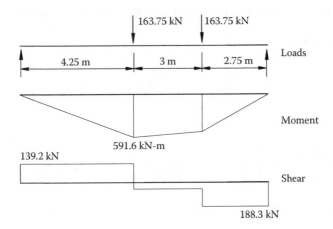

Figure 4.40 Unfactored internal forces due to wheel loads at x_1 = 4.25 m.

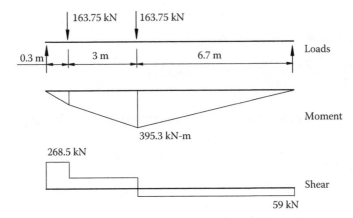

Figure 4.41 Unfactored internal forces due to wheel loads at x_1 = 0.3 m.

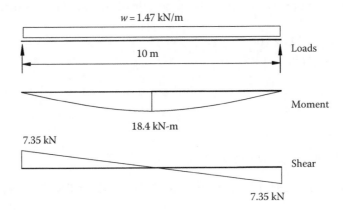

Figure 4.42 Unfactored internal forces due to self-weight.

The classification of cross sections is according to Section 5.5 of EN 1993-1-1.

$$\frac{c_f}{t_f} = 4.89 < 9\varepsilon = 8.5$$

$$\frac{c_w}{t_w} = 46.0 < 72\varepsilon = 67.7$$

The whole cross section is Class 1.

Transverse forces and stresses
The transverse force $F_{z,Ed}$ exerted on the crane runway beam is the design value of the maximum wheel load $Q_{r,max}$:

$$F_{z,Ed} = \gamma_Q Q_{r,max} = 1.35 \times 163.75 = 221\,\text{kN}$$

Taking wear into account, $h_{r,wear} = 50$ mm, the effective flange width is

$$b_{eff} = b_{fr} + h_{r,wear} + t_f \leq b$$

$$= 125 + 50 + 19.6 = 194.6\,\text{mm} < 229\,\text{mm}$$

The second moment of inertia of the flange is

$$I_{f,eff} = \frac{b_{eff}t_f^3}{12} = \frac{194.6 \times 19.6^3}{12} \times 10^{-4} = 146.5\,\text{cm}^4$$

The crane rail does not rigidly fix to flanges, and the effective loaded length becomes

$$l_{eff} = 3.25\left(\frac{I_r + I_{f,eff}}{t_w}\right)^{\frac{1}{3}}$$

$$= 3.25\left(\frac{90 \times 10^4 + 146.5 \times 10^4}{11.9}\right)^{\frac{1}{3}} = 189.7\,\text{mm}$$

The local vertical compressive stress at the underside of the top flange is determined by

$$\sigma_{z,Ed} = \frac{F_{z,Ed}}{l_{eff}t_w} = \frac{221 \times 10^3}{189.7 \times 11.9} = 97.9\,\text{N/mm}^2$$

Resistance to shear forces
For I sections,

$$A_v = A - 2bt_f + (t_w + 2r)t_f \geq \eta h_w t_w$$

$$= 15,900 - 2 \times 229 \times 19.6 + (11.9 + 2 \times 12.7) \times 19.6 \geq 1.0 \times 573 \times 11.9$$

$$= 7,654\,\text{mm}^2 \geq 6,819\,\text{mm}^2$$

The shear resistance of the section is given by

$$V_{pl,Rd} = \frac{A_v f_y}{\sqrt{3}\gamma_{M0}} \quad (\gamma_{M0} = 1.0)$$

$$= \frac{7654 \times 265}{\sqrt{3}} \times 10^{-3} = 1171 \text{ kN}$$

The maximum shear force under wheel load at $x_2 = 0.3$ m

$$V_{Q,x2} = \gamma_Q Q_{r,x2} = 1.35 \times 268.5 = 362.5 \text{ kN}$$

The shear forces under self-weight at $x_2 = 0.3$ m

$$V_{G,x2} = \gamma_G V_{x2} = 1.35 \times \left(7.35 \times \frac{5-0.3}{5}\right) = 9.3 \text{ kN}$$

Thus, the design value of shear forces is

$$V_{Ed} = V_{Q,x2} + V_{G,x2} = 371.8 \text{ kN}$$

$$V_{pl,Rd} = 1171 \text{ kN} > V_{Ed} = 371.8 \text{ kN}$$

Shear resistance is satisfied.
Check web-plate buckling from shear at the ultimate limit state:

$$\frac{h_w}{t_w} = \frac{573}{11.9} = 48.2$$

$$72\frac{\varepsilon}{\eta} = 72 \times \frac{0.94}{1.0} = 67.7 > 48.2$$

No shear buckling check is required.

Transverse force resistance check
The length of stiff bearing s_s on the surface of the top flange is given by

$$s_s = l_{eff} - 2t_f = 189.7 - 2 \times 19.6 = 150.5 \text{ mm}$$

Type of load application is type (a). The beam there is not stiffener, thus the ratio of h_w to the spacing a is zero.
The buckling coefficient for the load application is given by

$$k_F = 6 + 2\left(\frac{h_w}{a}\right) = 6$$

The critical force F_{cr} is given by

$$F_{cr} = \frac{0.9k_F Et_w^3}{h_w} = \frac{0.9 \times 6 \times 210{,}000 \times 11.9^3}{573} \times 10^{-3} = 3{,}335 \text{ kN}$$

The dimensionless parameters

$$m_1 = \frac{f_{yf} b_f}{f_{yw} t_w} = \frac{265 \times 229}{275 \times 11.9} = 18.5$$

Assuming $\bar{\lambda}_F > 0.5$,

$$m_2 = 0.02 \left(\frac{h_w}{t_f} \right)^2 = 0.02 \left(\frac{573}{19.6} \right)^2 = 17.1$$

The effective loaded length

$$l_y = s_s + 2t_f \left(1 + \sqrt{m_1 + m_2} \right)$$

$$= 150.5 + 2 \times 19.6\sqrt{18.5 + 17.1} = 384.4 \text{ mm}$$

$$\bar{\lambda}_F = \sqrt{\frac{l_y t_w f_{yw}}{F_{cr}}} = \sqrt{\frac{384.4 \times 11.9 \times 275}{3335 \times 10^3}} = 0.61 > 0.5 \text{ Satisfactory}$$

$$\text{Reduction factor } \chi_F = \frac{0.5}{\bar{\lambda}_F} = \frac{0.5}{0.61} = 0.82 < 1$$

The effective length L_{eff} is given by

$$L_{eff} = \chi_F l_y = 0.82 \times 384.4 = 315.2 \text{ mm}$$

The resistance to transverse force

$$F_{Rd} = \frac{f_{yw} L_{eff} t_w}{\gamma_{M1}} = \frac{275 \times 315.2 \times 11.9}{1.0} \times 10^{-3} = 1031.5 \text{ kN}$$

$$\eta_2 = \frac{F_{z,Ed}}{F_{Rd}} = \frac{221}{1031.5} = 0.214$$

The resistance to transverse force is satisfied.

Interaction checks
At location $x_1 = 4.25$ m, the resistance to transverse forces requires an interaction check with bending moment by the following interaction expression:

$$\eta_2 + 0.8\eta_1 \leq 1.4$$

$$M_{y,Ed} = 1.35 \times 591.6 + 1.35 \times 16.8 = 821.3 \text{ kN}$$

$$\eta_1 = \frac{M_{Ed}}{(W_{el,y}f_y)/\gamma_{M0}} = \frac{821.3 \times 10^3}{(3220 \times 265)/1.0} = 0.96$$

$$\eta_2 + 0.8\eta_1 = 0.21 + 0.8 \times 0.96 = 0.98 \leq 1.4$$

The resistance to interaction is satisfied.

LTB verification
Bending moment about z–z axis and torsion is ignored in this case; thus, the check expression is simplified as

$$\frac{M_{y,Ed}}{\chi_{LT}M_{y,Rk}/\gamma_{M1}} \leq 1$$

The design-factored bending moment diagram is shown in Figure 4.43. The lateral restraint is given by crane girder in each wheel-load location. The curve of moment diagram is nearly linear. Hence, Table 4.2 is used to determine the factor C_1.
    LTB check for segment AC
From Table 4.2,

$$C_1 = 1.88$$

$$L_{cr} = 4250 \text{ mm}$$

The critical moment for segment AC

$$M_{cr} = C_1 \frac{\pi^2 EI_z}{L_{cr}^2} \sqrt{\frac{I_W}{I_Z} + \frac{L_{cr}^2 GI_T}{\pi^2 EI_z}}$$

$$= 1.88 \frac{\pi^2 \times 210{,}000 \times 393 \times 10^5}{4{,}250^2} \sqrt{\frac{3.45 \times 10^{12}}{393 \times 10^5} + \frac{4{,}250^2 \times 81{,}000 \times 154 \times 10^4}{\pi^2 \times 210{,}000 \times 393 \times 10^5}} \times 10^{-6}$$

$$= 2{,}881 \text{ kN-m}$$

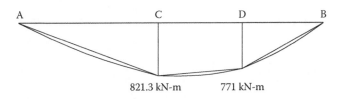

Figure 4.43 Design-factored bending moment.

Non-dimensional lateral-torsional slenderness for segment AC

$$\bar{\lambda}_{LT} = \sqrt{\frac{W_y f_y}{M_{cr}}} = \sqrt{\frac{3680 \times 265 \times 10^{-3}}{2881}} = 0.58$$

$$\bar{\lambda}_{LT} = 0.58 > \bar{\lambda}_{LT,0} = 0.4$$

So LTB must be checked.
Using Table 4.7,

$$3.1 > \frac{h}{b} = 2.67 > 2$$

For a rolled I section, use buckling curve c.

$$\alpha_{LT} = 0.49$$

$$\bar{\lambda}_{LT,0} = 0.4 \text{ (maximum value)}$$

$$\beta = 0.75 \text{ (maximum value)}$$

$$\phi_{LT} = 0.5\left[1 + \alpha_{LT}\left(\bar{\lambda}_{LT} - \bar{\lambda}_{LT,0}\right) + \beta\bar{\lambda}_{LT}^2\right]$$

$$= 0.5\left[1 + 0.49 \times (0.58 - 0.4) + 0.75 \times 0.58^2\right] = 0.67$$

Reduction factor for LTB χ_{LT}

$$\chi_{LT} = \frac{1}{\phi_{LT} + \sqrt{\phi_{LT}^2 - \beta\bar{\lambda}_{LT}^2}} \quad \text{but} \quad \chi_{LT} \leq \begin{cases} 1.0 \\ \dfrac{1}{\bar{\lambda}_{LT}^2} \end{cases}$$

$$= \frac{1}{0.67 + \sqrt{0.67^2 - 0.75 \times 0.58^2}} = 0.9 \leq \begin{cases} 1.0 \\ 2.97 \end{cases}$$

$$M_{b,y,Rd} = \chi_{LT}\frac{W_{pl,y}f_y}{\gamma_{M1}} = 0.9 \times \frac{3680 \times 265}{1.0} \times 10^{-3} = 878 \text{ kN-m}$$

$$M_{b,y,Rd} > M_{Ed} = 821.3 \text{ kN-m}$$

The LTB resistance for segment AC is satisfied.
LTB check for segment CD
From Table 4.2,

$$\psi = \frac{771}{821.3} = 0.94$$

$$C_1 = 1.88 - 1.4\psi + 0.52\psi^2 = 1.02$$

$$L_{cr} = 3000 \text{ mm}$$

The critical moment for segment AC

$$M_{cr} = C_1 \frac{\pi^2 E I_z}{L_{cr}^2} \sqrt{\frac{I_W}{I_Z} + \frac{L_{cr}^2 G I_T}{\pi^2 E I_z}}$$

$$= 1.02 \frac{\pi^2 \times 210,000 \times 393 \times 10^5}{3,000^2} \sqrt{\frac{3.45 \times 10^{12}}{393 \times 10^5} + \frac{3,000^2 \times 81,000 \times 154 \times 10^4}{\pi^2 \times 210,000 \times 393 \times 10^5}} \times 10^{-6}$$

$$= 2,942 \text{ kN-m}$$

Non-dimensional lateral-torsional slenderness for segment CD

$$\overline{\lambda}_{LT} = \sqrt{\frac{W_y f_y}{M_{cr}}} = \sqrt{\frac{3680 \times 265 \times 10^{-3}}{2942}} = 0.58$$

$$\overline{\lambda}_{LT} = 0.58 > \overline{\lambda}_{LT,0} = 0.4$$

So LTB must be checked.

The beam section of segment CD is same as segment AC, then the reduction factor for LTB $\chi_{LT} = 0.9$

$$M_{b,y,Rd} = \chi_{LT} \frac{W_{pl,y} f_y}{\gamma_{M1}} = 0.9 \times \frac{3680 \times 265}{1.0} \times 10^{-3} = 878 \text{ kN-m}$$

$$M_{b,y,Rd} > M_{Ed} = 821.3 \text{ kN-m}$$

The LTB resistance for segment CD is satisfied.

Deflection check

The maximum deflection due to the unfactored wheel load is at the location $x_1 = 4.25$ m:

$$\delta = \frac{FL^3}{48EI} \left(\frac{3(a+c)}{L} - \frac{4(a^3 + c^3)}{L^3} \right) + \frac{5wL^4}{384EI}$$

$$= \frac{163.75 \times 10^3 \times 10^{12}}{48 \times 210,000 \times 986 \times 10^6} \left[\frac{3(4.25 + 2.75)}{10} - \frac{4(4.25^3 + 2.75^3)}{10^3} \right]$$

$$+ \frac{5 \times 1.74 \times 10^{16}}{384 \times 210,000 \times 986 \times 10^6}$$

$$= 29.3 \text{ mm}$$

where

$$F = Q_{r,max} = 163.75 \times 10^3 \text{ kN} \quad a = 4251 \text{ mm} \quad c = 2750 \text{ mm}$$

Limiting deflection $\delta_{lim t} = \min\left(\frac{L}{600} : 25 \right) = \min\left(\frac{10,000}{600} : 25 \right) = 16.7 \text{ mm}$

$$\delta_{lim t} = 16.7 \text{ mm} < \delta = 29.3 \text{ mm}$$

Deflection check is not satisfied.

4.12 PURLINS AND SHEETING RAIL

4.12.1 General

Purlin is a beam and it supports roof decking on flat roofs or cladding on sloping roofs on industrial buildings.

Members used for purlins and sheeting rails are shown in Figure 4.44. These are cold-formed sections, angles, channels joists and structural hollow sections. Cold-formed sections are now used on most industrial buildings. Only cold-formed members are considered in this book.

In general, the conventions for member axes are as used in EN1993-1-1 (see Figure 4.45). For profiled sheets and liner trays, the following axis convention is used:

- y–y axis parallel to the plane of sheeting
- z–z axis perpendicular to the plane of sheeting

4.12.2 Purlins restrained by sheeting

The purlins were restrained by sheeting, which may be steel or aluminium corrugated or profile sheets or decking. On sloping roofs, sheeting is placed over insulation board or

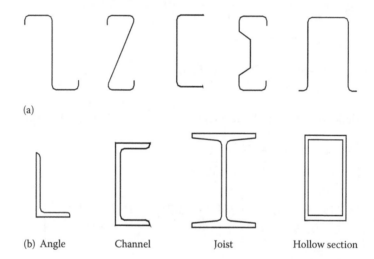

(a)

(b) Angle   Channel   Joist   Hollow section

Figure 4.44 Section used for purlins and sheeting rails: (a) cold-formed sections; (b) hot-rolled sections.

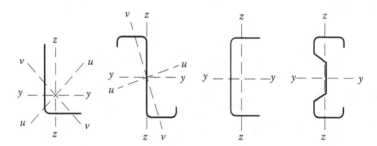

Figure 4.45 Axis convention.

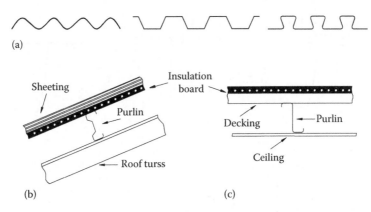

Figure 4.46 Roof material and constructions: (a) profiled sheeting; (b) slope roof construction; (c) flat roof construction.

glass wool. On flat roofs, insulation board, felt and bitumen are laid over the steel decking. Typical roof cladding and roof construction for flat and sloping roofs are shown in Figure 4.46.

The purlins discussed in this book of Z, C, Σ and Hat cross sections are all with the limits $h/t < 233$, $c/t \leq 20$ for single fold and $d/t \leq 20$ for double-edge fold. The purlins at the connection to trapezoidal sheeting or other profiled steel sheeting may be regarded as laterally restrained, if the following equation is fulfilled:

$$S \geq \left(EI_w \frac{\pi^2}{L^2} + GI_t + EI_z \frac{\pi^2}{L^2} 0.25h^2 \right) \frac{70}{h^2} \tag{4.42}$$

where

$S = 1000\sqrt{t^3}\left(50 + 10\sqrt[3]{b_{roof}}\right)s/h_w$ is the portion of the shear stiffness
t is the design thickness of sheeting
b_{roof} is the width of the roof
s is the distance between the purlins
h_w is the profile depth of sheeting
I_w is the warping constant of the purlin
I_t is the torsion constant of the purlin
I_z is the second moment of area of the cross section about the minor axis
L is the span of the purlin
h is the height of the purlin

4.12.3 Spring stiffness

The behaviour of a free flange under uplift loading is shown in Figure 4.47. The deformation can be spitted into in-plane bending and torsional and lateral bending due to cross-sectional distortion. The connection of the purlin to the sheeting may be assumed to partially restrain the twisting of the purlin, in which partial torsional restraint is represented by a rotational spring with a spring stiffness C_D. The lateral move can be treated as beam subject to a lateral load $q_{h,Ed}$. For using this method, the rotational spring should be replaced by an equivalent lateral linear spring of stiffness K (see Figure 4.47).

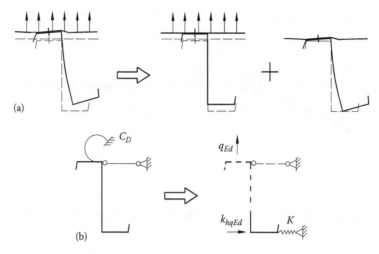

Figure 4.47 (a) Behaviour of purlins and (b) rotational and lateral spring.

4.12.3.1 Lateral spring stiffness

The lateral spring support given to the free flange of the purlin by the sheeting should be modelled as lateral spring acting at the free flange. The total lateral spring stiffness K per unit length should be determined from

$$\frac{1}{K} = \frac{4(1-v^2)h^2\left(h_d + b_{mod}\right)}{Et^3} + \frac{h^2}{C_D} \tag{4.43}$$

where
$\quad K$ is the lateral spring stiffness
$\quad v$ is Poisson's ratio of the steel
$\quad C_D$ is the total rotational spring stiffness

The dimension b_{mod} is depending on uplift or downwards actions

- For downwards actions, $b_{mod} = a$.
- For uplift actions, $b_{mod} = 2a + b$.

a, b, h and h_d are shown in Figure 4.48.

4.12.3.2 Rotational spring stiffness

The rotational restraint given to the purlin by the sheeting that is connected to its top flange should be modelled as rotational spring acting at the top flange of the purlin. The total rotational spring stiffness C_D should be determined from

$$C_D = \frac{1}{\left(1/C_{D,A} + 1/C_{D,C}\right)} \tag{4.44}$$

$$C_{D,A} = C_{100}k_{ba}k_t k_{bR}k_A k_{bT} \tag{4.45}$$

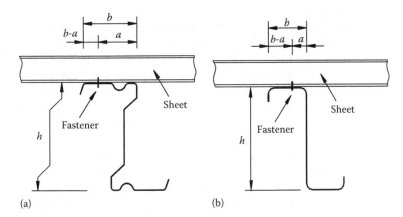

Figure 4.48 Purlin and attached sheeting: (a) Z-section sheeting with stiffeners; (b) typical Z-section sheeting.

$$C_{D,C} = \frac{kEI_{eff}}{s}$$ (4.46)

where

$C_{D,A}$ is the rotational stiffness of the connection between the sheeting and the purlin

$C_{D,C}$ is the rotational stiffness corresponding to the flexural stiffness of the sheeting

C_{100} is a rotation coefficient for trapezoidal sheeting (Table 10.3 of EN 1993-1-3)

k_{ba} is the width factor of purlin flange (Table 4.14)

k_t is the thickness factor (constant values, linear between given limits)

k_{bR} is the corrugation width factor of sheeting (Table 4.14)

k_A is the supporting force factor (Table 4.14)

k_{bT} is the connection factor (Table 4.14)

k is a numerical coefficient (see Figure 4.49)

I_{eff} is the effective second moment of area per unit width of the sheeting

s is the spacing of the purlins

4.12.4 Design of cold-formed beams to EN 1993-1-3

The design criteria about purlins are stated in Chapter 10 and Annex E of EN 1993-1-3. For single-span purlins, the cross-sectional resistance should be satisfied under gravity loading. If subject to axial compression, the purlin should satisfy the criteria for stability of the free flange. These criteria are given in Clause 10.1.3 and 10.1.4 in EN 1993-1-3. For double- or multiple-span continuous purlins, more details of internal force should be considered such as bending moment + support reaction in mid-support.

These design criteria may also be applied to cold-formed members used as side rails, floor beams and other similar types of beam that are similarly restrained by sheeting. Sheeting rail is designed on the basis that wind pressure has a similar effect on them to gravity loading on purlins and that wind suction acts on them in a similar way to uplift loading on purlins.

Table 4.14 Factors for $C_{D,A}$

Factors		Expressions	Conditions
k_{ba}		$k_{ba} = (b_a/100)^2$	$b_a < 125$ mm
		$k_{ba} = 1.25(b_a/100)$	125 mm $\leq b_a < 200$ mm
k_t		$k_t = (t_{nom}/0.75)^{1.1}$	$t_{nom} \geq 0.75$ mm; positive position
		$k_t = (t_{nom}/0.75)^{1.5}$	$t_{nom} \geq 0.75$ mm; negative position
		$k_t = (t_{nom}/0.75)^{1.5}$	$t_{nom} < 0.75$ mm
k_{bR}		$k_{bR} = 1.0$	$b_R \leq 185$ mm
		$k_{bR} = 185/b_R$	$b_R > 185$ mm
k_A	Gravity load	Not valid	$t_{nom} < 0.75$ mm
		$k_A = 1.0 + (A - 1.0)0.08$	$t_{nom} = 0.75$ mm; positive position
		$k_A = 1.0 + (A - 1.0)0.16$	$t_{nom} = 0.75$ mm; negative position
		Linear interpolation	0.75 mm $< t_{nom} < 1.00$ mm
		$k_A = 1.0 + (A - 1.0)0.095$	$t_{nom} \geq 1.00$ mm
	Uplift load	$k_A = 1.0$	All cases
k_{bT}		$k_{bT} = \sqrt{b_{T,max}/b_T}$	$b_T > b_{T,max}$
		$k_{bT} = 1.0$	$b_T \leq b_{T,max}$

where
  b_a is the width of the purlin flange
  b_R is the corrugation width
  b_T is the width of the sheeting flange
   through which it is fastened to
   the purlin
  $b_{T,max}$ is given in Table 10.3 in EC3-1-3

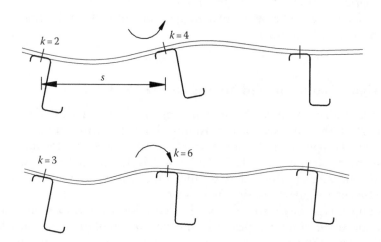

Figure 4.49 Model for calculating $C_{D,C}$.

Table 4.15 Limitations of width-to-thickness ratios

Sections	t (mm)	b/t	h/t	h/b	c/t	b/c	L/h
	≥ 1.25	≤ 55	≤ 160	≤ 3.43	≤ 20	≤ 4.0	≥ 15
	≥ 1.25	≤ 55	≤ 160	≤ 3.43	≤ 20	≤ 4.0	≥ 15

A simplified design method of purlins made of C-, Z- and Σ-cross sections is given in Annex E in EN 1993-1-3. The purlins with or without additional stiffeners in web or flange may be designed by the simplified method if the following conditions are fulfilled:

- The purlins are horizontally restrained by trapezoidal sheeting where the horizontal restraints fulfil the conditions of Equation 4.42.
- The cross-sectional dimensions are within the range of Table 4.15.
- The purlins are restrained rotationally by the trapezoidal sheeting.
- The purlins have equal spans and uniform loading.

However, there are some conditions where the method should not be used:

- For systems using antisag bars
- For sleeve or overlapping systems
- For application of axial force

The design value of the bending moment M_{Ed} should satisfy

$$\frac{M_{Ed}}{M_{LT,Rd}} \leq 1$$

where

$$M_{LT,Rd} = \left(\frac{f_y}{\gamma_{M1}}\right) W_{eff,y} \frac{\chi_{LT}}{k_d} \tag{4.47}$$

$W_{eff,y}$ is section modulus of the effective cross section

χ_{LT} is reduction factor for LTB in dependency of $\bar{\lambda}_{LT}$ due to Table 4.16, where α_{LT} is substituted by $\alpha_{LT,eff}$

$$\bar{\lambda}_{LT} = \sqrt{\frac{W_{eff,y} f_y}{M_{cr}}} \tag{4.48}$$

$$\alpha_{LT,eff} = \alpha_{LT} \sqrt{\frac{W_{el,y}}{W_{eff,y}}} \tag{4.49}$$

$k_d = (a_1 - a_2(L/h))$ but ≥ 1.0 is coefficient for consideration of the non-restraint part of the purlin

a_1, a_2 are coefficients from Table 4.17

Table 4.16 Appropriate buckling curve for various types of cross section

Type of cross section		Buckling about axis	Buckling curve
	If f_{yb} is used	Any	b
	If f_{ya} is used	Any	c
		y–y	a
		z–z	b
		Any	b
		Any	c

The average yield strength f_{ya} should not be used unless $A_{eff} = A_g$.

Table 4.17 Coefficients a_1, a_2

System	Z-Purlins		C-Purlins		Σ-Purlins	
	a_1	a_2	a_1	a_2	a_1	a_2
Single-span beam gravity load	1.0	0	1.1	0.002	1.1	0.002
Single-span beam uplift load	1.3	0	3.5	0.05	1.9	0.02
Continuous beam gravity load	1.0	0	1.6	0.02	1.6	0.02
Continuous beam uplift load	1.4	0.01	2.7	0.04	1.0	0

Table 4.18 Factors k_ϑ

Statical system	Gravity load	Uplift load
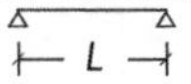 ⊢ L ⊣	—	0.21
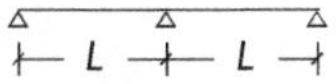 ⊢ L ⊣ L ⊣	0.07	0.029
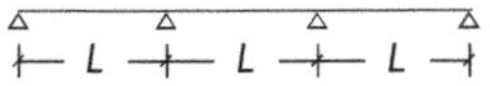 ⊢ L ⊣ L ⊣ L ⊣	0.15	0.066
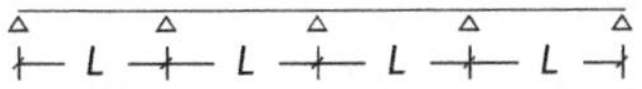 ⊢ L ⊣ L ⊣ L ⊣ L ⊣	0.1	0.053

The reduction factor χ_{LT} may be chosen by $\chi_{LT} = 1.0$, if a single-span beam under gravity load is present or if the following expression is met:

$$C_D \geq \frac{M_{el,u}^2}{EI_v} k_\vartheta \tag{4.50}$$

where

$M_{el,u} = W_{el,ufy}$ is the elastic moment of the gross cross section with regard to the major u–u axis

I_v is the moment of inertia of the gross cross section with regard to the minor v–v axis

k_ϑ is a factor for considering the static system of the purlin due to Table 4.18

It is noted that, for equal-flanged C- and Σ-purlins, $I_v = I_z$, $W_u = W_y$ and $M_{el,u} = M_{el,y}$.

If Equation 4.50 is not met, the reduction factor χ_{LT} should be calculated using $\bar{\lambda}_{LT}$ and $\alpha_{LT,eff}$. The elastic critical moment for LTB M_{cr} may be calculated by the following equation:

$$M_{cr} = \frac{k}{L} \sqrt{GI_t^* EI_v} \tag{4.51}$$

where

I_t^* is the fictitious St. Venant torsion constant considering the effective rotational restraint

k is the LTB coefficient due to Table 4.19

$$I_t^* = I_t + C_D \frac{L^2}{\pi^2 G} \tag{4.52}$$

Table 4.19 LTB coefficient k

Statical system	Gravity load	Uplift load
	$\propto$	10.3
	17.7	27.7
	12.2	18.3
	14.6	20.5

where I_t is the torsional constant of the purlin

$$\frac{1'}{C_D} = \frac{1}{C_{D,A}} + \frac{1}{C_{D,B}} + \frac{1}{C_{D,C}}$$ (4.53)

where $C_{D,B}$ is rotational stiffness due to distortion of the cross section of the purlin

$$C_{D,B} = K_B h^2$$ (4.54)

where
   h is the depth of the purlin
   K_B is the lateral stiffness due to distortion of the cross section of the purlin

Chapter 5

Plate girders

5.1 DESIGN CONSIDERATIONS

5.1.1 Uses and construction

Plate girders are used when the bending moments are larger than the standard rolled universal or compound beams can resist, usually because of a large span. These girders are invariably very deep, resulting in non-compact or slender webs. They are used in buildings and industrial structures for long-span floor girders, heavy crane girders and in bridges.

Plate girders are constructed by welding steel plates together to form I sections. Such welding is generally performed as a double pass one on either side of the girder as shown in Figure 5.1. It should be noted that this process may cause very high residual stresses to exist in the flanges and web. A closed section is termed a 'box girder'. Typical sections, including a heavy fabricated crane girder, are shown in Figure 5.2a.

The web of a plate girder is relatively thin, and stiffeners are required either to prevent buckling due to compression from bending and shear or to take advantage of tension field action, depending on the design method used. Stiffeners are also required at load points and supports. Thus, the side elevation of a plate girder has an array of stiffeners, as shown in Figure 5.2b.

5.1.2 Depth and breadth of girder

The depth of a plate girder may be fixed by headroom requirements, but it can often be selected by the designer. The depth is usually made from one-tenth to one-twelfth of the span. The breadth of flange plate is made about one-third of the depth.

The deeper the girder is made, the smaller the flange plates required. However, the web plate must then be made thicker, or additional stiffeners must be provided to meet particular design requirements. A method to obtain the optimum depth is given in Section 5.3.2. A shallow girder can be very much heavier than a deeper girder in carrying the same loads.

5.1.3 Variation in girder sections

Flange cover plates can be curtailed or single-flange plates can be reduced in thickness when reduction in bending moment permits. This is shown in Figure 5.3a. In the second case mentioned, the girder depth is kept constant throughout.

For simply supported girders, where the bending moment is maximum at the centre, the depth may be varied, as shown in Figure 5.3b. In the past, hog-back or fish-belly girders were commonly used. In modem practice with automatic methods of fabrication, it is more economical to make girders of uniform depth and section throughout.

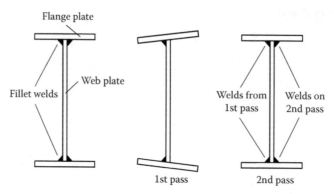

Figure 5.1 Plate girder fabrication.

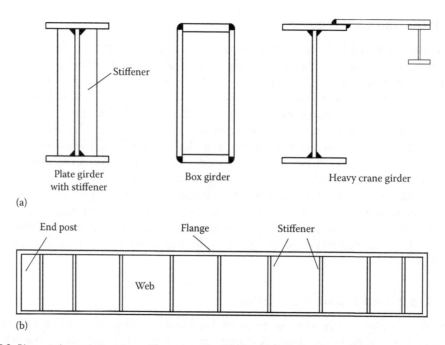

(a)

(b)

Figure 5.2 Plate girder constructions: (a) sections for fabricated girders; (b) plate girder.

In rigid frame construction and in continuous girders, the maximum moment occurs at the supports. The girders may be haunched to resist these moments, as shown in Figure 5.3c.

The design procedures for plate buckling that are based on the effective width method were developed for web or flange panels of uniform width. Usually these panels are stiffened or unstiffened plates between rigid transverse stiffeners. The panels may be considered as uniform when

- The shape of the panel is rectangular or almost rectangular. In the latter case, the angle α_{limit} should not exceed 10°.
- The diameter of any unstiffened hole or cut-out does not exceed $0.05b$, where b is the width of the panel (see Figure 5.4).

If $\alpha_{limit} > 10°$, the panel may be conservatively treated as rectangular with the width equal to the larger of widths on both ends of the panel as shown in Figure 5.5.

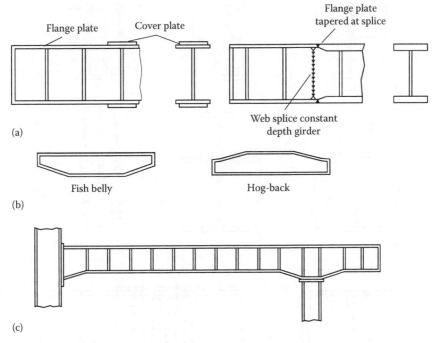

Figure 5.3 Variation in plate girder sections: (a) local reinforced girders; (b) variable depth girders; (c) haunched ends continuous girder.

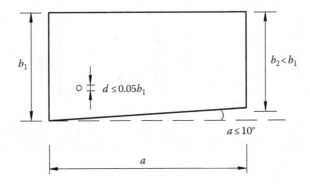

Figure 5.4 Nominally uniform panels.

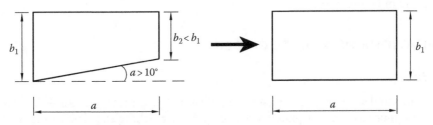

Figure 5.5 Non-uniform panel transformed to equivalent uniform panel.

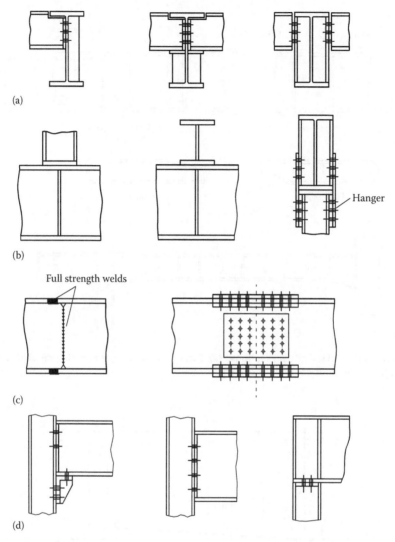

Figure 5.6 Plate girder connection and splices: (a) beam to girder connections; (b) loads from columns, beams and hangers; (c) welded and bolted solices; (d) plate girder end connections.

5.1.4 Plate girder connections and splices

Typical connections of beams and columns to plate girders are shown in Figure 5.6a and b. Splices are necessary in long girders. Bolted and welded splices are shown in Figure 5.6c and end supports in Figure 5.6d.

5.2 BEHAVIOUR OF A PLATE GIRDER

5.2.1 Basis

Slender plates under compression possess significant post-critical resistance that can be utilized in design procedures for plated structures. After reaching the elastic critical stress σ_{cr}, the resistance is not exhausted, but it increases further until plastic collapse occurs. In the

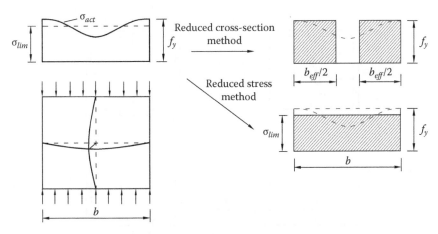

Figure 5.7 Actual stresses distribution and simplified methods.

post-critical state, the redistribution of compressive stresses takes place with the reduction of stresses in the middle buckled part, where axial stiffness is decreased, and with the increase of stresses near straight plate edges, as shown in Figure 5.7. The ultimate resistance is reached after the maximum edge stress has reached the plate yield strength, as in general slender plates do not have any ductility to redistribute stresses by developing zones of plastic strains. To deal with non-linear distribution of actual stresses, σ_{act}, which is not very practical, two simplified methods are given in EN 1993-1-5:

1. Effective width method (main method)
2. Reduced stress method

The effective width method (reduced cross-sectional method) is based on the appropriate reduction of cross section in the central buckled part of the plate, assuming effective width b_{eff} adjacent to edges as fully effective with stresses equal to f_y all over the effective width.

The reduced stress method is based on the average stress σ_{lim} of the actual stress distribution σ_{act} in the ultimate state.

5.2.2 Effective cross section

The effective width method is used to determine the resistance of Class 4 cross sections subject to direct stresses. Based on the effective width, effective cross-sectional area A_{eff}, effective second moment of inertia I_{eff} and effective section modulus W_{eff} are calculated. The effective cross section is then treated as an equivalent Class 3 cross section, and the linear elastic strain and stress distribution over the reduced section is assumed.

If axial loading and bending moment act simultaneously, the effective area A_{eff} should be determined assuming that the cross section is subject only to stresses due to uniform axial compression (see Figure 5.8), and the effective section modulus W_{eff} should be determined assuming the section is subject to stresses just due to pure bending (see Figure 5.9). The symmetric cross section has the same centroid in pure compression. For nonsymmetric cross sections, the possible shift e_N of the centroid of the effective area A_{eff} is relative to the centre of gravity of the gross cross section (see Figure 5.8). This shift results in an additional bending moment that should be taken into account in the cross-sectional verification.

For I girders, the maximum stress in a flange should be calculated using the elastic section modulus with reference to the mid-plane of the flange (Figure 5.9).

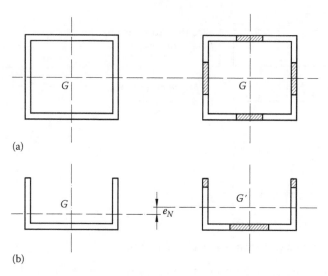

(a)

(b)

Figure 5.8 Class 4 cross-sectional axial force: (a) symmetric cross-section; (b) unsymmetric cross-sections.

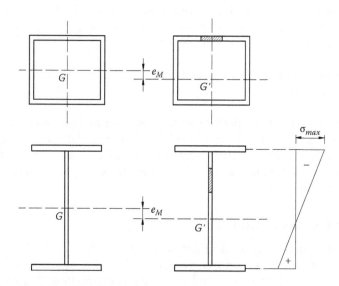

Figure 5.9 Class 4 cross-sectional bending moment.

5.2.3 Flange-induced buckling

Optimizing a girder for bending action is achieved by making the web of the plate girder more slender. When lateral supports are used to prevent lateral-torsional buckling (LTB), then flange-induced buckling will become the critical failure mechanism. With no web stiffeners, the web should be sized to avoid the flange undergoing local buckling due to the web being unable to support the flange. This is known as flange-induced buckling.

To prevent the compression flange buckling in the plane of the web, the following criterion should be met:

$$\frac{h_w}{t_w} \leq k\frac{E}{f_{yf}}\sqrt{\frac{A_W}{A_{fc}}}$$ (5.1)

where
    h_w is the depth of the web
    t_w is the thickness of the web
    f_{yf} is the yield strength of the compression flange
    A_w is the cross-sectional area of the web
    A_{fc} is the effective cross-sectional area of the compression flange

The value of the factor k should be taken as follows:

    $k = 0.3$, where plastic hinge rotation is utilized
    $k = 0.4$, where the plastic moment resistance is utilized
    $k = 0.55$, where the elastic moment resistance is utilized

Thus, for rigid or continuous design, $k = 0.3$, unless the analysis is elastic with no redistribution. For simply supported beams, k may be taken as 0.4.

5.2.4 Transverse loading and web buckling

The collapse behaviour of girders subjected to transverse loading is characterized by three failure modes: yielding, buckling and crippling of the web, as shown in Figure 5.10. There are many parameters that influence the ultimate load capacity, such as geometry dimensioning and steel strength of the composed plates. The rules cover of section 6 in EN 1993-1-5 cover

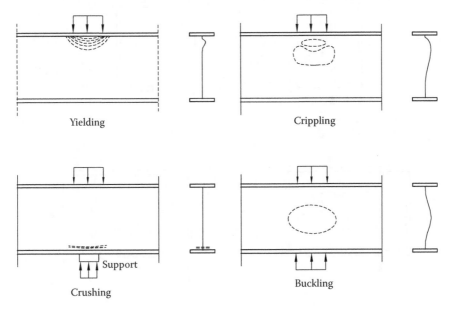

Figure 5.10 Failure modes of girders subjected to transverse loading.

both rolled and welded girders, with and without longitudinal stiffeners, up to steel grades of S690. And also assumed, the compression flange has fully lateral-torsional restraint, which needs to be taken into account in design.

5.2.5 Shear buckling and tension field action

The behaviour of plates under shear comprises two phenomena: the state of pure shear stress and the tension field. Prior to buckling, pure shear stresses occur in the plate. If these shear stresses τ are transformed into principal stresses, they correspond to principal tensile tresses σ_1 and principal compressive stresses σ_2 with equal magnitude and inclined by 45° with regard to the longitudinal axis of the girder. In this state, only constant shear stresses occur at the edges (see Figure 5.11a).

As for plates subjected to direct compressive stresses, slender plates under shear possess a post-critical reserve. After buckling, the plate reaches the post-critical stress state, whilst a shear buckling forms in the direction of the principal tensile stresses σ_1. Due to buckling, no significant increase of the stresses in the direction of the principal compressive stresses σ_2 is possible, whereas the principal tensile stresses can still increase. The compressive stresses can cause the web to buckle. As a result, the tensile stresses are larger than the compressive stresses, which lead to a rotation of the shear field and which are denoted tension field action (see Figure 5.11b).

At the point of impending buckling, the web loses its ability to support the diagonal compressive stresses, and this stress is shifted to the transverse stiffeners and the flanges. The stiffeners resist the vertical component of the diagonal compressive force, and the flanges resist the horizontal component. The web will need to resist only the diagonal tensile force. This behaviour can be likened to that of a Pratt truss, in which the vertical web members carry compression and the diagonals carry tension. When reaching ultimate load, a plastic hinge mechanism forms in the flange (see Figure 5.11c).

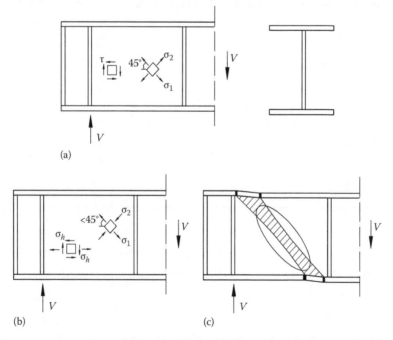

Figure 5.11 Stress states and collapse behaviour of a plate girder subjected to shear: (a) pure shear stress; (b) tension field section; (c) plastic hinge flange mechanism.

5.3 DESIGN TO EN 1993-1-5

5.3.1 Bending resistance and flange design

The section classification is determined in the same manner as rolled sections.

For compression flange restrained plate girder, the method is used for girder design: the flanges carrying the bending moment and the web carrying the shear force. This is probably best used where the maximum bending moment and the maximum shear force are not coincident, as the ability of the flange to contribute towards shear capacity may be utilized. If both the maximum loads are coincident in either a simply supported beam under point loading or a continuous beam at the internal support, then the flange capacity will not be able to be utilized to resist shear. With this method, the moment capacity is only dependent on the section classification of the flanges as the web does not carry compression.

An initial flange area may be established on the assumption that bending moment is carried by the flanges only, this can be shown to be conservative in most practical applications [22]; then

$$M_{Rd} = A_f \frac{f_y}{\gamma_{M0}} h_w \geq M_{Ed} \tag{5.2}$$

or

$$A_f = 2b_f t_f \geq \frac{M_{Ed}}{f_y h_w} \gamma_{M0} \tag{5.3}$$

Generally proportion flanges to remain non-slender. Maximum ratio flange outstand to flange thickness for Class 2 section is 10ε and for Class 3 section is 14ε. For Class 3 welded section,

$$\frac{c_f}{t_f} = \frac{(b - t_w - 2s)/2}{t_f} \leq 14\varepsilon$$

$$\varepsilon = \sqrt{\frac{235}{f_y}}$$

Then the dimension of flange can be determined.

5.3.2 Optimum depth

The optimum depth based on minimum area of cross section may be derived as follows. This treatment applies to a girder with restrained compression flange for a given web depth-to-thickness ratio. Define the following terms:

λ = ratio of web depth/thickness = h_w/t_w
A_f = area of one flange
W_{pl} = plastic modulus based on the flanges only

$$W_{pl} = A_f h_w = \frac{M_{Rd}}{f_y}$$

$$A_f = \frac{W_{pl}}{h_w}$$

$$A_w = \text{area of web} = h_w t_w = \frac{h_w^2}{\lambda}$$

$$A = \text{total area} = A_f + A_w = 2\frac{W_{pl}}{h_w} + \frac{h_w^2}{\lambda}$$

For an optimum solution, $\dfrac{dA}{dh_w} = 0$ so

$$\frac{dA}{dh_w} = -2\frac{W_{pl}}{h_w^2} + 2\frac{h_w}{\lambda} = 0$$

$$h_w = \sqrt[3]{W_{pl}\lambda}$$

The area of the web A_w is then given by

$$A_w = \frac{h_w^2}{\lambda} = \sqrt[3]{\frac{W_{pl}^2}{\lambda}} \tag{5.4}$$

The area of the flange A_f is given by

$$A_f = \frac{W_{pl}}{h_w} = \sqrt[3]{\frac{W_{pl}^2}{\lambda}} \tag{5.5}$$

Curves drawn for depth do against plastic modulus W_{pl} for values of λ of 150, 200 and 250 are shown in Figure 5.12b. For the required value of

$$W_{pl} = \frac{M_{Rd}}{f_y} = \frac{M_{Ed}}{f_y}$$

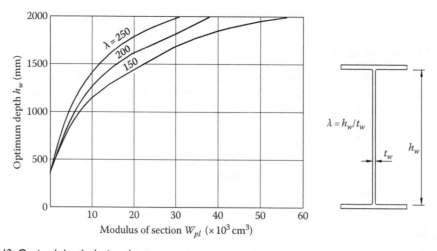

Figure 5.12 Optimal depth design chart.

The optimum depth do can be read from the chart for a given value of λ, where M_{Ed} is the applied moment. And

$$h_w = \sqrt[3]{\frac{\lambda M_{Ed}}{f_y}} \tag{5.6}$$

5.3.3 Shear capacity and web design

1. Minimum thickness of web
   With no web stiffeners, the web should be sized to avoid the flange undergoing local buckling due to the web being unable to support the flange. This is known as flange-induced buckling given in Section 8 of EN 1993-1-5.
   The optimum depth h_w can be determined by the aforementioned method. To avoid the flange buckling into web, the minimum web thickness should be satisfied for the following condition:

$$\frac{h_w}{t_w} \leq k \frac{E}{f_{yf}} \sqrt{\frac{A_W}{A_{fc}}}$$

2. Shear buckling resistance of slender web
   When the web depth-to-width h_w/t_w ratio of plate girder is satisfied, the following conditions can be determined as a stocky web. The design method is discussed in Chapter 4.

$$\text{For unstiffened webs,} \quad \frac{h_w}{t_w} \leq 72 \frac{\varepsilon}{\eta}$$

$$\text{For webs with intermediate stiffeners,} \quad \frac{h_w}{t_w} \leq 31 \frac{\varepsilon}{\eta} \sqrt{k_\tau}$$

where
   $\eta = 1.0$ for all steels (NA to BS EN 1993-1-1)
   k_τ is the shear buckling coefficient

The depth-to-width ratio beyond the aforementioned limits is slender web. The slender web should be checked for the resistance to shear buckling and should be provided with transverse stiffeners at the supports. In EN 1993-1-5, the shear resistance $V_{b,Rd}$ comprises contributions from the web $V_{bw,Rd}$ and from the flanges $V_{bw,Rd}$. However, the full shear resistance $V_{b,Rd}$ can never be larger than the plastic shear resistance of the web alone (see Equation 5.7):

$$V_{b,Rd} = V_{bw,Rd} + V_{bf,Rd} \leq h_w t_w \frac{\eta f_{yw}}{\sqrt{3} \gamma_{M1}} \tag{5.7}$$

where
   $V_{bw,Rd}$ is the resistance contribution from the web
   $V_{bf,Rd}$ is the resistance contribution from the flanges
   f_{yw} is the web yield strength
   $\gamma_{M1} = 1.0$

For unstiffened webs, the flange contribution is negligible, and webs carrying shear stress only. This states that webs with intermediate stiffeners may be designed with tension field action. After web buckling occurred, there was still a reserve of strength in the web. The additional reserve of strength is due to a diagonal tension field forming in the web including the flanges contribution.

3. Contribution from the web

The contribution of the web is determined as the following:

$$V_{bw,Rd} = \chi_w h_w t_w \frac{f_{yw}}{\sqrt{3}\gamma_{M1}} \tag{5.8}$$

where χ_w is the reduction factor for shear buckling.

The reduction factor χ_w considers components of pure shear and anchorage of membrane forces by transverse stiffeners due to tension field action. Since the axial and flexural stiffness of the transverse end stiffeners influence the post-critical reserve, the distinction between non-rigid and rigid end posts in the determination of the reduction factor χ_w is shown in Table 5.1 and Figure 5.13.

As shown in Figure 5.13, $\chi_w = \eta$ is defined for a small slenderness. The recommended η values according to EN 1993-1-5 are

For $f_y \leq 460 \text{ N/mm}^2$, $\eta = 1.2$

For $f_y > 460 \text{ N/mm}^2$, $\eta = 1.0$

Table 5.1 Contribution from the web χ_w to shear buckling resistance

	Rigid end post	Non-rigid end post
$\bar{\lambda}_w < 0.83/\eta$	η	η
$0.83/\eta \leq \bar{\lambda}_w < 1.08$	$0.83/\bar{\lambda}_w$	$0.83/\bar{\lambda}_w$
$\bar{\lambda}_w \geq 1.08$	$1.37/(0.7 + \bar{\lambda}_w)$	$0.83/\bar{\lambda}_w$

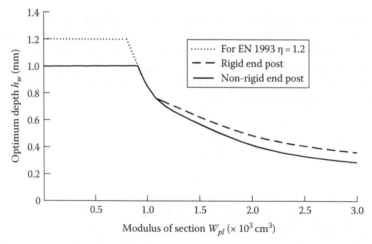

Figure 5.13 Reduction curves for shear buckling.

The reduction curves according to Table 5.1 apply for the verification of both unstiffened and stiffened webs. They are based on the plate slenderness $\bar{\lambda}_w$. The non-dimensionalized web slenderness $\bar{\lambda}_w$ is defined as

$$\bar{\lambda}_w = \sqrt{\frac{f_{yw}/\sqrt{3}}{\tau_{cr}}} = 0.76\sqrt{\frac{f_{yw}}{\tau_{cr}}} \tag{5.9}$$

Equation 5.9 can be simplified as follows:

a. For webs with transverse stiffeners at the supports and either intermediate transverse or longitudinal stiffeners or both,

$$\bar{\lambda}_w = \frac{h_w}{37.4\varepsilon t\sqrt{k_\tau}} \tag{5.10}$$

b. For webs with transverse stiffeners at supports only, $a/h_w \propto \infty$, hence with little loss in accuracy $k_\tau = 5.34$ (see the following),

$$\bar{\lambda}_w = \frac{h_w}{37.4\varepsilon t\sqrt{5.34}} = \frac{h_w}{86.4\varepsilon t} \tag{5.11}$$

where k_τ is the shear buckling coefficient dependent upon the aspect ratio of a web panel.

Annex A.3 of EN 1993-1-5 gives the calculation of the shear buckling coefficient for plates as follows:

c. For panels with rigid transverse stiffeners only,

$$k_\tau = 5.34 + 4.00\left(\frac{a}{h_w}\right)^2 \quad \text{for } \frac{a}{h_w} \geq 1.0$$

$$k_\tau = 4.00 + 5.34\left(\frac{a}{h_w}\right)^2 \quad \text{for } \frac{a}{h_w} < 1.0 \tag{5.12}$$

d. For stiffened panels with one or two longitudinal stiffeners and $\alpha = a/h_w < 3.0$,

$$k_\tau = 4.1 + \frac{6.3 + 0.18(I_{sl}/t^3h_w)}{\alpha^2} + 2.2 \cdot \sqrt[3]{\frac{I_{sl}}{t^3h_w}} \tag{5.13}$$

e. For stiffened panels with one or two longitudinal stiffeners and $\alpha = a/h_w \geq 3.0$ or for stiffened panels with more than two longitudinal stiffeners,

$$k_\tau = 5.34 + 4.00\left(\frac{a}{h_w}\right)^2 + k_{\tau sl} \quad \text{for } \frac{a}{h_w} \geq 1.0$$

$$k_\tau = 4.00 + 5.34\left(\frac{a}{h_w}\right)^2 + k_{\tau sl} \quad \text{for } \frac{a}{h_w} < 1.0 \tag{5.14}$$

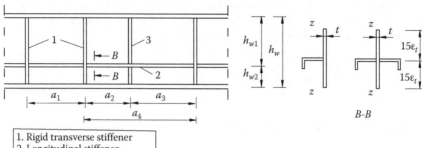

1. Rigid transverse stiffener
2. Longitudinal stiffener
3. Non-rigid transverse stiffener

Figure 5.14 Web with transverse and longitudinal stiffeners.

with

$$k_{\tau sl} = 9 \cdot \left(\frac{a}{h_w}\right)^2 \cdot \sqrt[4]{\left(\frac{I_{sl}}{t^3 h_w}\right)^3} > \frac{2.1}{t} \cdot \sqrt[3]{\frac{I_{sl}}{h_w}} \tag{5.15}$$

where
  a is the distance between transverse stiffeners (see Figure 5.14)
  I_{sl} is the second moment of area of the longitudinal stiffener about z–z axis (see Figure 5.14). For stiffened panels with two or more longitudinal stiffeners, I_{sl} is the sum of all individual stiffeners

4. Contribution from the flanges
When the flange resistance is not completely utilized in resisting the bending moment ($M_{Ed} < M_{f,Rd}$), the contribution from the flanges should be accounted for according to Equation 5.18, which assumes the formation of four plastic hinges in the flanges at a distance c (see Figure 5.11):

$$V_{bf,Rd} = \frac{b_f t_f^2 f_{yf}}{c\gamma_{M1}}\left(1 - \left(\frac{M_{Ed}}{M_{f,Rd}}\right)^2\right) \tag{5.16}$$

where c is the width of the portion of the web between the plastic hinges

$$c = a\left(0.25 + \frac{1.6 b_f t_f^2 f_{yf}}{t_w h_w^2 f_{yw}}\right) \tag{5.17}$$

$M_{f,Rd}$ is the design moment of resistance of the cross section determined using the effective flanges only

$$M_{f,Rd} = \frac{M_{f,k}}{\gamma_{M0}} = \frac{A_f f_{yf} h_f}{\gamma_{M0}} \tag{5.18}$$

where h_f is the distance between mid-planes of flanges.

For areas with a high utilization level due to normal force and bending moment, the determination of $V_{bf,Rd}$ can be neglected, because the main resistance to shear comes from the web contribution $V_{bw,Rd}$.

5.3.4 Stiffener design

Two main types of stiffeners used in plate girders are

1. Intermediate stiffeners: These carry concentrated transverse forces and divide the web into panels and prevent it from buckling. They resist direct forces from tension field action if utilized and possible external loads. They are usually designed as rigid stiffeners, and consequently, the panels between two rigid transverse stiffeners may be designed independently without an interaction with adjacent panels.
2. Load-bearing stiffeners: These are required at all points where substantial loads are applied and at supports to prevent local buckling and crushing of the web. They are usually designed to be the most effective. This is achieved when a further increase of the stiffener cross section does not significantly increase the resistance of the stiffened plate.

The stiffeners at the supports are termed 'end posts'. They are always double sided to avoid eccentricity at the introduction of large reaction forces.

Cross sections of stiffeners may be open or closed and single or double sided, as shown in Figure 5.15.

5.3.4.1 Transverse stiffener

Transverse stiffeners should preferably provide a rigid support up to the ultimate limit state for a plate with or without longitudinal stiffeners. Hence, they should be designed for both strength and stiffness. Due to inevitable geometrical imperfections, transverse stiffeners carry deviation forces from the adjacent compressed panels including out-of-plane bending (see Figure 5.16).

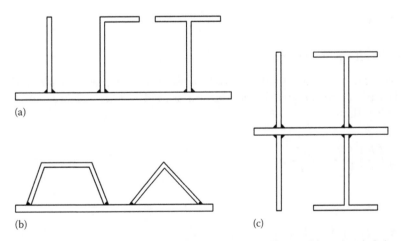

(a)

(b)   (c)

Figure 5.15 Typical cross section of stiffeners: (a) single sided open stiffeners; (b) single sided closed stiffeners; (c) double sided stiffeners.

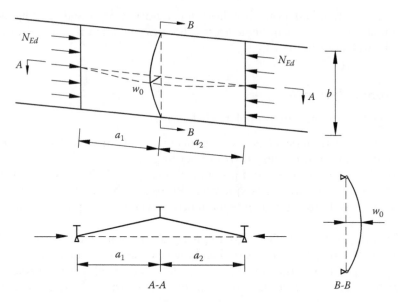

Figure 5.16 Mechanical model for transverse stiffener.

In principle, based on the second-order elastic analysis, both the following criteria should be satisfied at the ultimate limit state:

- The maximum stress in the stiffener under design load should not exceed the yield strength:

$$\sigma_{max} \leq \frac{f_y}{\gamma_{M1}} \tag{5.19}$$

- The additional deflection should not exceed $b/300$:

$$w \leq \frac{b}{300} \tag{5.20}$$

where b is the plate width.

In the absence of transverse loads or axial forces in the stiffeners, the strength and deflection criteria earlier are satisfied if they have a second moment of area of the transverse stiffener not less than

$$I_{st} = \frac{\sigma_m}{E}\left(\frac{b}{\pi}\right)^4\left(1 + w_0\frac{300}{b}u\right) \tag{5.21}$$

where

$$\sigma_m = \frac{\sigma_{cr,c}}{\sigma_{cr,p}}\frac{N_{Ed}}{b}\left(\frac{1}{a_1} + \frac{1}{a_2}\right)$$

$$u = \frac{\pi^2 E e_{max}}{f_y 300 b / \gamma_{M1}} \geq 1.0$$

w_0 is the initial imperfection and $w_0 \leq (1/300)\min(a_1, a_2, b)$
a_1 and a_2 are the panel lengths either side if the stiffener is under consideration
b is the height of the stiffener. They are defined in Figure 5.16
e_{max} is the distance from the edge of the stiffener to the centroid of the stiffener
N_{Ed} is the maximum compressive force in the adjacent panels
$\sigma_{cr,c}$, $\sigma_{cr,p}$ are the elastic critical stresses for column-like buckling and plate-like buckling

For unstiffened plates, the critical column buckling stress is given by

$$\sigma_{cr,c} = \frac{\pi^2 E t^2}{12(1 - v^2) a^2} \tag{5.22}$$

The critical plate buckling stress is given by

$$\sigma_{cr,p} = k_\sigma \frac{\pi^2 E}{12(1 - v^2)} \left(\frac{t}{b}\right)^2 = 190,000 k_\sigma \left(\frac{t}{b}\right)^2 \tag{5.23}$$

where k_σ is the buckling factor corresponding to the stress ratio ψ and boundary. For long plates ($\alpha = a/b \geq 1.0$), k_σ is given in Table 5.2 or Table 5.3 as appropriate; for plates with $\alpha = a/b < 1.0$, and subject to uniform compression, k_σ is given as

$$k_\sigma = \left(\alpha + \frac{1}{\alpha}\right)^2$$

For other details about $\sigma_{cr,c}$ and $\sigma_{cr,p}$, see Section 4 and Annex A in EN 1993-1-5.

Table 5.2 Internal compression elements

Stress distribution (compression positive)	Effective width b_{eff}
σ_1 [uniform compression] σ_2 b_{e1} b_{e2} b	$\psi = 1$ $b_{eff} = \rho b$ $b_{e1} = 0.5 b_{eff}$ $b_{e2} = 0.5 b_{eff}$
σ_1 [tapered compression] σ_2 b_{e1} b_{e2} b	$1 > \psi \geq 0$ $b_{eff} = \rho b$ $b_{e1} = \dfrac{2}{5 - \psi} b_{eff}$ $b_{e2} = b_{eff} - b_{e1}$
b_c b_t σ_1 σ_2 b_{e1} b_{e2} b	$\psi < 0$ $b_{eff} = \rho b_c = \rho b (1 - \psi)$ $b_{e1} = 0.4 b_{eff}$ $b_{e2} = 0.6 b_{eff}$

$\psi = \sigma_2 / \sigma_1$	1	$1 > \psi > 0$	0	$0 > \psi > -1$	-1	$-1 > \psi > -3$
Buckling coefficient k_σ	4.0	$8.2/(1.05 + \psi)$	7.8	$7.81 - 6.29\psi + 9.78\psi^2$	23.9	$5.98(1 - \psi)^2$

Table 5.3 Internal compression elements

Stress distribution (compression positive)	Effective width b_{eff}

$1 > \psi \geq 0$
$b_{eff} = \rho b$

$\psi < 0$
$b_{eff} = \rho b_c = \rho b(1 - \psi)$

$\psi = \sigma_2/\sigma_1$	1	0		-1	$-1 \geq \psi \geq -3$
Buckling coefficient k_σ	0.43	0.57		0.85	$0.57 - 0.21\psi + 0.07\psi^2$

$1 > \psi \geq 0$
$b_{eff} = \rho b$

$\psi < 0$
$b_{eff} = \rho b_c = \rho b(1 - \psi)$

$\psi = \sigma_2/\sigma_1$	1	$1 > \psi > 0$	0	$0 > \psi > -1$	-1
Buckling coefficient k_σ	0.43	$0.578/(0.34 + \psi)$	1.7	$1.7 - 5\psi + 17.1\psi^2$	23.8

When u is less than 1.0, a displacement check is decisive and u is taken as 1.0. Otherwise, a strength check governs.

To avoid LTB of the stiffeners with open cross sections, the following criterion should be satisfied:

$$\frac{I_T}{I_P} \geq 5.3 \frac{f_y}{E} \tag{5.24}$$

where
  I_p is the polar second moment of area of the stiffener alone around the edge fixed to the plate (see Figure 5.17)
  I_T is the St. Venant torsional constant for the stiffener alone

5.3.4.2 Rigid end posts

Besides acting as a bearing stiffener resisting the reaction force at the support, a rigid end post should be able to provide adequate anchorage for the longitudinal component of the membrane tension stresses in the web. Anchorage may be provided in different ways, as

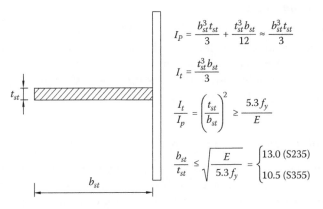

$$I_p = \frac{b_{st}^3 t_{st}}{3} + \frac{t_{st}^3 b_{st}}{12} \approx \frac{b_{st}^3 t_{st}}{3}$$

$$I_t = \frac{t_{st}^3 b_{st}}{3}$$

$$\frac{I_t}{I_p} = \left(\frac{t_{st}}{b_{st}}\right)^2 \geq \frac{5.3 f_y}{E}$$

$$\frac{b_{st}}{t_{st}} \leq \sqrt{\frac{E}{5.3 f_y}} = \begin{cases} 13.0 \ (S235) \\ 10.5 \ (S355) \end{cases}$$

Figure 5.17 Preventing torsional buckling of flat stiffeners.

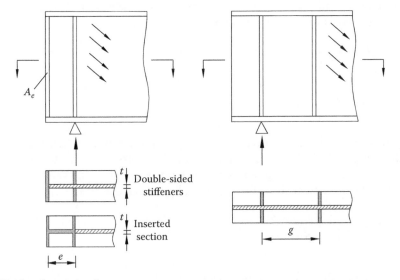

Figure 5.18 Rigid end-post details.

shown in Figure 5.18. A short I beam can be formed at the end of the plate girder by providing two double-sided stiffeners or by inserting a hot-rolled section. This short beam resists longitudinal membrane stresses by its bending strength. The other possibility is to limit the length g of the last panel, so that the panel resists shear loading for the non-rigid end-post conditions.

To assure adequate stiffness and strength of the end post, the centre to centre distance between the stiffeners e should be fulfilled:

$$e > 0.1 h_w$$

$$A_e > \frac{4 h_w t_w^2}{e} \tag{5.25}$$

The other flange of an end post with cross section Au should be checked also as a bearing stiffener to carry reaction force R.

When a rolled section other than flats is used for end post, the section modulus should be not less than $4 h_w t_w^2$ for bending around the horizontal axis perpendicular to the web.

5.3.4.3 Non-rigid end posts

When design criteria for rigid end posts are not fulfilled, the end post should be considered as non-rigid. The reduced shear resistance of the end panels shall be calculated accordingly.

Examples of non-rigid end posts are shown in Figure 5.19.

Generally, single- and double-sided stiffeners may be used as non-rigid end posts. If the end post acts as a bearing stiffener resisting the reaction at the girder support, they should be double sided (see Figure 5.19).

5.3.4.4 Intermediate transverse stiffeners

Intermediate transverse stiffeners that act as a rigid support at the boundary of interior web panels shall be checked for strength and stiffness. If the relevant requirements are not met, transverse stiffeners are considered flexible. Their actual stiffness may be considered in the calculation of the shear buckling coefficient k_τ.

The minimum second moment of area of the effective section of intermediate stiffeners acting as rigid supports for web panels should fulfilled that

$$I_{st} \geq 1.5 \frac{h_w^3 t^3}{a^3} \quad \text{for } \frac{a}{h_w} < \sqrt{2}$$

$$I_{st} \geq 0.75 h_w^3 t^3 \quad \text{for } \frac{a}{h_w} \geq \sqrt{2}$$

$$(5.26)$$

The strength is checked for the axial force $N_{st,ten}$ coming from the tension field action in the two adjacent panels. EN 1993-1-5 gives a simplified procedure for the determination of $N_{st,ten}$. The axial force $N_{st,ten}$ is taken as the difference between the shear force V_{Ed} in the panels and the elastic critical shear force carried by the tension field action:

$$N_{st,ten} = V_{Ed} - \frac{1}{\overline{\lambda}_w^2} t h_w \frac{f_{yw}}{\sqrt{3}\gamma_{M1}}$$

$$(5.27)$$

where
V_{Ed} is the design shear force in the adjacent panels
$\overline{\lambda}_w$ is the slenderness of the panel adjacent to the stiffener assuming the stiffener under consideration is removed:

$$\overline{\lambda}_w = \frac{h_w}{37.4\varepsilon t \sqrt{k_\tau}}$$

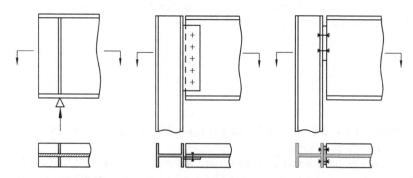

Figure 5.19 Non-rigid end posts.

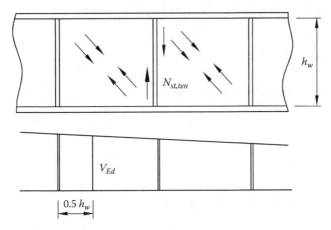

Figure 5.20 Axial force in intermediate transverse stiffener.

The axial force $N_{st,ten}$ is larger than the actual force induced in the transverse stiffeners. A reduced shear force V_{Ed} may be taken at the distance $0.5h_w$ from the edge of the panel with the largest shear force, as shown in Figure 5.20.

If $N_{st,ten}$ is negative, it is taken as 0.

To determine the buckling resistance of the stiffener, a portion of the web may be taken into account. A section of the web in length equal to $15\varepsilon t$ on either side of the stiffener may be considered (Figure 5.21).

For a symmetric stiffener, the effective area $A_{eff,st}$ is given by

$$A_{eff,st} = \sum A_{st} + (30\varepsilon t_w + t_s)t_w \tag{5.28}$$

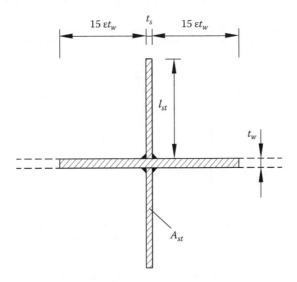

Figure 5.21 Effective cross section of stiffener.

And the effective second moment of the area $I_{eff,st}$ is given by

$$I_{eff,st} = \frac{1}{12}t_s\left(\sum l_{st,eff} + t_w\right)^3 \tag{5.29}$$

where A_{st} is the effective area of the stiffener

$$A_{st} = l_{st,eff} \times t_s$$

$$l_{st,eff} = \min\left(l_{st}, 14\varepsilon t_s\right)$$

I_{st} is the second moment of area of the stiffener

$$I_{st} = \frac{1}{12}t_s l_{st,eff}^3$$

The out-of-plane buckling resistance of the transverse stiffener under transverse loads and shear force should be determined from Clause 6.3.3 or 6.3.4 of EN 1993-1-1, using buckling curve c. The effective length of the stiffener may be taken as $0.75h_w$ when both ends are fixed.

5.3.4.5 Longitudinal stiffeners

Longitudinal stiffeners may be continuous or discontinuous. They are continuous when they pass through the opening made in the transverse stiffeners or when they pass connected to either side of the transverse stiffener. In all other cases, stiffeners have to be considered as discontinuous.

Longitudinal stiffeners are usually installed on the same side of the web as single-sided intermediate transverse stiffeners (see Figure 5.22a and b). Easier fabrication and better fatigue details may be achieved by putting single-sided transverse stiffeners on one side of the web and longitudinal stiffeners on the other side, as shown in Figure 5.22c.

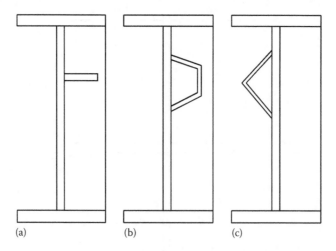

Figure 5.22 Position of longitudinal stiffeners: (a) stiffener with plates; (b) stiffener with channels; (c) stiffener with angles.

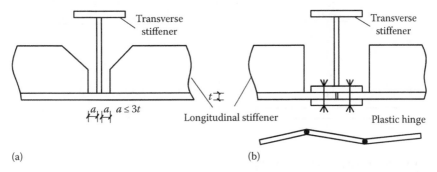

Figure 5.23 Discontinuity of longitudinal stiffeners: (a) appropriate detail; (b) unappropriate detail.

EN 1993-1-5 imposes the following limitations to discontinuous stiffeners:

- To be used only for webs (i.e. not allowed in flanges)
- To be neglected in global analysis
- To be neglected in the calculation of stresses
- To be considered in the calculation of the effective widths of web subpanels
- To be considered in the calculation of the elastic critical stresses $\sigma_{cr,p}$ and $\sigma_{cr,p}$

In other words, discontinuous stiffeners should be taken into account only to increase bending stiffness of the stiffened plates and in the calculation of effective widths of subpanels but should be excluded from transferring forces from one stiffened panel to another. To avoid undesirable local failure modes in the plating, it is important that discontinuous stiffeners terminate close to the transverse stiffeners, as shown in Figure 5.23.

5.3.4.6 Structural details

1. Welded plates

   When the plate thickness is changed, the transverse weld in the plate should be sufficiently close to the transverse stiffener so that the effect of eccentricity may be disregarded. The distance of the weld from the transverse stiffener should fulfil the requirement given in Figure 5.24. Otherwise the effect of eccentricity should be accounted for in the design of the plate.

2. Cut-outs in stiffeners

   Cut-outs in longitudinal stiffeners should be limited in length and depth to prevent plate buckling and to control the net section resistance, as shown in Figure 5.25.

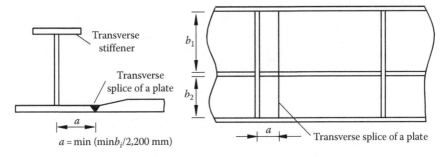

Figure 5.24 Transverse weld in the plate.

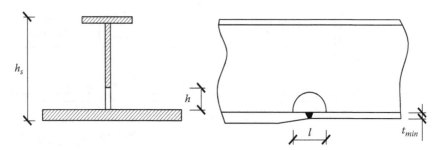

Figure 5.25 Cut-outs in longitudinal stiffeners.

The length l and depth h of the cut-out should not exceed

$l \leq 6t_{min}$ for flat stiffeners in compression
$l \leq 8t_{min}$ for other stiffeners in compression
$l \leq 15t_{min}$ for stiffeners without compression

$$h \leq \min\left(\frac{h_s}{4}, 40 \text{ mm}\right)$$

When $\sigma_{x,Ed} < \sigma_{x,Rd}$ and $l \leq 15t_{min}$, the limiting value of the hole length l for stiffeners in compression may be increased by the factor $\sqrt{\sigma_{x,Rd} / \sigma_{x,Ed}}$ up to $15t_{min}$.

The dimensions of cut-outs in transverse stiffeners should be shown in Figure 5.26.

Additionally, the gross cross section of the web a–a adjacent to the cut-outs should resist a shear force V_{Ed}:

$$V_{Ed} = \frac{I_{net}}{e} \frac{f_{yk}}{\gamma_{M0}} \frac{\pi}{b_G} \tag{5.30}$$

where
$\quad I_{net}$ is the second moment of area for the net section of the transverse stiffener b–b (see Figure 5.26)
$\quad e$ is the maximum distance from the neutral axis of the net section to the outmost edge of the stiffener
$\quad b_G$ is the length of the transverse stiffener between the flanges for web stiffeners

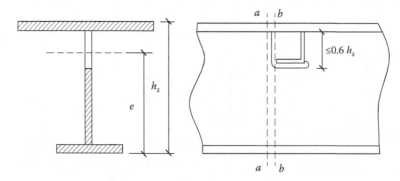

Figure 5.26 Cut-outs in transverse stiffeners.

5.4 DESIGN OF A PLATE GIRDER

A simply supported plate girder has a span of 18 m and carries two concentrated loads on the top flange at the third points, consisting of 400 kN permanent action and 300 kN variable load. In addition, it carries a uniformly distributed permanent load of 20 kN/m, which includes an allowance for self-weight and an imposed load of 10 kN/m. The compression flange is fully restrained laterally. The girder is supported on a heavy stiffened bracket at each end. The materials of both web and flanges are Grade S275 steel. The web will be designed with rigid end post.

5.4.1 Loads, shears and moments

The factored loads are

Concentrated actions = $(1.35 \times 400) + (1.5 \times 300) = 990$ kN
Distributed action = $(1.35 \times 20) + (1.5 \times 10) = 42$ kN/m

The actions and reactions are shown in Figure 5.27a and the shear force diagram in Figure 5.27b. The moments are

$M_C = (1368 \times 6) - (42 \times 6 \times 3) = 7452$ kN-m
$M_E = (1368 \times 6) - (990 \times 3) - (42 \times 9 \times 4.5) = 7641$ kN-m

The bending moment diagram is shown in Figure 5.27c.

5.4.2 Girder section

The critical slenderness ratio for the web can be controlled by flange-induced buckling with $k = 0.4$ as plastic rotation is not utilized. If the flanges resist the bending moment, from Equations 5.4 and 5.5 $A_w = A_{cf}$, the critical h_w/t ratio is given by Equation 5.1.

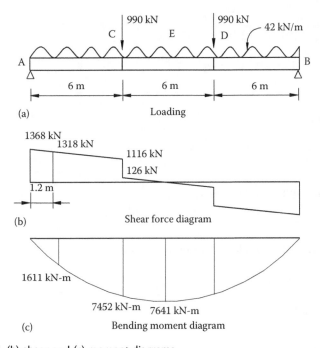

Figure 5.27 (a) Load, (b) shear and (c) moment diagrams.

Assuming the thickness of the flange is larger than 16 mm, then the yield strength $f_{yf} = 265 \ N/mm^2$.

$$\frac{h_w}{t} \leq k \frac{E}{f_{yf}} \sqrt{\frac{A_W}{A_{fc}}} = 0.4 \frac{210,000}{265} \sqrt{1} = 317$$

The optimum depth h_w is given by Equation 5.6.

$$h_w = 3\sqrt{\frac{\lambda M_{Ed}}{f_{yf}}} = 3\sqrt{\frac{317 \times 7642 \times 10^6}{265}} = 2090 \ mm$$

Then the thickness t is given as

$$t = \frac{h_w}{\lambda} = \frac{2090}{317} = 6.6 \ mm$$

Use $t = 8$ mm.

Assuming the flange is Class 2, the maximum ratio flange outstand to flange thickness is 10ε where

$$\varepsilon = \sqrt{\frac{235}{275}} = 0.924$$

So for the maximum flange outstand, $c_f = 10\varepsilon t_f = 10 \times 0.924 \times t_f = 9.24 t_f$

Use $s = t_w$, so the flange width $b_f = 2c_f + 2s + t_w = 18.48 t_f + 24$ mm, and the flange area

$$A_f = (18.48 t_f + 24) t_f$$

From Equation 5.5,

$$A_f = \frac{W_{pl}}{h_w} = \frac{M_{Ed}}{h_w (f_y / \gamma_{M0})} = \frac{7.641 \times 10^9}{(2,090 \times 265)/1.0} = (18.48 t_f + 24) t_f = 13,796 \ mm^2$$

So

$$t_f = 26.6 \ mm$$

$$b_f = 515.6 \ mm$$

Use a 28 × 500 mm plate.

Overall depth

$$h = h_w + 2 t_f = 2090 + 2 \times 28 = 2146 \ mm$$

Use $h = 2000$ mm:

$$h_w = h - 2 t_f = 2000 - 2 \times 28 = 1944 \ mm$$

Actual web slenderness

$$\frac{h_w}{t_w} = \frac{1944}{8} = 243$$

The web is Class 4.

Maximum h_w/t ratio

$$\frac{h_w}{t} = k\frac{E}{f_{yf}}\sqrt{\frac{A_w}{A_{cf}}} = 0.4\frac{210,000}{265}\sqrt{\frac{1,944\times8}{500\times28}} = 321 > 243$$

The actual value is below the allowable and is therefore satisfactory.
  Plastic moment of resistance of the flanges

$$M_{pl,Rd} = A_f\frac{f_y}{\gamma_{M0}}\left(h-t_f\right) = 28\times500\times\frac{265}{1.0}\times1972 = 7316 \text{ kM-m}$$

5.4.3 Design of web

An arrangement for the stiffeners is set out in Figure 5.28. The design strength of web f_{yw} is 275 kN/m; the design strength of flange f_{yf} is 265 kN/m.

  1. Panel I
      Web contribution
      Intermediate stiffener is 1.2 m from the support, so $a = 1.2$ m:

$$\frac{h_w}{a} = \frac{1944}{1200} = 1.62$$

$$\frac{a}{h_w} = 0.62 < 1.0$$

For $a/h_w < 1.0$,

$$k_\tau = 4 + 5.34\left(\frac{h_w}{a}\right)^2 = 18.01$$

$$\frac{h_w}{t_w} = \frac{1944}{8} = 243 > \frac{31}{1.0}\varepsilon\sqrt{k_\tau} = \frac{31}{1.0}0.924\sqrt{18.01} = 121.6$$

Verification of shear buckling is necessary.

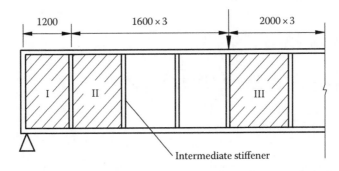

Figure 5.28 Stiffener arrangement.

Determine the normalized web slenderness ratio $\bar{\lambda}_w$:

$$\bar{\lambda}_w = \frac{h_w}{37.4\varepsilon t\sqrt{k_\tau}} = \frac{1944}{37.4\times0.924\times8\times\sqrt{18.01}} = 1.657$$

As $\bar{\lambda}_w > 1.08$

$$\chi_w = \frac{1.37}{0.7+\bar{\lambda}_w} = 0.58$$

The shear resistance contribution from the web

$$V_{bw,Rd} = \chi_w h_w t_w \frac{f_{yw}}{\sqrt{3}\gamma_{M1}} = 0.58\times\frac{275\times1944\times8}{\sqrt{3}\times1.0} = 1432 \text{ kN}$$

Flange contribution

$$c = a\left(0.25 + \frac{1.6b_f t_f^2 f_{yf}}{t_w h_w^2 f_{yw}}\right) = 1200\left(0.25 + \frac{1.6\times500\times28^2\times265}{8\times1944^2\times275}\right) = 325 \text{ mm}$$

$$M_{Ed} = 1368\times1.2 - 42\times\frac{1.2^2}{2} = 1611 \text{ kN-m}$$

$$M_{f,Rd} = M_{pl,Rd} = 7316 \text{ kN-m}$$

The shear resistance contribution from the flanges

$$V_{bf,Rd} = \frac{b_f t_f^2 f_{yf}}{c\gamma_{M1}}\left(1 - \left(\frac{M_{Ed}}{M_{f,Rd}}\right)^2\right) = \frac{500\times28^2\times265}{325\times1}\left(1 - \left(\frac{1611}{7316}\right)^2\right) = 304 \text{ kN}$$

Verification for shear resistance

$$V_{b,Rd} = V_{bw,Rd} + V_{bf,Rd} \le \frac{\eta f_{yw} h_w t_w}{\sqrt{3}\gamma_{M1}}$$

$$= 1432 + 304 \le \frac{1\times275\times1944\times8}{\sqrt{3}\gamma_{M1}}\times10^{-3}$$

$$= 1736 \le 2469 \text{ kN}$$

$$V_{Ed} = 1368 \text{ kN} < V_{b,Rd} = 1736 \text{ kN}$$

2. Panel II

$$V_{Ed} = 1368 - 1.2 \times 42 = 1318 \text{ kN} \quad \text{(at the edge of panel II, } x = 1.2 \text{ m)}$$

$a = 1.6$ m

$$\frac{h_w}{a} = \frac{1944}{1600} = 1.22$$

$$\frac{a}{h_w} = 0.82 < 1.0$$

For $a/h_w < 1.0$,

$$k_\tau = 4 + 5.34\left(\frac{h_w}{a}\right)^2 = 11.88$$

$$\frac{h_w}{t_w} = \frac{1944}{8} = 243 > \frac{31}{1.0}\varepsilon\sqrt{k_\tau} = \frac{31}{1.0}0.924\sqrt{11.88} = 98.7$$

Verification of shear buckling is necessary.
Determine the normalized web slenderness ratio $\bar{\lambda}_w$:

$$\bar{\lambda}_w = \frac{h_w}{37.4\varepsilon t\sqrt{k_\tau}} = \frac{1944}{37.4 \times 0.924 \times 8 \times \sqrt{11.88}} = 2.04$$

As $\bar{\lambda}_w > 1.08$

$$\chi_w = \frac{1.37}{0.7 + \bar{\lambda}_w} = 0.5$$

The shear resistance contribution from the web

$$V_{bw,Rd} = \chi_w h_w t_w \frac{f_{yw}}{\sqrt{3}\gamma_{M1}} = 0.5 \times \frac{275 \times 1944 \times 8}{\sqrt{3} \times 1.0} = 1234 \text{ kN}$$

Flange contribution

$c = 325$ mm

$$M_{Ed} = 1368 \times 1.8 - 42 \times \frac{1.8^2}{2} = 3666 \text{ kN-m}$$

The shear resistance contribution from the flanges

$$V_{bf,Rd} = \frac{b_f t_f^2 f_{yf}}{c\gamma_{M1}}\left(1 - \left(\frac{M_{Ed}}{M_{f,Rd}}\right)^2\right) = \frac{500 \times 28^2 \times 265}{325 \times 1}\left(1 - \left(\frac{3666}{7316}\right)^2\right) = 240 \text{ kN}$$

Verification for shear resistance

$$V_{b,Rd} = V_{bw,Rd} + V_{bf,Rd} \leq \frac{\eta f_{yw} h_w t_w}{\sqrt{3}\gamma_{M1}}$$

$$= 1474 \leq 2469 \text{ kN}$$

$$V_{Ed} = 1318 \text{ kN} < V_{b,Rd} = 1474 \text{ kN}$$

3. Panel III

$$V_{Ed} = 126 \text{ kN} \quad (\text{at the edge of panel III}, x = 6 \text{ m})$$

$$M_{Ed} = 7452 \text{ kN-m} > M_{f,Rd} = 7316 \text{ kN-m}$$

So the contribution of flanges should not be accounted.

$$V_{b,Rd} = V_{bw,Rd} = 1110 \text{ kN} > V_{Ed} = 126 \text{ kN}$$

The girder is satisfactory for the stiffener arrangement assumed.

5.4.4 Stiffener design

1. Rigid end post
   The design will be made using twin stiffener end post as shown in Figure 5.29. The end post is designed to be rigid. Therefore, it must fulfil the following criteria:

$$e \geq 0.1 h_w = 0.1 \times 1.944 \text{ m} = 0.1944 \text{ m}$$

Use $e = 0.2$ m

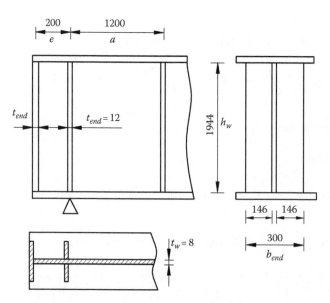

Figure 5.29 Rigid end post.

End-plate dimensions: $\dfrac{b_{end}}{t_{end}} = \dfrac{300}{12}$ (double sided)

$$A_{end} \geq 4\frac{b_w t_w^2}{e}$$

$$= 300 \times 12 \geq 4\frac{1944 \times 8^2}{200}$$

$$= 3600 \text{ mm}^2 \geq 2488 \text{ mm}^2$$

$$\frac{\left(b_{end} - (t_w/2)\right)}{t_{end}} = \frac{146}{12} = 12.2 < 14\varepsilon = 12.9$$

The end-post plate is Class 3 and, therefore, fully effective.

2. *Intermediate stiffeners*

The axial force from the tension field action can usually occur in the first intermediate stiffener near support, where the shear force is the maximum.

The axial force $N_{st,ten}$ is given by

$$N_{st,ten} = V_{Ed} - \frac{1}{\bar{\lambda}_w^2} t_w h_w \frac{f_{yw}}{\sqrt{3}\gamma_{M1}}$$

V_{Ed} is determined at $0.5h_w$ from the stiffener with the higher shear force:

$$V_{Ed} = 1368 - 42(1.2 - 0.5 \times 1.944) = 1358 \text{ kN}$$

The parameter of plate slenderness $\bar{\lambda}_w$ is calculated assuming the stiffener is removed; thus, $a = 1200 + 1600 = 2800$:

$$\frac{h_w}{a} = \frac{1944}{2800} = 0.69$$

$$\frac{a}{h_w} = 1.44 > 1.0$$

For $a/h_w > 1.0$,

$$k_\tau = 5.34 + 4\left(\frac{h_w}{a}\right)^2 = 7.27$$

$$\bar{\lambda}_w = \frac{h_w}{37.4\varepsilon t\sqrt{k_\tau}} = \frac{1944}{37.4 \times 0.924 \times 8 \times \sqrt{7.27}} = 2.61$$

$$N_{st,ten} = 1358 - \frac{1}{2.61^2}\frac{1.944 \times 8 \times 275}{\sqrt{3}} = 995 \text{ kN}$$

The minimum stiffness requirement for all but two panels either side of the central stiffener is given by the case $a < \sqrt{2}h_w$.

So design on the least value of a

$$I_s = 1.5\frac{h_w^3 t^3}{a^2} = 1.5\frac{1944^3 \times 8^3}{1200^2} = 3.93 \times 10^6 \text{ mm}^4$$

Use a 12 mm thick plate; then the total breath of the stiffener l_{st} is given by

$$b = \sqrt[3]{\frac{12 \times 3.92 \times 10^6}{12}} = 157.7 \text{ mm}$$

Use l_{st} = 160 mm.

Strength check
Effective area

$$\frac{l_{st}}{t_s} = \frac{160}{12} = 13.3 > 14\varepsilon = 12.9$$

$$l_{st,eff} = 14\varepsilon t = 14 \times 0.924 \times 12 = 155 \text{ mm}$$

$$A_{eff,st} = 2A_{st} + (30\varepsilon t_w + t_s)t_w = 2 \times 155 \times 12 + (30 \times 0.924 \times 8 + 12) \times 8 = 5590 \text{ mm}^2$$

The axial force that can be carried by the stiffener N_{Rd} is given as

$$N_{Rd} = A_{eff,st}\frac{f_y}{\gamma_{M0}} = 5590 \times \frac{275}{1.0} \times 10^{-3} = 1537 \text{ kN}$$

$$N_{Rd} > N_{st,ten} = 995 \text{ kN}$$

And resistance N_{Rd} exceeds the values of N_{Ed} of all stiffeners.

Buckling check
Effective length $l_e = 0.75h_w = 0.75 \times 1944 = 1458$ mm

Effective second moment of area

$$I_{eff,st} = \frac{1}{12}t_s\left(\sum l_{st,eff} + t_w\right)^3 = \frac{12}{12} \times (2 \times 155 + 8)^3 = 32,157,432 \text{ mm}^4$$

$$i_{eff,st} = \sqrt{\frac{I_{eff,st}}{A_{eff,st}}} = \sqrt{\frac{32,157,432}{5,590}} = 75.8 \text{ mm}$$

$$\bar{\lambda} = \frac{l_{cr}}{i_{eff,st}\lambda_1} = \frac{1458}{75.8 \times 93.9 \times 0.924} = 0.22 > 0.2$$

Use the buckling curve c and the imperfect factor $\alpha = 0.49$:

$$\phi = 0.5\left(1 + \alpha\left(\bar{\lambda} - 0.2\right) + \bar{\lambda}^2\right) = 0.5\left(1 + 0.49(0.22 - 0.2) + 0.22^2\right) = 0.53$$

$$\chi = \frac{1}{\phi + \sqrt{\phi^2 - \bar{\lambda}^2}} = \frac{1}{0.53 + \sqrt{0.53^2 - 0.22^2}} = 0.99$$

$$N_{b,Rd} = \chi N_{Rd} = 0.99 \times 1537 = 1521 \text{ kN}$$

$$N_{b,Rd} > N_{st,ten} = 995 \text{ kN}$$

The stiffener is satisfactory.

Chapter 6

Tension members

6.1 USES, TYPES AND DESIGN CONSIDERATIONS

6.1.1 Uses and types

A tension member transmits a direct axial tensile force between two points in a structural frame. An axially loaded tension member is subjected to uniform, normal tensile stresses at all cross sections along its length. Ropes supporting a load or cables in a suspension bridge are obvious examples. In building frames, tension members occur as

1. Tension chords and internal ties in trusses
2. Tension bracing members
3. Hangers supporting floor beams

Examples of these members are shown in Figure 6.1.
  The main sections used for tension members are

1. Open sections such as angles, channels, tees, joists, universal beams and columns.
2. Closed sections: circular, square and rectangular hollow sections.
3. Compound and built-up sections: Double angles and double channels are common compound sections used in trusses. Built-up sections are used in bridge trusses.

Round bars, flats and cables can also be used for tension members where there is no reversal of load. These elements as well as single angles are used in cross bracing, where the tension diagonal only is effective in carrying a load, as shown in Figure 6.1d. Common tension member sections are shown in Figure 6.2.

6.1.2 Design considerations

Theoretically, the tension member is the most efficient structural element, but its efficiency may be seriously affected by the following factors:

1. The end connections: For example, bolt holes reduce the member section.
2. The member may be subject to reversal of load, in which case it is liable to buckle because a tension member is more slender than a compression member.
3. Many tension members must also resist moment as well as axial load. The moment is due to eccentricity in the end connections or to lateral load on the member.

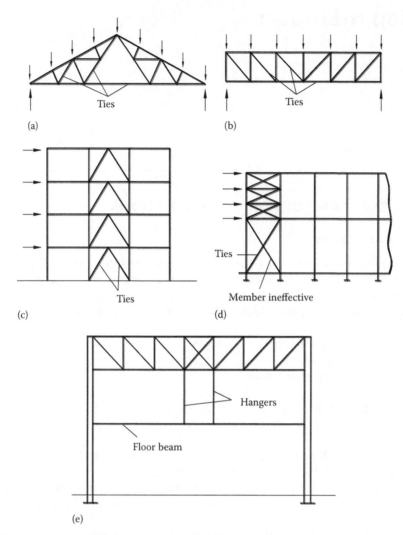

Figure 6.1 Tension members in building: (a) roof truss; (b) lattice girder; (c) multi-storey building; (d) industrial building; (e) hangers supporting floor beam.

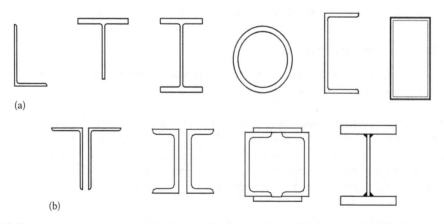

Figure 6.2 Tension member sections: (a) rolled and formed sections; (b) compound and built-up sections.

6.2 END CONNECTIONS

Some common end connections for tension members are shown in Figure 6.3a and b. Comments on the various types are

1. Bolt or threaded bar: The strength is determined by the tensile area at the threads.
2. Single angle connected through one leg: The outstanding leg is not fully effective, and if bolts are used, the connected leg is also weakened by the bolt hole.

Full-strength joints can be made by welding. Examples occur in lattice girders made from hollow sections. However, for ease of erection, most site joints are bolted and welding is normally confined to shop joints.

Site splices are needed to connect together large trusses that have been fabricated in sections for convenience in transport. Shop splices are needed in long members or where the member section changes. Examples of bolted and welded splices in tension members are shown in Figure 6.2c and d.

6.3 STRUCTURAL BEHAVIOUR OF TENSION MEMBERS

6.3.1 Stress redistribution and connection

A tension member is often connected to the main or other members by bolts or welds. It is usual to assume that the distribution of tensile stresses in the tension member is uniform. When connected using bolts, stress concentrations would occur at holes as shown in Figure 6.4a. Connection details may affect the validity of this assumption in two ways.

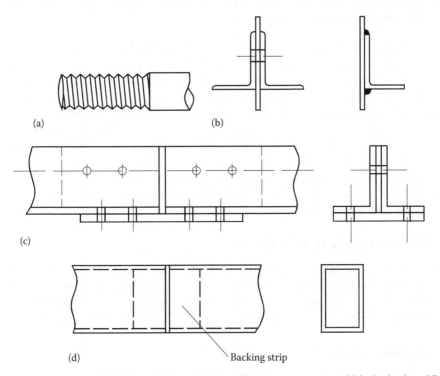

Figure 6.3 End connections and splices: (a) threaded bar; (b) angle connections; (c) bolted splice; (d) welded splice.

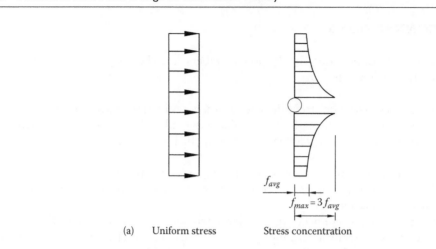

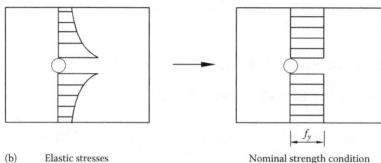

Figure 6.4 Distribution of stress of tension member with hole: (a) stress concentration due to hole; (b) equivalent stress distribution.

Firstly, if bolting is used, the cross-sectional area is reduced at bolt holes and the stresses around the holes are increased locally. From the theory of elasticity, the tensile stress adjacent to a hole will be about two or three times the average stress on the net area. The high local stress concentrations will cause favourable stress redistribution under static loading situations. Secondly, some eccentricity in connections is often inevitable, and secondary moments are therefore induced. These problems may be accounted for by the use of an effective net area rather than the gross area in the calculation of the design plastic resistance.

When a tension member with a hole is loaded statically, the point adjacent to the hole reaches the yield stress f_y first. The further loading, the stress at that point, remains constant at yield stress, and each fibre away from the hole progressively reaches the yield stress, as shown in Figure 6.4b. Deformations continue with an increase load; finally, rupture of the member occurs.

6.3.2 Concentrically tensile force

A tension member can fail by reaching one of two limit states: excessive deformation or fracture. To prevent excessive deformation, initiated by yielding, the load on the gross section must be small enough that the stress on the gross section is than the yield stress f_y. To prevent fracture, the stress on the net section must be less than the ultimate stress f_u.

6.3.2.1 Member without holes

A steel member of constant cross-sectional area A and length L subjected to a concentrically tensile load N is shown in Figure 6.5. When loaded, the member elongates by an amount e.

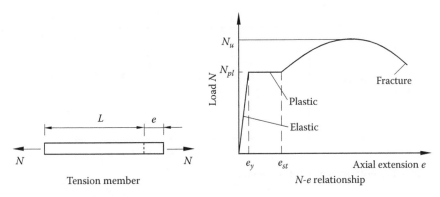

Figure 6.5 Load–extension behaviour of a perfect tension member.

The load–extension relationship for the member is similar to the material stress–strain relationship, as shown in Figure 6.5. The initial portion of the curve shows a linear elastic response typical of a ductile material. That is, unloading of the member anywhere in that region causes the member to return to its original undeformed state.

The tensile force N, the strain ε and the extension e are related in the elastic region by the expressions

$$e = \varepsilon L = \frac{NL}{EA} \tag{6.1}$$

where E is the Young's modulus of elasticity. Linear behaviour continues until the stress reaches the yield stress f_y of the material.

The yield load of the member in tension is given by

$$N_{pl} = Af_y \tag{6.2}$$

When the applied load reaches the yield load, the extension increases suddenly with little or no increase in load until fibres begin to strain harden. After strain hardening commences, the load increases slowly until it reaches the ultimate strength of the tension member:

$$N_u = Af_u \tag{6.3}$$

Here, f_u is the ultimate tensile stress of the material.

Beyond the ultimate strength, a local cross section of the member necks down, and the applied load N decreases until fracture occurs.

6.3.2.2 Member with holes

For members connected by bolting, the section resistance is weakened by the reduction of the cross-sectional area due to the presence of the holes, and an additional check is required. The tension member may fail before the gross yield load N_{pl} is reached by fracturing at s hole. The fracture load is given by

$$N_u = A_{net}f_u \tag{6.4}$$

where A_{net} is the net area of the cross section.

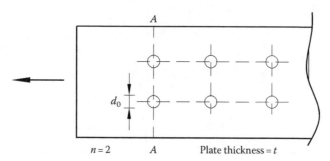

Figure 6.6 Non-staggered arrangement of fasteners.

The net area of the cross section is the gross area less the deductions for bolt holes and other openings. For each fastener hole, the deduction is the gross cross-sectional area of the hole.

When fasteners are not staggered, the total area to be deducted from any cross section perpendicular to the member axis is the maximum sum of the sectional areas of the holes as shown in Figure 6.6. The net area of the cross section is given by

$$A_{net} = A - nd_0t \tag{6.5}$$

where n is the number of holes in line A–A.

When fasteners are staggered, the total area to be deducted should be taken as the greater of

1. The sum of the sectional area of the holes on any line perpendicular to the member axis
2. $t\left(nd_0 - \Sigma s^2/4p\right)$, which measured on any diagonal or zigzag line across the member

   where
   > s is the staggered pitch parallel to the member axis
   > p is the spacing of the centres of the same holes measured perpendicular to the member axis
   > n is the number of holes extending in any diagonal or zigzag line progressively across the section t
   > d_0 are shown in Figure 6.7

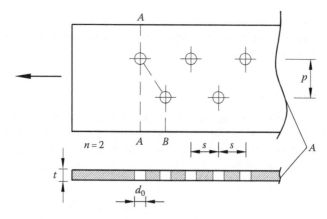

Figure 6.7 Staggered arrangement of fasteners.

An example is taken in Figure 6.7.
On section A–A,

Net area = $A - d_0 t$ – the minimum area is taken as A_{net}.

On section A–B,

Net area = $A - 2d_0 t + \dfrac{s^2 t}{4p}$ – the minimum area is taken as A_{net}.

6.3.3 Eccentrically tensile force

Where members are connected unsymmetrically or the member itself is unsymmetrical (angles, tees, channels), the eccentricity of the connection should be accounted for. The effect of eccentric connections is to induce bending moments in the member, whilst the effect of connecting to some but not all elements in the cross section is to cause those regions most remote from the connection points to carry fewer loads. In the particular case of an angle connected by a single row of bolts in one leg, the member may be treated as concentrically loaded, as shown in Figure 6.8.

The effect of induced bending moment can be approximated by reducing the cross-sectional area of the member to an effective net area, and design it as concentric load. The effective net area $A_{net,eff}$ is dependent on the number of bolts and the pitch p_1. The expression of $A_{net,eff}$ is given by

- For one bolt, $A_{net,eff} = 2(e_2 - 0.5d_0)t$ (6.6)
- For two bolts, $A_{net,eff} = \beta_2 A_{net}$ (6.7)
- For three or more bolts, $A_{net,eff} = \beta_3 A_{net}$ (6.8)

where A_{net} is the net area of the angle.

Values of reduction factors β_2 and β_3 can be found in Table 6.1.

For an unequal angle connected by its smaller leg, A_{net} is taken as the net section area of an equivalent equal angle of leg length equal to the smaller leg of the unequal leg, as shown in Figure 6.9a.

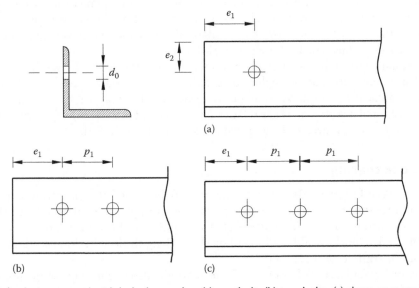

Figure 6.8 Angles connected with bolts by one leg: (a) one bolt; (b) two bolts; (c) three or more bolts.

Table 6.1 Reduction factors β_2 and β_3

Pitch p_1	$\leq 2.5\ d_0$	$\geq 5.0\ d_0$
β_2	0.4	0.7
β_3	0.5	0.7

Note: For intermediate values of pitch p_1, values of β may be determined by linear interpolation.

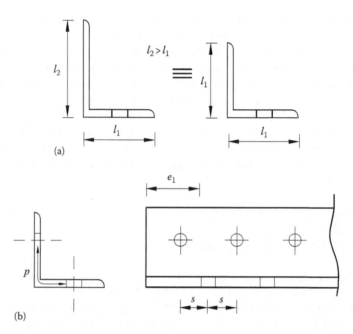

Figure 6.9 Net area of angle with holes: (a) equivalent equalangle; (b) measure of spacing p.

For angles and other members with holes in more than one plane, the spacing p should be measured along the centre of thickness of the material as shown in Figure 6.9b. From Figure 6.9b, the spacing p comprises two straight portions and one curved portion of radius equal to the root radius plus half the material thickness.

For an equal angle or an unequal angle connected by the larger leg by welding rather than bolting, the eccentricity may be neglected, and the effective area A_{net} may be taken as equal to the gross area A. For an unequal angle connected by the small leg by welding, the effective cross-sectional area taken as the relative equal angle uses the small leg. Examples of the effective cross-sectional area to be adopted are shown in Figure 6.10.

6.3.4 Block tearing

Block tearing is a failure mode in which one or more blocks of plate material tear out from the end of a tension member or a gusset plate. For example, the connection of the single angle tension member shown in Figure 6.11 is susceptible to this failure. For the figure illustrated, the shaded block would tend to fail by shear along the longitudinal section ab and by tension on the transverse section bc.

For certain arrangements of bolts, block shear would also occur in gusset plates as shown in Figure 6.12. The blocks are generally rectangular in shape when bolts are arranged

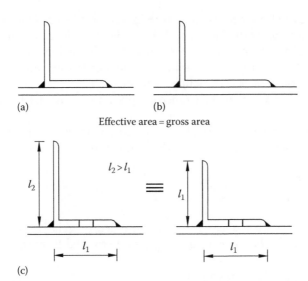

Effective area = gross area

(a)   (b)

(c)

Figure 6.10 Effective area of welding connected angle: (a) equal angle; (b) unequal angle; (c) effective area taken as gross area of a smaller angle.

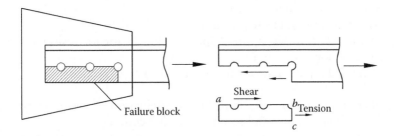

Figure 6.11 Block tearing subject to concentric load.

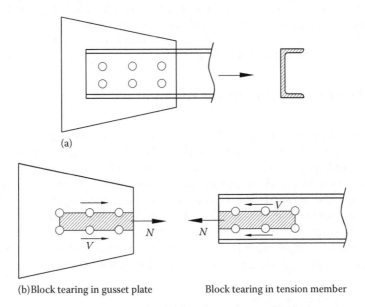

(a)

(b) Block tearing in gusset plate   Block tearing in tension member

Figure 6.12 Block tearing subject to eccentric load: (a) end connection; (b) block tearing.

without stagger and are bounded by the centrelines of bolt holes. The tensile failure occurs along the horizontal limit of the block, and shear plastic yielding occurs along the vertical limit of the block. Thus, block tearing resistance may be defined as the sum of the shear resistance provided by the shear area parallel to the tensile force and the tensile resistance provided by the tension area perpendicular to the load.

Design criteria for block tearing resistance are given in Clause 3.10.2 of EN 1993-1-8.

- For symmetric bolt groups subject to concentric loading (see Figure 6.12), the design resistance for block tearing $V_{eff,1,Rd}$ is determined from the following equation:

$$V_{eff,1,Rd} = \frac{f_u A_{nt}}{\gamma_{M2}} + \frac{(1/\sqrt{3})f_y A_{nv}}{\gamma_{M0}} \tag{6.9}$$

- For member end with a shear force acting eccentric relative to the bolt group, the design resistance for block tearing $V_{eff,2,Rd}$ is determined from the following equation:

$$V_{eff,2,Rd} = 0.5\frac{f_u A_{nt}}{\gamma_{M2}} + \frac{(1/\sqrt{3})f_y A_{nv}}{\gamma_{M0}} \tag{6.10}$$

where
A_{nt} is the net area subjected to tension
A_{nv} is the net area subjected to shear
γ_{M2} is the 1.1 from Singapore Annex (partial factor for resistance to fracture of cross sections in tension)
γ_{M0} is the 1.00 (partial factor for resistance of cross sections)

Equations earlier consider block shear to be a combination of shear fracture and tensile yield, as evidenced from the first and second term, respectively.

Block tearing failure occurs when high tensile forces are transmitted through relatively thin material and a short connection length. Thus, connections detailed using relatively few large diameter bolts and minimum values specified for bolt pitch p_1, end distance e_1 and spacing p should be checked for adequate block shear strength.

Block tearing failure can also occur in welded connections, especially in gusset plates, and is bounded by the centrelines of fillet welds.

6.3.5 Bending moment in tension member

1. Elastic analysis

   Tension members often have bending actions caused by eccentric connections, eccentric loads or transverse loads, including self-weight. These bending actions, which interact with the tensile loads, reduce the ultimate strengths of tension members and must therefore be accounted for in design. Figure 6.13 illustrates that the transverse deflections w of a tension member caused by the axial tension N induce restoring moment M_N that oppose the bending action. It is a common and conservative practice to ignore these restoring moments that are small when either the deflection w or the tensile force N is small.

   When the tension member is subjected to axial tension N and moments M_y and M_z about the y–y and z–z axes, respectively, the individual stresses can be given as

   $$\text{Direct tensile stress } f_t = \frac{N}{A} \tag{6.11}$$

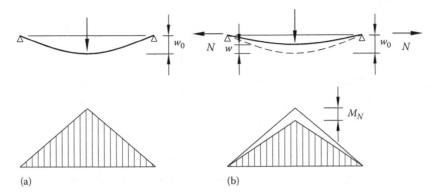

Figure 6.13 Bending moment diagrams of a tension member: (a) transverse load only; (b) transverse load with axial tension.

Tensile bending stress caused by the $y–y$ axis bending moment

$$f_{by} = \frac{M_y}{W_{el,y}}$$ (6.12)

Tensile bending stress caused by the $z–z$ axis bending moment

$$f_{bz} = \frac{M_z}{W_{el,z}}$$ (6.13)

where $W_{el,y}$ and $W_{el,z}$ are elastic modulus for the $y–y$ and $z–z$ axes, respectively.

In this case, the maximum stress in the member can be safely approximated by

$$f_{max} = f_t + f_{by} + f_{bz}$$ (6.14)

The nominal first yield of the member therefore occurs when $f_{max} = f_y$.

The interaction expression to give permissible combinations of stress is given by

$$f_t + f_{by} + f_{bz} = f_y$$ (6.15)

2. Plastic analysis

For a section with two axes of symmetry (as shown in Figure 6.14a), the moment is resisted by two equal areas extending inwards from the extreme fibres. The central core resists the axial tension. The stress distribution is shown in Figure 6.14a for the case where the tension area lies in the web. At higher loads, the area needed to support tension spreads to the flanges, as shown in Figure 6.14b.

For design strength f_y, the tension and bending capacity of the section can be given as

$$N_{Rd} = f_y A$$ (6.16)

$$M_{y,pl,Rd} = f_y W_{pl,y}$$ (6.17)

$$M_{z,pl,Rd} = f_y W_{pl,z}$$ (6.18)

where
$W_{y,pl}$ is the plastic modulus of the section about $y–y$ axis
$W_{z,pl}$ is the plastic modulus of the section about $z–z$ axis

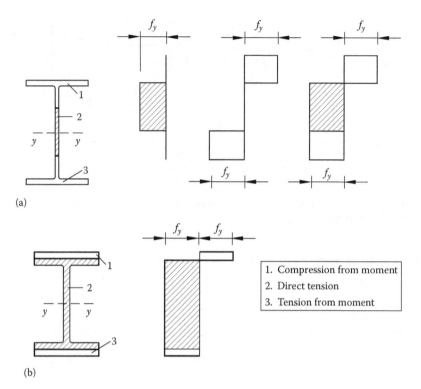

(a)

(b)

1. Compression from moment
2. Direct tension
3. Tension from moment

Figure 6.14 Plastic analysis: tension and moment about one axis: (a) Tension area in web; (b) tension area spread to flanges.

For the section subjected to an axial tension N that is less than N_{Rd} and bending moment about the major y–y axis, if the tension area is in the web (as shown in Figure 6.14a), the length l_N of web supporting N is

$$l_N = \frac{N}{f_y t_w}$$ (6.19)

The reduced moment capacity in the presence of axial load is

$$M_{ry} = \left(W_{pl,y} - \frac{t_w l_N^2}{4} \right) f_y$$ (6.20)

A more complicated formula is needed where the tension area enters the flanges, as shown in Figure 6.14c.

In this case, the applied moment must be less than the reduced moment capacity.

Solutions can be found for sections subject to axial tension and moments about both axes at full plasticity. I sections with two axes of symmetry have been found to give a convex failure surface, which are constructed in terms of N/N_{Rd}, $M_{ry}/M_{y,pl,Rd}$ and $M_{rz}/M_{z,pl,Rd}$, as shown in Figure 6.15, where M_{rz} is the reduced moment capacity for the z–z axis in the presence of axial tension.

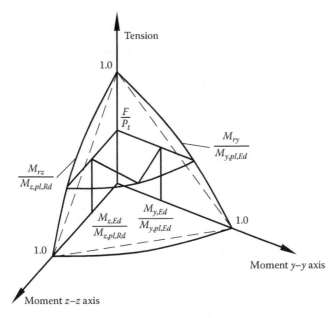

Figure 6.15 Plastic analysis: tension and moment about two axes.

Any point **P** on the failure surface gives the permissible combination of axial load and moments the section can support. A plane may be drawn through the terminal points on the failure surface. This can be used to give a simplified and conservative linear interaction expression:

$$\frac{N}{N_{Rd}} + \frac{M_{y,Ed}}{M_{y,pl,Rd}} + \frac{M_{z,Ed}}{M_{z,pl,Rd}} = 1 \qquad\qquad (6.21)$$

where
   $M_{y,Ed}$ is the applied moment about the y–y axis
   $M_{z,Ed}$ is the applied moment about the z–z axis

6.4 DESIGN OF TENSION MEMBERS

Basic design of a tension member is essentially simple; it can be summarized as a provision of sufficient cross-sectional area to resist the applied force. EN 1993 states that tension resistance should be verified as follows:

$$\frac{N_{Ed}}{N_{t,Rd}} \le 1.0$$

or

$$N_{Ed} \le N_{t,Rd}$$

where
   N_{Ed} is the design value of tension force
   $N_{t,Rd}$ is the resistance of the tension member

6.4.1 Axially loaded tension members

The design tension resistance $N_{t,Rd}$ is limited either by the lesser of yielding resistance of the gross cross section $N_{pl,Rd}$ that can prevent excessive deformation of the member or ultimate (or fracture) resistance of the net cross section.

The yielding resistance of the section is given as

$$N_{pl,Rd} = \frac{Af_y}{\gamma_{M0}}$$

(6.22)

The ultimate resistance of the net cross section is given as

$$N_{u,Rd} = \frac{0.9A_{net}f_u}{\gamma_{M2}}$$

(6.23)

where

A is the area of the cross section
A_{net} is the net area of the section
f_u is the ultimate stress of steel
$\gamma_{M0} = 1.0$
$\gamma_{M2} = 1.1$ from Singapore Annex

For capacity design, the design plastic resistance of the gross cross section should be less than the design ultimate resistance of the net cross section. The factor 0.9 was deduced from a statistical evaluation of a large number of test results.

Inclusion of the 0.9 factor enabled the partial γ_M factor to be harmonized with that applied to the resistance of other connection parts.

6.4.2 Simple tension members

If a member is connected in such a way as to eliminate any moments due to connection eccentricity, the member may be designed as a simple tension member. The method of strength design of simple tension member is similar to the method of concentrically loaded members. The '$0.9A_{net}$' used in Equation 6.23 can be replaced by the effective net area $A_{net,eff}$. For two or more bolts connected member, the factor 0.9 can be replaced by the reduction factors β_2 or β_3. The design ultimate resistance of simple tension member should be determined as follows:

$$\text{With 1 bolt, } N_{u,Rd} = \frac{2.0(e_2 - 0.5d_0)f_u}{\gamma_{M2}}$$

(6.24)

$$\text{With 2 bolts, } N_{u,Rd} = \frac{\beta_2 A_{net}f_u}{\gamma_{M2}}$$

(6.25)

$$\text{With 3 or more bolts, } N_{u,Rd} = \frac{\beta_3 A_{net}f_u}{\gamma_{M2}}$$

(6.26)

If there are two or more bolts across the angle leg, the net area of the connected leg is determined using the method in Section 6.3.2, and the reduction factor-β is determined by the number of bolts along the length of the leg. This arrangement is often used to minimize the size of the connection components, such as the gusset plate. For more details, see Section 6.6.2.

For welded-angle tension member, the effective area discussed in Section 6.3.3 should be used in Equation 6.23 to determine the design resistance.

6.4.3 Tension members with moments

Clause 6.2.1 (7) of EN 1993 gives a conservative approximation for all cross sections of a tension member subjected to combined bending and axial force. The bending may be about one or both principal axes, and the axial force may be tension or compression. The linear interaction expression is given by

$$\frac{N_{Ed}}{N_{Rd}} + \frac{M_{y,Ed}}{M_{y,Rd}} + \frac{M_{z,Ed}}{M_{z,Rd}} \leq 1 \tag{6.27}$$

where
$M_{y,Ed}$ and $M_{z,Ed}$ are the design bending moments
$M_{y,Rd}$ and $M_{z,Rd}$ are the design values of the resistance depending on the cross-sectional classification and including any necessary reduction due to shear effects

Equation 6.27 is rather conservative, and its purpose is to determine the initial member sizing. EN1993 also prescribes refined methods for designing Class 1 and 2 and Class 3 and 4 cross sections.

Class 1 or 2 sections with mono-axial bending and axial tension
For Class 1 and 2 cross sections, the applied bending moment M_{Ed} is less than the reduced plastic moment resistance $M_{N,Rd}$, as follows:

$$M_{Ed} \leq M_{N,Rd} \tag{6.28}$$

where $M_{N,Rd}$ is the design plastic moment resistance reduced due to the axial force N_{Ed}.

1. For a rectangular solid section without fastener holes, the reduced plastic moment resistance is given by

$$M_{N,Rd} = M_{pl,Rd}(1 - n^2) \tag{6.29}$$

where

$$n = \frac{N_{Ed}}{N_{pl,Rd}}$$

2. For small axial loads, the reduction in plastic moment capacity is essentially offset by material strain hardening and may be neglected. For doubly symmetrical I and H sections or other flanges sections subjected to axial force and major (y–y) axis bending moment, no reduction in the major axis plastic moment resistance is necessary provided that both of the following criteria are satisfied:

$$N_{Ed} \leq 0.25 N_{pl,Rd} \tag{6.30}$$

$$N_{Ed} \leq \frac{0.5 h_w t_w f_y}{\gamma_{M0}} \tag{6.31}$$

And similarly, for doubly symmetrical I and H sections subjected to axial force and major (z–z) axis bending moment, no reduction in the minor axis plastic moment resistance is necessary when

$$N_{Ed} \leq \frac{h_w t_w f_y}{\gamma_{M0}} \tag{6.32}$$

If the aforementioned criteria are not met, a reduced plastic moment resistance must be calculated, which is provided in Clause 6.2.9.1(5) of EN 1993.

- For rolled or welded I and H sections with equal flanges,

$$M_{N,y,Rd} = M_{pl,y,Rd}\frac{1-n}{1-0.5a} \quad \text{but } M_{N,y,Rd} \leq M_{pl,y,Rd} \tag{6.33}$$

$$M_{N,z,Rd} = \begin{cases} M_{pl,z,Rd} & \text{for } n \leq a \\ M_{pl,z,Rd}\left[1-\left(\frac{n-a}{1-a}\right)^2\right] & \text{for } n > a \end{cases} \tag{6.34}$$

where a is the ratio of the area of web to the total area and

$$a = \frac{A-2bt_f}{A} \quad \text{but } a \leq 0.5$$

- For rectangular hollow sections of uniform thickness and welded box sections with equal flanges and equal webs,

$$M_{N,y,Rd} = M_{pl,y,Rd}\frac{1-n}{1-0.5a_w} \quad \text{but } M_{N,y,Rd} \leq M_{pl,y,Rd} \tag{6.35}$$

$$M_{N,z,Rd} = M_{pl,z,Rd}\frac{1-n}{1-0.5a_f} \quad \text{but } M_{N,y,Rd} \leq M_{pl,y,Rd} \tag{6.36}$$

where

For hollow sections, $a_w = \dfrac{A-2bt}{A}$ but $a_w \leq 0.5$

For welded box sections, $a_w = \dfrac{A-2bt_f}{A}$ but $a_w \leq 0.5$

For hollow sections, $a_f = \dfrac{A-2bt}{A}$ but $a_f \leq 0.5$

For welded box sections, $a_f = \dfrac{A-2bt_w}{A}$ but $a_f \leq 0.5$

Class 1 or 2 sections with mono-axial bending and axial tension
EN 1993 also provides an alternative to Equation 6.27 for the section resistance to axial tension and bending about both principal axes of Class 1or 2 cross sections. Equation 6.37 represents a more sophisticated convex interaction expression as follows:

$$\left(\frac{M_{y,Ed}}{M_{N,y,Rd}}\right)^\alpha + \left(\frac{M_{z,Ed}}{M_{N,z,Rd}}\right)^\beta \leq 1 \tag{6.37}$$

The interaction expression was discussed in Section 4.4.2(2).

Class 3 or 4 sections

As an overview, for Class 3 sections, the maximum longitudinal stress due to bending moment and axial tension must be less than the yield stress of the steel. For Class 4 sections, the maximum longitudinal stress calculated using the effective cross sections due to combined actions must be less than the yield stress of the steel.

6.5 DESIGN EXAMPLES

6.5.1 Angle connected through one leg

Design a single angle to carry an axial permanent action of 70 kN and an imposed load of 35 kN.

1. Bolted connection

   Design load $= 1.35 \times 70 + 1.5 \times 35 = 147$ kN

   Try an $80 \times 60 \times 7$ angle of S275 steel connected through the long leg by a single low of two 20 mm bolts in 22 mm holes at 80 mm centres, as shown in Figure 6.16a. Design strength

   $t = 7$ mm, $\quad f_y = 275$ N/mm^2, $\quad f_u = 430$ N/mm^2

Net area

   $A = (56.5 + 76.5) \times 7 = 931 \, \text{mm}^2$

   $A_{net} = (56.5 + 76.5 - 22) \times 7 = 777 \, \text{mm}^2$

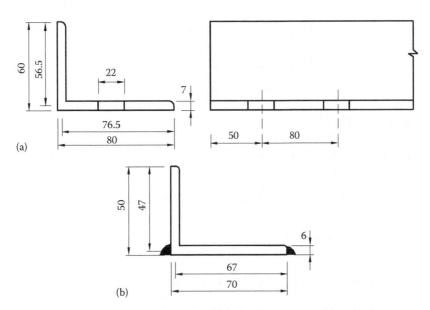

Figure 6.16 Single angle connected through one leg: (a) bolted connection; (b) welded connection.

The spacing

$$p_1 = 80 \text{ mm}$$

$$\frac{p_1}{d_0} = \frac{80}{22} = 3.64$$

Using Table 6.1, for intermediate values of pitch p_1, values of β may be determined by linear interpolation

$$\beta = 0.4 + (3.64 - 2.5)\left(\frac{0.7 - 0.4}{5 - 2.5}\right) = 0.54$$

The ultimate resistance of the net cross section

$$N_{u,Rd} = \frac{\beta A_{net} f_u}{\gamma_{M2}} = \frac{0.54 \times 777 \times 430}{1.1} \times 10^{-3} = 164 \text{ kN}$$

The yielding resistance of the section

$$N_{pl,Rd} = \frac{A f_y}{\gamma_{M0}} = \frac{931 \times 275}{1.0} \times 10^{-3} = 256 \text{ kN} > 164 \text{ kN}$$

$$N_{t,Rd} = N_{u,Rd} = 164 \text{ kN} > 147 \text{ kN}$$

Design resistance in block tearing considering rows as staggered:

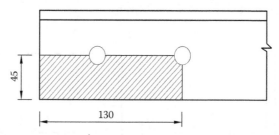

$$A_{nt} = (45 - 11) \times 7 = 238 \text{ mm}^2$$

$$A_{nv} = (130 - 33) \times 7 = 679 \text{ mm}^2$$

$$V_{eff,2,Rd} = 0.5 \frac{f_u A_{nt}}{\gamma_{M2}} + \frac{(1/\sqrt{3}) f_y A_{nv}}{\gamma_{M0}}$$

$$= \frac{0.5 \times 430 \times 238}{1.1} + \frac{275 \times 679}{\sqrt{3}}$$

$$= 46.5 + 107.8 = 154.3 \text{ kN} > 147 \text{ kN}$$

The angle is satisfactory.

2. Welded connection

Try 70 × 50 × 6 L connected through the long leg (see Figure 6.16b):

The net area equal to the cross-sectional area

$$A_{net} = A = (67 + 47) \times 6 = 684 \, \text{mm}^2$$

The yielding resistance of the section

$$N_{pl,Rd} = \frac{A f_y}{\gamma_{M0}} = \frac{684 \times 275}{1.0} \times 10^{-3} = 188 \, \text{kN}$$

The ultimate resistance of the net cross section

$$N_{u,Rd} = \frac{0.9 A_{net} f_u}{\gamma_{M2}} = \frac{0.9 \times 684 \times 430}{1.1} \times 10^{-3} = 241 \, \text{kN}$$

$$N_{t,Rd} = N_{pl,Rd} = 188 \, \text{kN} > 147 \, \text{kN}$$

The angle is satisfactory.

6.5.2 Unequal angle connected to a gusset plate

A 125 × 75 × 10 unequal angle loaded in tension is to be connected to a gusset plate of 10 mm thickness. S275 steel is used for the angle and the gusset plate. Six 20 mm diameter bolts in 22 mm holes are used in a staggered line to connect one leg of the angle to the gusset plate, as shown in Figure 6.17. Check the member for a design tension force of $N_{t,Ed} = 300$ kN.

Design strength

$$t = 10 \, \text{mm}, \quad f_y = 275 \, \text{N/mm}^2, \quad f_u = 430 \, \text{N/mm}^2$$

Gross cross-sectional area of the angle

$$A = 1910 \, \text{mm}^2$$

The yielding resistance of the section

$$N_{pl,Rd} = \frac{A f_y}{\gamma_{M0}} = \frac{1910 \times 275}{1.0} \times 10^{-3} = 525 \, \text{kN}$$

For staggered holes, the net area should be taken as the lesser of

- Gross area minus the deduction for non-staggered holes

- $A - t \left(n d_0 - \sum \dfrac{s^2}{4p} \right)$

Deductions for non-staggered holes

$$A - t d_0 = 1910 - 10 \times 22 = 1690 \, \text{mm}^2$$

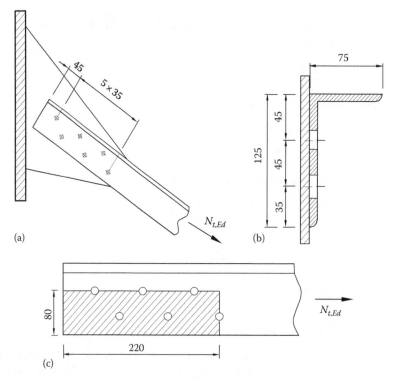

Figure 6.17 Angle connected to a gusset plate: (a) gusset plate connection; (b) connection detail; (c) block tearing.

Net area through two staggered holes

$n = 2$

$s = 35$ mm

$p = 45$ mm

$$A - t\left(nd_0 - \sum \frac{s^2}{4p} \right) = 1910 - 10\left(2 \times 22 - \frac{35^2}{4 \times 45} \right)$$

$$= 1538 \text{ mm}^2 < 1690 \text{ mm}^2$$

Therefore, $A_{net} = 1538$ mm²

The spacing

$p_1 = 2s = 70$ mm

$$\frac{p_1}{d_0} = \frac{70}{22} = 3.18 \text{ mm}$$

By interpolation of three bolts in one row, using Table 6.1, for intermediate values of pitch p_1, values of β_3 may be determined by linear interpolation

$$\beta_3 = 0.4 + (3.18 - 2.5)\left(\frac{0.7 - 0.4}{5 - 2.5} \right) = 0.48$$

The ultimate resistance of the net cross section

$$N_{u,Rd} = \frac{\beta_3 A_{net} f_u}{\gamma_{M2}} = \frac{0.48 \times 1538 \times 430}{1.1} \times 10^{-3} = 288.6\,\text{kN} < 525\,\text{kN}$$

$$N_{t,Rd} = N_{u,Rd} = 288.6\,\text{kN} > 275\,\text{kN}$$

Design resistance in block tearing considering rows as staggered, as shown in Figure 6.16c

$$V_{eff,2,Rd} = 0.5 \frac{f_u A_{nt}}{\gamma_{M2}} + \frac{(1/\sqrt{3})f_y A_{nv}}{\gamma_{M0}}$$

$$= \frac{0.5 \times 430 \times (80 - 22) \times 10}{1.1} + \frac{275 \times (220 - 3 \times 22) \times 10}{\sqrt{3}}$$

$$= 113.4 + 244.5 = 357.9\,\text{kN} > 270\,\text{kN}$$

The angle is satisfactory.

6.5.3 Hanger supporting floor beams

A high-strength Grade S450 steel hanger consisting of a 203 × 203 UC 52 carries the factored loads from beams framing into it and from the floor below, as shown in Figure 6.18a. Check the hanger at the main floor beam connection.

The properties of the section

$$h = 206.2\,\text{mm}$$

$$b = 204.3\,\text{mm}$$

$$t_w = 7.9\,\text{mm}$$

$$t_f = 12.5\,\text{mm}$$

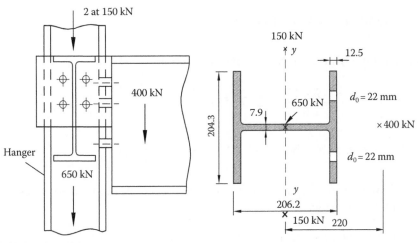

Figure 6.18 High-strength hanger.

$$A = 6630\,\text{mm}^2$$

$$\frac{c_f}{t_f} = 7.04$$

$$\frac{c_w}{t_w} = 20.4$$

$$W_{el,y} = 510\,\text{cm}^3$$

$$W_{pl,y} = 567\,\text{cm}^3$$

$$f_y = 450\,\text{N/mm}^2$$

$$f_u = 550\,\text{N/mm}^2$$

$$\varepsilon = \sqrt{\frac{235}{f_y}} = \sqrt{\frac{235}{450}} = 0.72$$

The net section is shown in Figure 6.18b.

$$A_{net} = 6630 - (2 \times 22) \times 9.5 = 6212\,\text{mm}^2$$

Check the limiting proportions of the section

$$\frac{c_f}{t_f} = 7.04 < 10\varepsilon = 7.2$$

$$\frac{c_w}{t_w} = 20.02 < 72\varepsilon = 51.84$$

The section is Class 2.
    The applied axial load

$$N_{Ed} = 2 \times 150 + 650 + 400 = 1350\,\text{kN}$$

The design plastic resistance of reduction cross section

$$N_{pl.Rd} = \frac{A_{net}f_y}{\gamma_{M0}} = \frac{6212 \times 450}{1.0} = 2794.5\,\text{kN}$$

$$0.25N_{pl.Rd} = 698.5\,\text{kN} < N_{Ed}$$

Therefore, allowance for the effect of axial force on the plastic moment resistance of the cross section must be made.

$$M_{Ed} = 400 \times 0.22 = 88\ \text{kN-m}$$

$$n = \frac{N_{Ed}}{N_{pl.Rd}} = 0.48$$

$$a = \frac{A - 2bt_f}{A} = \frac{6630 - 2 \times 204.3 \times 12.5}{6630} = 0.23$$

$$M_{pl,y,Rd} = \frac{W_{pl,y}f_y}{\gamma_{M0}} = \frac{567 \times 450}{1.0} \times 10^{-3} = 255\,\text{kN-m}$$

$$M_{N,y,Rd} = M_{pl,y,Rd} \frac{1-n}{1-0.5a} = 255 \times \frac{1-0.48}{1-0.5 \times 0.23} = 150.5\,\text{kN-m} < 255\,\text{kN-m}$$

$$M_{Ed} = 88\,\text{kN-m} < M_{N,y,Rd} = 150.5\,\text{kN-m}$$

The hanger is satisfactory.

Chapter 7

Compression members

7.1 TYPES AND USES

7.1.1 Types of compression members

Compression members are one of the basic structural elements and are described by the terms 'columns', 'stanchions' or 'struts', all of which primarily resist axial load.

Columns are vertical members supporting floors, roofs and cranes in buildings. Though internal columns in buildings are essentially axially loaded and are designed as such, most columns are subjected to axial load and moment. The term 'strut' is often used to describe other compression members such as those in trusses, lattice girders or bracing. Some types of compression members are shown in Figure 7.1. Building columns will be discussed in this chapter and trusses and lattice girders are dealt with in Chapter 8.

7.1.2 Compression member sections

Compression members must resist buckling, so they tend to be stocky with square sections. The tube is the ideal shape, as will be shown in the following. These are in contrast to the slender and more compact tension members and deep beam sections.

Rolled, compound and built-up sections are used for columns. Universal columns are used in buildings where axial load predominates, and universal beams are often used to resist heavy moments that occur in columns in industrial buildings. Single angles, double angles, tees, channels and structural hollow sections are the common sections used for struts in trusses, lattice girders and bracing. Compression member sections are shown in Figure 7.2.

7.1.3 Construction details

Construction details for columns in buildings are

1. Beam-to-column connections
2. Column cap connections
3. Column splices
4. Column bases

1. Beam-to-Column Cap Connections
   Typical beam-to-column connections and column cap connections are shown in Figure 7.3a and b, respectively.

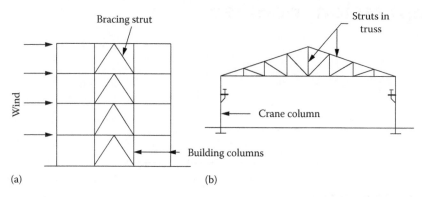

Figure 7.1 Types of compression members: (a) multi-storey building; (b) industrial building.

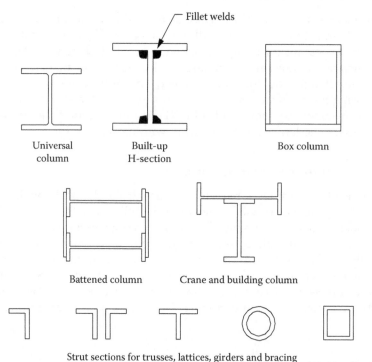

Figure 7.2 Compression member sections.

2. Column splices

   Splices in compression members are discussed in Section 6.2.7 of BS EN1993-1-8. Where the members are not prepared for full contact in bearing, the splice should be designed to transmit all the moments and forces to which the member is subjected at that point. Where the members are prepared for full contact, the splice should provide continuity of stiffness about both axes and resist any tension caused by bending.

   In multistorey buildings, splices are usually located just above floor level. If butted directly together, the ends are usually machined for bearing. Fully bolted splices and combined bolted and welded splices are used. If the axial load is high and the

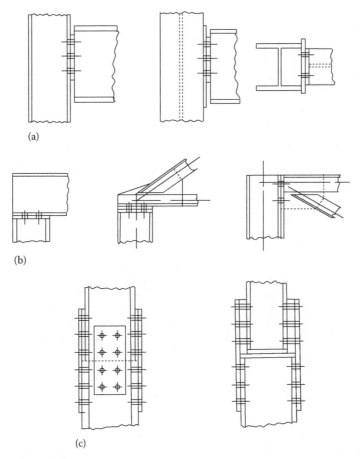

Figure 7.3 Column construction details: (a) flexible beam to column connections; (b) column cap connections; (c) column splices.

moment does not cause tension, the splice holds the columns' lengths in position. Where high moments have to be resisted, high strength or friction-grip bolts or a full-strength welded splice may be required. Some typical column splices are shown in Figure 7.3c.

3. Column bases
   Column bases are discussed in Section 7.10.

7.2 LOADS ON COMPRESSION MEMBERS

Axial loading on columns in buildings is due to loads from roofs, floors and walls transmitted to the column through beams and to self-weight (see Figure 7.4a). Floor beam reactions are eccentric to the column axis, as shown, and if the beam arrangement or loading is asymmetrical, moments are transmitted to the column. Wind loads on multistorey buildings designed to the simple design method are usually taken to be applied at floor levels and to be resisted by the bracing and so do not cause moments.

In industrial buildings, loads from cranes and wind cause moments in columns, as shown in Figure 7.4b. In this case, the wind is applied as a distributed load to the column through the sheeting rails.

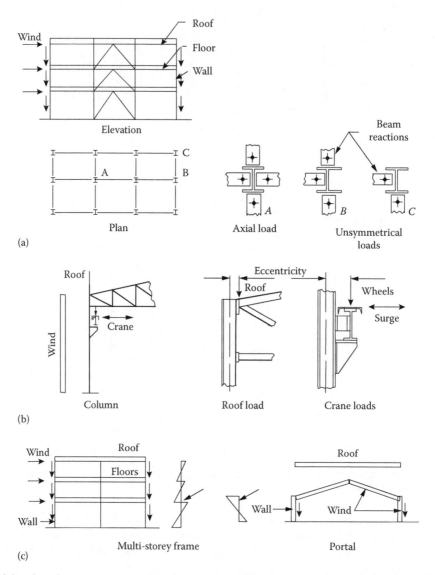

Figure 7.4 Loads and moments on compression members: (a) column in multi-storey buildings; (b) column in an industrial building; (c) rigid frame buildings.

In rigid frame construction, moments are transmitted through the joints from beams to column, as shown in Figure 7.4c. Rigid frame design is outside the scope of this book.

7.3 CLASSIFICATION OF CROSS SECTIONS

The same classification that was set out for beams in Section 5.3 is used for compression members. That is, to prevent local buckling, limiting proportions for flanges and webs in axial compression are given in Table 5.2, BS EN 1993-1-1. The proportions for rolled and welded column sections are shown in Figure 7.5.

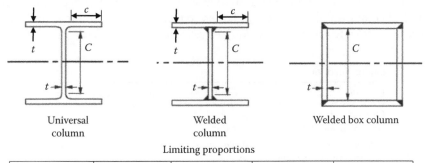

Universal
column

Welded
column

Welded box column

Limiting proportions

Element	Section type	Class 1	Class 2	Class 3
Outstand element of compression flange	Rolled $c/t \le$	9ε	10ε	14ε
	Welded $c/t \le$			
Web subject to compression throughout	Rolled $c/t \le$	33ε	38ε	42ε
	Welded $c/t \le$			

All elements in compression due to axial load: $\varepsilon = \sqrt{235/f_y}$

Figure 7.5 Limiting proportions for rolled and welded column sections.

7.4 AXIALLY LOADED COMPRESSION MEMBERS

7.4.1 General behaviour

Compression members may be classified by length. A short column, post or pedestal fails by crushing or squashing, as shown in Figure 7.6a. The squash load $N_{c,Rd}$ in terms of the design strength is

$$N_{c,Rd} = \frac{f_y A}{\gamma_{M0}}$$

where A is the area of cross section.

A long or slender column fails by buckling, as shown in Figure 7.6b.

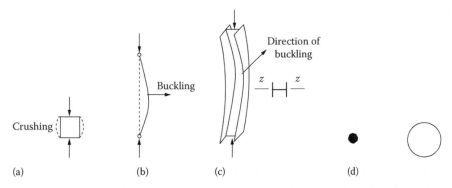

(a) (b) (c) (d)

Figure 7.6 Behaviour of members in axial compression: (a) short column; (b) slender column; (c) universal column; (d) bar and tube of same area.

The failure load is less than the squash load and depends on the degree of slenderness. Most practical columns fail by buckling. For example, a universal column under axial load fails in flexural buckling about the weaker z–z axis (see Figure 7.6c).

Limiting proportions

Element	Section type	Class 1	Class 2	Class 3
Outstand element of compression flange	Rolled $c/t \leq$ Welded $c/t \leq$	9ε	10ε	14ε
Web subject to compression throughout	Rolled $c/t \leq$ Welded $c/t \leq$	33ε	38ε	42ε

All elements in compression due to axial load: $\varepsilon = \sqrt{235/f_y}$

The strength of a column depends on its resistance to buckling. Thus, the column of tubular section shown in Figure 7.6d will carry a much higher load than the bar of the same cross-sectional area.

This is easily demonstrated with a sheet of A4 paper. Open or flat, the paper cannot be stand on edge to carry its own weight, but rolled into a tube, it will carry a considerable load. The tubular section is the optimum column section having equal resistance to buckling in all directions.

7.4.2 Basic strut theory

1. Euler load

Consider a pin-ended straight column. The critical value of axial load P is found by equating disturbing and restoring moments when the strut has been given a small deflection z, as shown in Figure 7.7a. The equilibrium equation is

$$EI_z \frac{d^2z}{dx^2} = -N_z$$

This is solved to give the Euler or lowest critical load:

$$N_{cr} = \pi^2 EI_z/L^2$$

In terms of stress, the equation is

$$N_{cr} = \frac{\pi^2 E}{(L/i_y)^2} = \frac{\pi^2 E}{\lambda^2}$$

where
I_z is the moment of inertia about the minor axis z–z
L is the length of the strut
N is the axial load
i_z is the radius of gyration for the minor axis z–$z = (I_z/A)^{0.5}$
$\sigma_{cr} = N_{cr}/A$ is the Euler critical stress
λ is the slenderness ratio $= L/i_z$

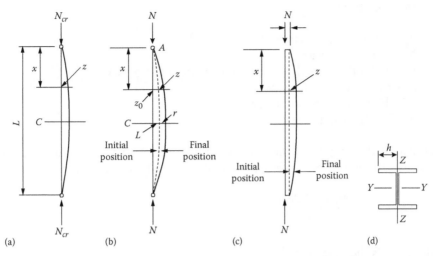

Figure 7.7 Load cases for struts: (a) initially straight strut Euler load; (b) strut with initial curvature; (c) strut with end eccentricity; (d) column section.

The slenderness λ is the only variable affecting the critical stress. At the critical load, the strut is in neutral equilibrium. The central deflection is not defined and may be of unlimited extent. The curve of Euler stress against slenderness for a universal column section is shown in Figure 7.9.

2. Strut with initial curvature

   In practice, columns are generally not straight, and the effect of out of straightness on strength is studied in this section. Consider a strut with an initial curvature bent in a half sine wave, as shown in Figure 7.7b. If the initial deflection at x from A is z_0 and the strut deflects z further under load N, the equilibrium equation is

$$EI_z \frac{d^2z}{dx^2} = N(z + z_0)$$

where deflection $z = \sin(\pi x/L)$. If δ_0 is the initial deflection at the centre and δ the additional deflection caused by N, then it can be shown by solving the equilibrium equation that

$$\frac{\delta}{\delta_0} = \frac{N/N_{cr}}{1 - N/N_{cr}}$$

The maximum stress at the centre of the strut is given by

$$\sigma_{max} = \frac{N}{A} + \frac{N(\delta_0 + \delta)h}{I_Z}$$

where h is shown in Figure 7.7d.

In the earlier equation,

$\sigma_{max} = f_y$ is the yield strength
N_{cr} is the Euler load
$I_z = Ai_z^2$ is the second moment of area about the z–z axis
A is the area of cross section
i_z is the radius of gyration for the z–z axis
h is half the flange breath

If the limited strength is taken as the yield strength, f_y, then the limiting axial buckling load, N_b, for which the earlier elastic analysis is valid, can be rewritten as

$$N_b = Af_y - \frac{N_b(\delta + \delta_0)hA}{I_z}$$

where Af_y is the squash load

$$N_b = Af_y - \frac{\delta_0 h}{i_z^2}\frac{N_b}{(1 - N_b/N_{cr})}$$

which can be solved for the dimensionless limiting factor, $\chi = N_b/Af_y$

$$\frac{N_b}{Af_y} = \left[\frac{1+(1+\eta)N_{cr}/Af_y}{2}\right] - \left\{\left[\frac{1+(1+\eta)N_{cr}/Af_y}{Af_y(1-N_b/N_{cr})}\right]^2 - \frac{N_{cr}}{Af_y}\right\}^{1/2}$$

where $\eta = \delta_0 h/i^2$ is the imperfection parameter.

Alternatively, the earlier equation can be rearranged to give the dimensionless reduction factor, $\chi = N_b/Af_y$:

$$\chi = \frac{1}{\Phi + \sqrt{\Phi^2 - \bar{\lambda}^2}}$$

in which

$$\Phi = \frac{1+\eta+\bar{\lambda}^2}{2} \quad \text{and} \quad \bar{\lambda} = \sqrt{\frac{Af_y}{N_{cr}}}$$

and η can be rewritten as $\eta = \alpha(\bar{\lambda} - 0.2)$ in which α is an imperfection factor corresponding to the appropriate buckling curve in Figure 6.4 and should be obtained using Tables 6.1 and 6.2 from BS EN1993-1-1. For the slenderness, $\bar{\lambda} \leq 0.2$, the buckling effects may be ignored and only cross-sectional checks apply.

The imperfection factor, α, can be obtained from Table 6.1 from BS EN1993-1-1; $\alpha = 0.13, 0.21, 0.34, 0.49$ and 0.76 represent the compressive curves a_0, a, b, c and d, respectively. The reduction factor, χ, can also be obtained from Figure 6.4 using the non-dimensional slenderness, $\bar{\lambda}$.

7.4.3 Practical strut behaviour and design strength

1. Residual stresses

   As noted earlier, in general, practical struts are not straight and the load is not applied concentrically. In addition, rolled and welded strut sections have residual stresses, which are locked in when the section cools.

   A typical pattern of residual stress for a hot-rolled H section is shown in Figure 7.8. If the section is subjected to a uniform load, the presence of these stresses causes yielding to occur first at the ends of the flanges. This reduces the flexural rigidity of the section, which is now based on the elastic core, as shown in Figure 7.8b. The effect on buckling about the z–z axis is more severe than for the y–y axis. Theoretical studies and tests show that the effect of residual stresses can be taken into account by adjusting the Perry factor η.

2. Column tests and design strengths

   An extensive column-testing programme has been carried out, and this has shown that different design curves are required for

   a. Different column sections

   b. The same section buckling about different axes

3. Sections with different thicknesses of metal

   For example, H sections have high residual compressive stresses at the ends of the flanges, and these affect the column strength if buckling takes place about the minor axis.

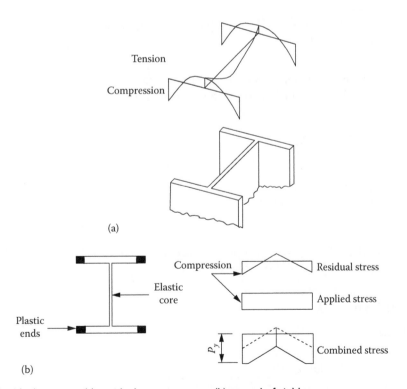

Figure 7.8 Residual stresses: (a) residual stress pattern; (b) spread of yield.

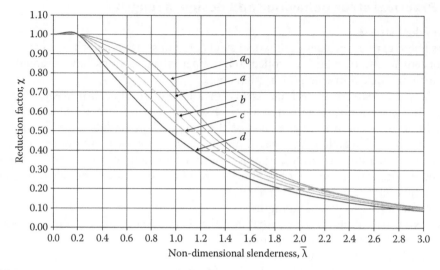

Figure 7.9 Buckling curves.

The total effect of the imperfections discussed earlier (initial curvature, end eccentricity and residual stresses on strength) is combined into the imperfection factor, α.
The non-dimensional slenderness $\overline{\lambda}$ is given by

$$\overline{\lambda} = \sqrt{\frac{Af_y}{N_{cr}}} = \frac{L_{cr}}{i}\frac{1}{\lambda_1} \quad \text{for Class 1, 2 and 3 cross sections}$$

$$\overline{\lambda} = \sqrt{\frac{A_{eff}f_y}{N_{cr}}} = \frac{L_{cr}}{i}\frac{\sqrt{A_{eff}/A}}{\lambda_1} \quad \text{for Class 4 cross sections}$$

where
   L_{cr} is the buckling length in the buckling plane considered
   i is the radius of gyration about the relevant axis

$$\lambda_1 = \pi\sqrt{\frac{E}{f_y}} = 93.9\varepsilon$$

$$\varepsilon = \sqrt{\frac{235}{f_y}} \quad (f_y \text{ in N/mm}^2)$$

For flexural buckling, the appropriate buckling curve given in Figure 7.9 should be determined from Table 6.2 of BS EN1993-1-1.

7.4.4 Slenderness, λ

The slenderness λ is defined as

$$\lambda = \frac{\text{Buckling length}}{\text{Radius of gyration about relevant axis}} = \frac{L_{cr}}{i}$$

The non-dimensional slenderness $\bar{\lambda}$ is given by

$$\bar{\lambda} = \frac{\lambda}{\lambda_1} \quad \text{for Class 1, 2 and 3 cross sections}$$

$$\bar{\lambda} = \frac{\lambda}{\lambda_1}\sqrt{\frac{A_{eff}}{A}} \quad \text{for Class 4 cross sections}$$

7.4.5 Buckling lengths

1. Theoretical considerations

   The actual length of a compression member on any plane is the distance between effective positional or directional restraints in that plane. A positional restraint should be connected to a bracing system, which should be capable of resisting 1% of the axial force in the restrained member.

   The actual column is replaced by an equivalent pin-ended column of the same strength that has a critical length for buckling:

   $$L_{cr} = KL$$

   where

       L is the actual length

       K is the buckling length ratio and K is to be determined from the end conditions

   An alternative method is to determine the distance between points of contraflexure in the deflected strut. These points may lie within the strut length or they may be imaginary points on the extended elastic curve. The distance so defined is the buckling length, L_{cr}.

   The theoretical buckling lengths for standard cases are shown in Figure 7.10. Note that for the cantilever and sway case, the point of contraflexure is outside the strut length.

2. Code definitions and rules

   Unlike the BS 5950: Part 1, no comprehensive guidance on buckling lengths for compression members with different end conditions is provided in the Eurocode 3.

Effective lengths (buckling length) for compression members are set out in Section 4.7.2 of the BS 5950: Part 1. This states that for members other than angles, channels and T sections,

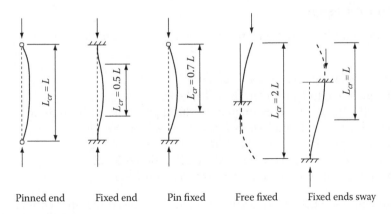

Pinned end   Fixed end   Pin fixed   Free fixed   Fixed ends sway

Figure 7.10 Buckling lengths.

the effective length should be determined from the actual length and conditions of restraint in the relevant plane. The code specifies

1. That restraining members, which carry more than 90% of their moment capacity after reduction for axial load, shall be taken as incapable of providing directional restraint.
2. Table 22 is used for standard conditions of restraint.
3. Appendix D1 is used for stanchions in single-storey buildings of simple construction.
4. Appendix E is used for members forming part of a frame with rigid joints.

The normal buckling lengths L_{cr} are given in Table 22 of the BS5950: Part 1. Some values from this table for various end conditions where L is the actual length are

1. Effectively held in position at both ends
   a.   Restrained in direction at both ends, $L_{cr} = 0.7L$
   b.   Partially restrained in direction at both ends, $L_{cr} = 0.85L$
   c.   Not restrained in direction at either end, $L_{cr} = L$
2. One end effectively held in position and restrained in direction. Other end not held in position
   a.   Partially restrained in direction, $L_{cr} = 1.5L$
   b.   Not restrained in direction, $L_{cr} = 2.0L$

In the absence of Eurocode 3 guidance, readers could consult Table 22 in the BS5950: Part 1 for other cases.

   Note the case for the fixed end strut, where the buckling length is given as 0.7L, is to allow for practical ends where true fixity is rarely achieved. The theoretical value shown in Figure 7.10 is 0.5L.

7.4.6 Buckling resistance

The buckling resistance of a strut is defined in Section 6.3.1 of BS EN1993-1-1 as

1. $N_{b,Rd} = \dfrac{\chi A f_y}{\gamma_{M1}}$ for Class 1, 2 and 3 cross sections

2. $N_{b,Rd} = \dfrac{\chi A_{eff} f_y}{\gamma_{M1}}$ for Class 4 cross sections

7.4.7 Column design

Column design is indirect, and the process is as follows (the tables referred to are in the BS EN1993-1-1):

1. The steel grade and section are selected.
2. The design strength f_y, is taken from Table 3.1.
3. The buckling length L_{cr} is estimated for the appropriate end conditions.
4. The non-dimensional slenderness $\bar{\lambda}$ is calculated for the relevant axis.
5. The imperfection factor is selected from Tables 6.1 and 6.2.
6. The reduction factor, χ, is calculated using Section 6.3.1.2.
7. The buckling resistance is calculated (see Section 7.4.6).

For a safe design, the buckling resistance should just exceed the applied load, and successive trials are needed to obtain an economical design. Load tables can be formed to give the

Table 7.1 Compression resistance of S275 steel UC sections

Serial size (mm)	Mass/m (kg)	Compression resistances (kN) for buckling lengths (m)								
		2	3	4	5	6	7	8	9	10
254 × 254	167	5260	4760	4210	3630	3060	2550	2130	1790	1510
Universal	132	4140	3740	3290	2820	2370	1970	1640	1370	1160
column	107	3340	3010	2640	2260	1880	1560	1300	1080	915
	89	2770	2490	2190	1860	1560	1290	1070	891	752
	73	2360	2110	1840	1550	1290	1060	873	728	613
203 × 203	86	2580	2230	1840	1480	1170	934	755	621	518
Universal	71	2120	1820	1510	1200	952	758	613	504	420
column	60	1840	1570	1280	1010	791	627	506	414	345
	52	1590	1360	1110	871	683	541	436	357	297
	46	1410	1200	970	762	596	472	380	311	259
152 × 152	37	1020	782	568	412	308	237	187	151	125
Universal	30	826	630	455	330	246	189	149	121	99.7
column	23	620	465	331	238	177	136	107	86.5	71.3

compression resistance for various sections for different values of buckling length. Table 7.1 gives compression resistances for some universal column sections. Column sizes may be selected from tables in the *P363 Steel Building Design: Design Data, In accordance with Eurocodes and the UK National Annexes*, Steel Construction Institute.

7.4.8 Example: Universal column

A part plan of an office floor and the elevation of internal column stack A are shown in Figure 7.11a and b. The roof and floor loads are as follows:

Roof

Dead load (total) = 5 kN/m²

Imposed load = 1.5 kN/m².

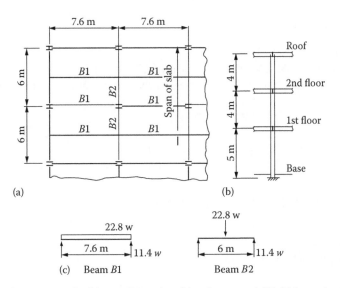

Figure 7.11 Column design example: (a) part floor plan; (b) column stack 'A'; (c) beam loads.

Floors

Dead load (total) = 7 kN/m²

Imposed load = 3 kN/m²

Design column *A* for axial load only. The self-weight of the column, including fire protection, may be taken as 1 kN/m. The roof and floor steels have the same layout. Use Grade S275 steel.

The roof is regarded as a floor for reckoning purposes.

The slabs for the floor and roof are precast one-way spanning slabs. The dead and imposed loads are calculated separately.

1. Loading

   Four floor beams are supported at column A. These are designated as B1 and B2 in Figure 7.11a. The reactions from these beams in terms of a uniformly distributed load are shown in Figure 7.11c:

   Load on beam B1 = 7.6 × 3 × 10 = 22.8 *w* (kN)

   where *w* is the uniformly distributed load. The dead and imposed loads must be calculated separately in order to introduce the different load factors.

   The self-weight of beam B2 is included in the reaction from beam B1.

   The design loading on the column can be set out as shown in Figure 7.12. The design loads are required just above the first floor, the second floor and the base.

2. Column design

   a. Top length = Roof to second floor

      Design load = 415.8 kN

      Try 152 × 152 universal column (UC) 30

      A = 38.3 cm²; i_y = 3.83 cm

      Yield strength f_y = 275 N/mm² (Table 3.1) where section thickness is less than 40 mm.

      If the beam connections are the simple joints, where end rotation is permitted, the buckling length for the column is taken as 0.85*L*:

      $$L_{cr} = 0.85 \times 4000 = 3400 \text{ mm}$$

      Slenderness, λ = 3400/38.3 = 88.8

      From Table 6.2 of EN1993-1-1, for an S275 rolled H section, t_f less than 100 mm, $h/b \leq 1.2$, buckling about the minor z–z axis, imperfection factor, α = 0.49,

      $$\lambda_1 = 93.3\varepsilon = 93.3 \times 0.92 = 85.8$$

      $$\bar{\lambda} = \frac{\lambda}{\lambda_1} = \frac{88.8}{85.8} = 1.035$$

      $$\Phi = 0.5\,[1 + 0.49(1.035 - 0.2) + 1.035^2] = 1.24$$

      $$\chi = \frac{1}{1.24 + \sqrt{1.24^2 - 1.035^2}} = 0.52$$

      Buckling resistance, $N_{b,Rd}$ = 0.52 × 38.3 × 275/10 = 547.7 kN

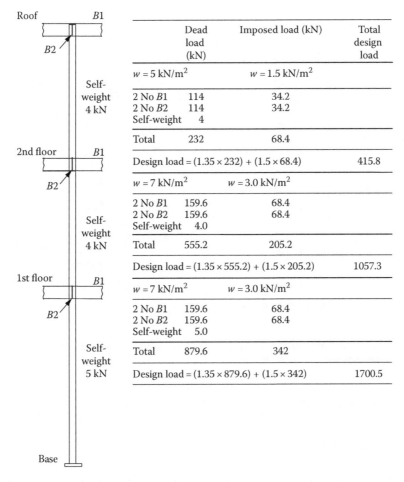

	Dead load (kN)	Imposed load (kN)	Total design load
	$w = 5$ kN/m^2	$w = 1.5$ kN/m^2	
2 No B1	114	34.2	
2 No B2	114	34.2	
Self-weight	4		
Total	232	68.4	
Design load = $(1.35 \times 232) + (1.5 \times 68.4)$			415.8
	$w = 7$ kN/m^2	$w = 3.0$ kN/m^2	
2 No B1	159.6	68.4	
2 No B2	159.6	68.4	
Self-weight	4.0		
Total	555.2	205.2	
Design load = $(1.35 \times 555.2) + (1.5 \times 205.2)$			1057.3
	$w = 7$ kN/m^2	$w = 3.0$ kN/m^2	
2 No B1	159.6	68.4	
2 No B2	159.6	68.4	
Self-weight	5.0		
Total	879.6	342	
Design load = $(1.35 \times 879.6) + (1.5 \times 342)$			1700.5

Figure 7.12 Column design loads.

The column splice and floor beam connections at the second-floor level are shown in Figure 7.13a. The net section at the splice is shown in Figure 7.13b with 4 No. 22 mm diameter holes. The section is satisfactory.

b. Intermediate length – first floor to second floor
Design load = 1057.2 kN
Try 203 × 203 UC 46
$A = 58.7$ cm^2; $i_y = 5.13$ cm
$f_y = 275$ N/mm^2
$\lambda = 3400/51.3 = 66.3$
From Table 6.2 of EN1993-1-1, for an S275 rolled H section, t_f less than 100 mm, $h/b \leq 1.2$, buckling about the minor z–z axis, imperfection factor, $\alpha = 0.49$,

$$\lambda_1 = 93.3\varepsilon = 93.3 \times 0.92 = 85.8$$

$$\bar{\lambda} = \frac{\lambda}{\lambda_1} = \frac{66.3}{85.8} = 0.77$$

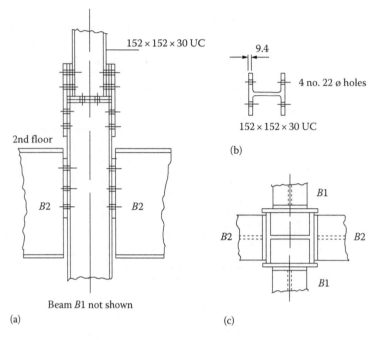

152 × 152 × 30 UC

9.4

4 no. 22 ⌀ holes

152 × 152 × 30 UC

(b)

2nd floor

B2   B2

Beam B1 not shown

(a)

B1

B2 ∶∶∶∶∶   ∶∶∶∶∶ B2

B1

(c)

Figure 7.13 Column connection details: (a) spice and floor beam connections; (b) net section at splice; (c) section at floor beam connections.

$$\Phi = 0.5\,[1 + 0.49(0.77 - 0.2) + 0.77^2] = 0.93$$

$$\chi = \frac{1}{0.93 + \sqrt{0.93^2 - 0.77^2}} = 0.69$$

Buckling resistance, $N_{b,Rd} = 0.69 \times 58.7 \times 275/10 = 1113.8$ kN
   The section is satisfactory.
c. Bottom length – base to first floor
Design load = 1700.5 kN
Try 254 × 254 UC 73
$A = 93.1$ cm^2; $i_y = 6.48$ cm
$f_y = 275$ N/mm^2
The beam connections do restrain the column in direction at the first-floor level. The base is considered pinned. The buckling length is taken as 0.85L:

$$L_{cr} = 0.85 \times 5000 = 4250 \text{ mm}$$

$$\lambda = 4250/64.8 = 65.6$$

From Table 6.2 of EN1993-1-1, for an S275 rolled H section, t_f less than 100 mm, $h/b \leq 1.2$, buckling about the minor z–z axis, imperfection factor, $\alpha = 0.49$,

$$\lambda_1 = 93.3\varepsilon = 93.3 \times 0.92 = 85.8$$

$$\bar{\lambda} = \frac{\lambda}{\lambda_1} = \frac{65.6}{85.8} = 0.76$$

$$\Phi = 0.5 \left[1 + 0.49(0.76 - 0.2) + 0.76^2 \right] = 0.93$$

$$\chi = \frac{1}{0.93 + \sqrt{0.93^2 - 0.76^2}} = 0.68$$

Buckling resistance, $N_{b,Rd} = 0.68 \times 93.1 \times 275/10 = 1741$ kN

The section is satisfactory. The same sections could have been selected from Table 7.1.

7.4.9 Built-up column: Design

The two main types of columns built up from steel plates are the H and box sections shown in Figure 7.2. The classification for cross sections is given in Figure 7.5.

For Class 1, 2 and 3 cross sections, the local compression capacity is based on the gross section.

Class 4 cross sections are dealt with in BS EN1993-1-5. The capacity of these sections is limited by local buckling and the design should be based on the effectively cross-sectional area. The buckling resistance of the compression member should be evaluated from Clause 6.3.1.1.

7.4.10 Example: Built-up column

Determine the compression resistance of the column section shown in Figure 7.14. The weld size is 8 mm. The buckling length of the column is 8 m and the steel is S275.

1. Flanges
   The design strength from Table 3.1 $f_y = 275$ N/mm²
   $t_f = 30$ mm, $t_w = 15$ mm, $\varepsilon = 0.92$
   $c/t_f = [(900/2 - 15/2) - 8]/30 = 14.48 \geq 14\varepsilon$ Class 4

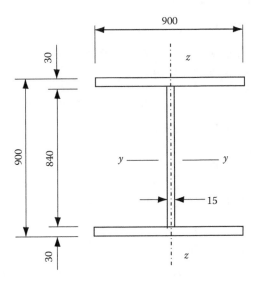

Figure 7.14 Built-up H column.

From BS EN1993-1-5, Table 4.2,

$k_\sigma = 0.43$

$$\overline{\lambda_p} = \frac{c/t}{28.4\varepsilon\sqrt{k_\sigma}} = \frac{14.48}{28.4 \times 0.92\sqrt{0.43}} = 0.845$$

$$\rho = \frac{\overline{\lambda_p} - 0.188}{\lambda_p^2} = 0.92$$

$A_{eff,f} = 0.92 \times 4 \times (900/2 - 15/2 - 8) \times 30 + (15 + 2 \times 8) \times 30 \times 2 = 49829$ mm²

2. Web
    This is an internal element in axial compression:
    $t_w = 15$ mm, $\varepsilon = 0.92$
    $c/t_w = (840 - 2 \times 8)/15 = 54.93 \geq 42\varepsilon$ Class 4
        From BS EN1993-1-5, Table 4.1,

$k_\sigma = 4.0$

$$\overline{\lambda_p} = \frac{b/t}{28.4\varepsilon\sqrt{k_\sigma}} = \frac{840/15}{28.4 \times 0.92\sqrt{4.0}} = 1.071$$

$$\rho = \frac{\overline{\lambda_p} - 0.188}{\lambda_p^2} = 0.77$$

$A_{eff,w} = 0.77 \times 840 \times 15 = 9702$ mm²

3. Properties of the gross section and effective section
    Gross area = $(2 \times 30 \times 900) + (840 \times 15) = 66\,600$ mm²

$$I_z = 2 \times (30 \times 9003/12) + (840 \times 153/12) = 3.645 \times 109 \text{ mm}^4$$

$$i_z = [3.645 \times 10^9/6.66 \times 10^4]^{0.5} = 233.9 \text{ mm}$$

$$\lambda_z = L_{cr}/i_z = 8000/233.9 = 34.2$$

Effective sectional area = $49829 + 9702 = 59531$ mm².
4. Compressive resistance of the column
    From Table 6.2 of EN1993-1-1, for an S275 welded I section, t_f less than 40 mm, buck-ling about the minor z–z axis, imperfection factor, $\alpha = 0.49$,

$$\lambda_1 = 93.3\varepsilon = 93.3 \times 0.92 = 85.8$$

$$\overline{\lambda} = \frac{\lambda}{\lambda_1} = \frac{34.2}{85.8} = 0.40$$

$$\Phi = 0.5\,[1 + 0.49(0.40 - 0.2) + 0.40^2] = 0.63$$

$$\chi = \frac{1}{0.63 + \sqrt{0.63^2 - 0.40^2}} = 0.89$$

Buckling resistance, $N_{b,Rd} = 0.89 \times 59531 \times 275/10^{-3} = 14570$ kN

7.5 BEAM COLUMNS

7.5.1 General behaviour

1. Behaviour classification

As already stated at the beginning of this chapter, most columns are subjected to bending moment in addition to axial load. These members, termed 'beam columns', represent the general load case of an element in a structural frame. The beam and axially loaded columns are limiting cases.

Consider a plastic or compact H section column as shown in Figure 7.15a. The behaviour depends on the column length, how the moments are applied and the lateral support, if any, provided. The behaviour can be classified into the following five cases:

Case 1: A short column subjected to axial load and uniaxial bending about either axis or biaxial bending. Failure generally occurs when the plastic capacity of the section is reached. Note limitations set in (2).

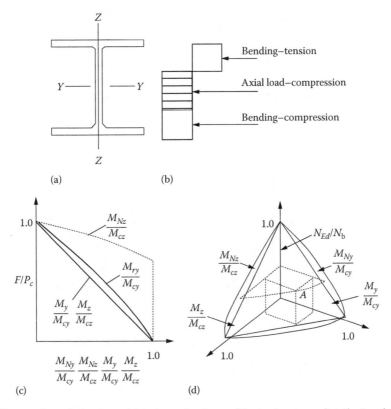

Figure 7.15 Short-column behaviour: (a) universal column; (b) plastic stress distribution bending about YY axis; (c) interaction curves for universal bending YY and ZZ axes; (d) interaction surface biaxial bending full plasticity.

Case 2: A slender column subjected to axial load and uniaxial bending about the major axis *y–y*. If the column is supported laterally against buckling about the minor axis *z–z* out of the plane of bending, the column fails by buckling about the *y–y* axis. This is not a common case (see Figure 7.16a). At low axial loads or if the column is not very slender, a plastic hinge forms at the end or point of maximum moment.

Case 3: A slender column subjected to axial load and uniaxial bending about the minor axis *z–z*. The column does not require lateral support and there is no buckling out of the plane of bending. The column fails by buckling about the *z–z* axis. At very low axial loads, it will reach the bending capacity for *z–z* axis (see Figure 7.16b).

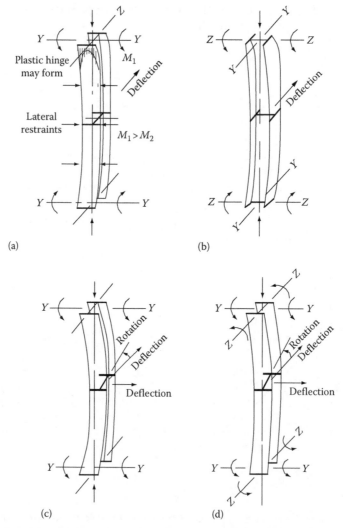

Figure 7.16 Slender columns subjected to axial load and moment: (a) moments about YY axis buckling restrained about ZZ axis; (b) moments about ZZ axis no restraint; (c) moments about YY axis no restraint; (d) moments about YY axis and ZZ axis no restraint.

Case 4: A slender column subjected to axial load and uniaxial bending about the major axis y–y. This time the column has no lateral support. The column fails due to a combination of column buckling about the y–y axis and lateral torsional buckling where the column section twists as well as deflecting in the y–y and z–z planes (see Figure 7.16c).

Case 5: A slender column subject to axial load and biaxial bending. The column has no lateral support. The failure is the same as in Case 4 earlier, but minor axis buckling will usually have the greatest effect. This is the general loading case (see Figure 7.16d).

Some of these cases are discussed in more detail in the following.

2. Short-column failure

The behaviour of short columns subjected to axial load and moment is the same as for tension members subjected to identical loads. This was discussed in Section 7.3.3.

The plastic stress distribution for uniaxial bending is shown in Figure 7.15b. The moment capacity for Class 1 or 2 section in the absence of axial load is given by

$$M_{c,Rd} = M_{pl,Rd} = \frac{W_{pl}f_y}{\gamma_{M0}}$$

where W_{pl} is the plastic modulus for the relevant axis.

The interaction curves for axial load and bending about the two principal axes separately are shown in Figure 7.17a. Note the effect of the limitation of bending capacity for the y–y axis.

These curves are in terms of $N_{Ed}/N_{c,Rd}$ against M_{Ny}/M_{cy} and M_{Nz}/M_{cz}, where N_{Ed} is the applied axial load, $N_{c,Rd}$ is the $f_y A/\gamma_{M0}$ the squash load, M_{Ny} is the reduced moment capacity about the y–y axis in the presence of axial load, M_{cy} is the moment capacity about the y–y axis in the absence of axial load, M_{Nz} is the reduced moment capacity about the z–z axis in the presence of axial load and M_{cz} is the moment capacity about the z–z axis in the absence of axial load.

Values for M_{Ny} and M_{Nz} are calculated using equations for reduced plastic modulus given in BS EN1993-1-1.

For Class 1 and 2 sections subjected to axial and bending,

$$M_{N,y,Rd} = M_{pl,y,Rd}\,\frac{(1-n)}{(1-0.5a)} \quad \text{but} \quad M_{N,y,Rd} \le M_{pl,y,Rd}$$

for $n \le a$: $M_{N,z,Rd} = M_{pl,z,Rd}$

for $n \ge a$: $M_{N,z,Rd} = M_{pl,z,Rd}\left[1 - \left(\frac{n-a}{1-a}\right)^2\right]$

where $n = N_{Ed}/N_{pl,Rd}$

$a = (A - 2bt_f)/A$ but $a \le 0.5$

A plane drawn through the terminal points of the surface gives a linear interaction expression:

$$\frac{N_{Ed}}{N_{c,Rd}} + \frac{M_y}{M_{cy}} + \frac{M_z}{M_{cz}} = 1.0$$

This results in a conservative design.

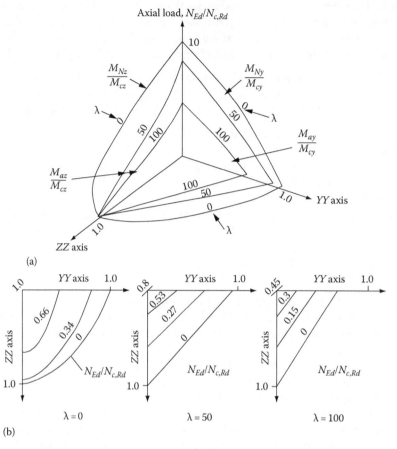

Figure 7.17 Failure surface for slender beam column: (a) failure surfaces; (b) failure contours.

3. Failure of slender columns

   With slender columns, buckling effects must be taken into account. These are minor axis buckling from axial load and lateral torsional buckling from moments applied about the major axis. The effect of moment gradient must also be considered.

   All the imperfections, initial curvature, eccentricity of application of load and residual stresses affect the behaviour. The end conditions have to be taken into account in estimating the effective length.

   Theoretical solutions have been derived and compared with test results. Failure surfaces for H section columns plotted from the more exact approach given in the code are shown in Figure 7.17a for various values of slenderness. Failure contours are shown in Figure 7.17b. These represent lower bounds to exact behaviour.

   The failure surfaces are presented in the following terms:

   Slenderness $\lambda = 0$ \qquad $N_{Ed}/N_{b,Rd}$; M_y/M_{cy}; M_z/M_{cz}
   \qquad\qquad $\lambda = 50, 100$ \quad $N_{Ed}/N_{b,Rd}$; M_{ay}/M_{cy}; M_{az}/M_{cz}

   Some of the terms were defined in Section 7.5.1(2). New terms used are

   M_{ay} is the maximum buckling moment about the x–x axis in the presence of axial load.
   M_{az} is the maximum buckling moment about the y–y axis in the presence of axial load.

The following points are to be noted:

a. M_{cy}, the moment capacity about the y–y axis.
b. At zero axial load, slenderness does not affect the bending strength of an H section about the y–y axis.
c. At high values of slenderness, the buckling resistance moment $M_{b,Rd}$ about the y–y axis controls the moment capacity for bending about that axis.
d. As the slenderness increases, the failure curves in the $N_{Ed}/N_{b,Rd}$, y–y axis plane change from convex to concave, showing the increasing dominance of minor axis buckling.

For design purposes, the results are presented in the form of an interaction expression, and this is discussed in Section 7.5.2.

7.5.2 Code design procedure

The code design procedure for compression members with moments is set out in Section 6.3.3 of BS EN1993-1-1. This requires the following to be satisfied:

$$\frac{N_{Ed}}{\chi_y N_{Rk}/\gamma_{M1}} + k_{yy}\frac{M_{y,Ed}}{\chi_{LT}M_{y,Rk}/\gamma_{M1}} + k_{yz}\frac{M_{z,Ed}}{M_{z,Rk}/\gamma_{M1}} \leq 1$$

$$\frac{N_{Ed}}{\chi_z N_{Rk}/\gamma_{M1}} + k_{zy}\frac{M_{y,Ed}}{\chi_{LT}M_{y,Rk}/\gamma_{M1}} + k_{zz}\frac{M_{z,Ed}}{M_{z,Rk}/\gamma_{M1}} \leq 1$$

where

N_{Ed}, $M_{y,Ed}$ and $M_{z,Ed}$ are the design values of the compression force and the maximum moments about the y–y and z–z axis along the member

χ_y and χ_z are the reduction factors due to flexural buckling from Cl. 6.3.1 from BS EN1993-1-1

χ_{LT} is the reduction factor due to lateral torsional buckling from Cl. 6.3.2 from BS EN1993-1-1

k_{yy}, k_{yz}, k_{zy}, k_{zz} are the interaction factors in accordance to Annex A or B in the BS EN1993-1-1

7.5.3 Example of beam column design

A braced column 4.5 m long is subjected to the factored end loads and moments about the x–x axis, as shown in Figure 7.18a. The column is held in position but only partially restrained in direction at the ends. Check that a 203 × 203 UC 52 in Grade S275 steel is adequate.

1. Column-section classification
   Yield strength from Table 3.1 f_y = 275 N/mm^2.
   Factor $\varepsilon = (235/f_y)0.5 = 0.92$ (see Figure 7.18b).
   Flange $c/t_f = 88/12.5 = 7 < 9.0\varepsilon$.
   Web $c/t_w = 160.8/8.0 = 20.1 < 33\varepsilon$.
   Referring to Table 5.2, the Section is Class 1.
   Section properties for 203 × 203 UC 52 are

   $A = 66.4$ cm^2
   $i_y = 8.91$ cm
   $i_z = 5.16$ cm

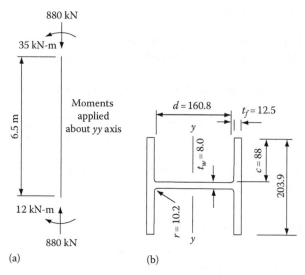

Figure 7.18 Beam column design example: (a) column length and loads; (b) trial section.

$W_{pl,y} = 567$ cm^3
$W_{el,y} = 510$ cm^3
Yield strength $f_y = 275$ N/mm^2
Buckling lengths $L_{cr,y} = 0.85 \times 4500 = 3825$ mm
Buckling lengths $L_{cr,z} = 0.85 \times 4500 = 3825$ mm

Maximum slenderness ratio

$$\lambda_y = L_{cr}, \ y/i_y = 3825/89.1 = 43$$
$$\lambda_z = L_{cr}, \ z/i_z = 3825/51.6 = 74.1$$

From Table 6.2 of EN1993-1-1, for an S275 rolled H section, t_f less than 100 mm, $b/b \le 1.2$, buckling about the major y–y axis, curve b, imperfection factor, $\alpha = 0.34$, buckling about the minor z–z axis, curve c, imperfection factor, $\alpha = 0.49$,

$$\lambda_1 = 93.9\varepsilon = 93.9 \times 0.92 = 86.4$$

$$\overline{\lambda_y} = \frac{\lambda_y}{\lambda_1} = \frac{43}{86.4} = 0.50$$

$$\Phi = 0.5 \ [1 + 0.34(0.50 - 0.2) + 0.50^2] = 0.68$$

$$\chi_y = \frac{1}{0.68 + \sqrt{0.68^2 - 0.5^2}} = 0.88$$

$$\overline{\lambda_z} = \frac{\lambda_z}{\lambda_1} = \frac{74.1}{86.4} = 0.86$$

$$\Phi = 0.5 \ [1 + 0.49(0.86 - 0.2) + 0.86^2] = 1.03$$

$$\chi_z = \frac{1}{1.03 + \sqrt{1.03^2 - 0.86^2}} = 0.63$$

Buckling lengths $L_{cr} = 0.85 \times 4500 = 3825$ mm

$$M_{cr} = C_1 \frac{\pi^2 E I_z}{L_{cr}^2} \left(\frac{I_w}{I_z} + \frac{L_{cr}^2 \, G I_T}{\pi^2 E I_z} \right)^{0.5}$$

For ratio of end moment, $\psi = 12/35 = 0.34$; $C_1 = 1.443$

$$M_{cr} = 1.443 \times \frac{\pi^2 \times 210{,}000 \times 17{,}800{,}000}{3825^2} \left(\frac{167 \times 10^9}{14{,}500{,}000} + \frac{3825^2 \times 81{,}000 \times 318{,}000}{\pi^2 \times 210{,}000 \times 17{,}800{,}000} \right)^{0.5}$$

$M_{cr} = 536.4$ kN-m

$$\overline{\lambda_{LT}} = \sqrt{\frac{W_y f_y}{M_{cr}}} = \sqrt{\frac{567 \times 10^3 \times 275}{536.4 \times 10^6}} = 0.54$$

$$\chi_{LT} = \frac{1}{\Phi_{LT} + \sqrt{\Phi_{LT}^2 - \overline{\lambda}_{LT}^2}}$$

$$\Phi_{LT} = 0.5 \left[1 + \alpha_{LT} \left(\overline{\lambda}_{LT} - 0.2 \right) + \overline{\lambda}_{LT}^2 \right]$$

For $h/b \leq 2$, curve a, $\alpha_{LT} = 0.21$

$$\Phi_{LT} = 0.5 \times [1 + 0.21(0.54 - 0.2) + 0.54^2] = 0.68$$

$$\chi_{LT} = \frac{1}{0.68 + \sqrt{0.68^2 - 0.54^2}} = 0.915$$

Determination of interaction factors k_{ij} (Annex B in BS EN1993-1-1)
    Table B.3. Equivalent uniform moment factors, C_m

For $\psi = 0.4$, C_{my}, $C_{mz} = 0.6 + 0.4(0.34) = 0.74 \geq 0.4$

$$k_{yy} = C_{my} \left(1 + \left(\overline{\lambda}_y - 0.2 \right) \frac{N_{Ed}}{\chi_y N_{Rk}/\gamma_{M1}} \right) \leq C_{my} \left(1 + 0.8 \frac{N_{Ed}}{\chi_y N_{Rk}/\gamma_{M1}} \right)$$

$$k_{yy} = 0.74 \left(1 + (0.5 - 0.2) \frac{880}{0.88 \times 66.4 \times 10^{-3} \times 275/1.0} \right) = 0.86$$

$$k_{zy} = 0.6, \ k_{yy} = 0.52$$

$$k_{zz} = C_{mz}\left(1+\left(\bar{\lambda}_z - 0.2\right)\frac{N_{Ed}}{\chi_z N_{Rk}/\gamma_{M1}}\right) \le C_{mz}\left(1+0.8\frac{N_{Ed}}{\chi_z N_{Rk}/\gamma_{M1}}\right)$$

$$k_{zz} = 0.74\left(1+(0.86-0.2)\frac{880}{0.63\times 66.4\times 10^{-3}\times 275/1.0}\right) = 1.11$$

$$k_{zz} = 1.11, \quad k_{yz} = 0.6, \quad k_{zz} = 0.67$$

$$\frac{N_{Ed}}{\chi_y N_{Rk}/\gamma_{M1}} + k_{yy}\frac{M_{y,Ed}}{\chi_{LT}M_{y,Rk}/\gamma_{M1}} + k_{yz}\frac{M_{z,Ed}}{M_{z,Rk}/\gamma_{M1}} \le 1$$

$$\frac{888}{0.88\times 1826/1.0} + 0.86\frac{35}{0.915\times 156.2/1.0} = 0.76$$

$$\frac{N_{Ed}}{\chi_z N_{Rk}/\gamma_{M1}} + k_{zy}\frac{M_{y,Ed}}{\chi_{LT}M_{y,Rk}/\gamma_{M1}} + k_{zz}\frac{M_{z,Ed}}{M_{z,Rk}/\gamma_{M1}} \le 1$$

$$\frac{888}{0.63\times 1826/1.0} + 0.25\frac{35}{0.915\times 156.2/1.0} = 0.833$$

The section is satisfactory.

7.6 ECCENTRICALLY LOADED COLUMNS IN BUILDINGS

7.6.1 Eccentricities from connections

The eccentricities to be used in column design in simple construction for beam, and truss reactions are given as follows:

1. For a beam supported on a cap plate, the load should be taken as acting at the face of the column or edge of the packing.
2. For a roof truss on a cap plate, the eccentricity may be neglected provided that simple connections are used.
3. In all other cases, the load should be taken as acting at a distance from the face of the column equal to 100 mm or at the centre of the stiff bearing, whichever gives the greater eccentricity.

The eccentricities for the various connections are shown in Figure 7.19.

7.6.2 Moments in columns of simple construction

For columns in simple construction, the moments are calculated using eccentricities given in Section 7.6.1. For multistorey columns effectively continuous at splices, the following interaction equation could be used for the overall buckling check:

$$\frac{N_{Ed}}{N_{b,z,Rd}} + \frac{M_{y,Ed}}{M_{b,Rd}} + 1.5\frac{M_{z,Ed}}{M_{z,Rd}} \le 1.0$$

where $M_{b,Rd}$ is the buckling resistance moment for a simple column calculated using non-dimensional slenderness for lateral torsional buckling, $\bar{\lambda}_{LT} = 0.9\bar{\lambda}_z$.

Other terms are defined in Section 7.5.2.

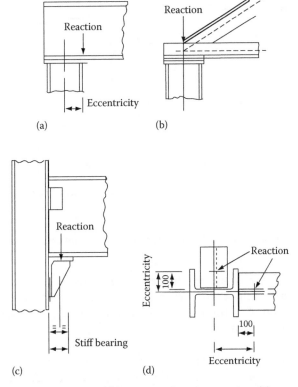

Figure 7.19 Eccentricities for end reactions: (a) beam to column connection; (b) truss to column connection; (c) beam supported on bracket; (d) eccentricities for beam-column connections.

7.6.3 Example: Corner column in a building

The part plan of the floor and roof steel for an office building is shown in Figure 7.20a and an elevation of the corner column is shown in Figure 7.20b. The roof and floor loading is as follows:

Roof

Total dead load = 5 kN/m²
Imposed load = 1.5 kN/m²

Floors

Total dead load = 7 kN/m²
Imposed load = 3 kN/m²

The self-weight of the column, including fire protection, is 1.5 kN/m. The external beams carry the following loads due to brick walls and concrete casing (they include self-weight):

Roof beams – parapet and casing = 2 kN/m
Floor beams – walls and casing = 6 kN/m

The reinforced concrete slabs for the roof and floors are one-way slabs spanning in the direction shown in the figure.

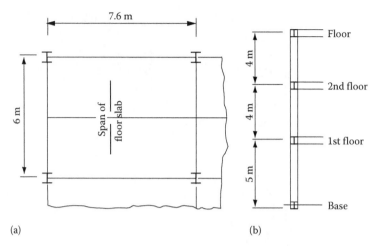

Figure 7.20 Corner-column design example: (a) roof and floor plan; (b) column stack.

Design the corner column of the building using S275 steel.

The roof is counted as a floor. Note that the reduction is only taken into account in the axial load on the column. The full imposed load at that section is taken in calculating the moments due to eccentric beam reactions.

1. Loading and reaction floor beams

   Mark numbers for the floor beams are shown in Figure 7.21a. The end reactions are calculated in the following:

   Roof

   > B1 dead load = $(5 \times 3.8 \times 1.5) + (2 \times 3.8) = 36.1$ kN
   > Imposed load = $1.5 \times 3.8 \times 1.5 = 8.55$ kN
   > B2 dead load = $5 \times 3.8 \times 3 = 57.0$ kN
   > Imposed load = $1.5 \times 3.8 \times 3 = 17.1$ kN
   > B3 dead load = $(0.5 \times 57.0) + (2 \times 3) = 34.5$ kN
   > Imposed load = $0.5 \times 17.1 = 8.55$ kN

   Floors

   > B1 dead load = $(7 \times 3.8 \times 1.5) + (6 \times 3.8) = 62.7$ kN
   > Imposed load = $3 \times 3.8 \times 1.5 = 17.1$ kN
   > B2 dead load = $7 \times 3.8 \times 3 = 79.8$ kN
   > Imposed load = $3 \times 3.8 \times 3 = 34.2$ kN
   > B3 dead load = $(0.5 \times 79.8) + (6 \times 3) = 57.9$ kN
   > Imposed load = $0.5 \times 34.2 = 17.1$ kN

   The roof and floor beam reactions are shown in Figure 7.21b.

2. Loads and moments at roof and floor levels

   The loading at the roof, second floor, first floor and base is calculated from values shown in Figure 7.21b. The values for imposed load are calculated separately, so that reductions permitted can be made and the appropriate load factors for dead and imposed load introduced to give the design loads and moments.

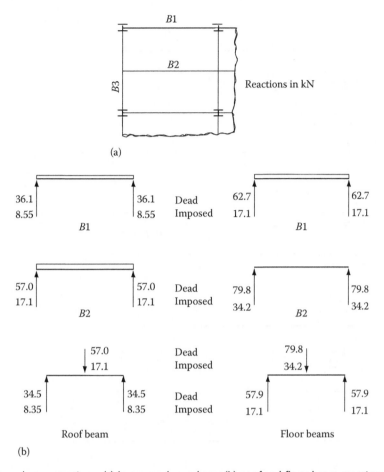

Figure 7.21 Floor–beam reactions: (a) beam mark numbers; (b) roof and floor beam reactions.

The moments due to the eccentricities of the roof and floor beam reactions are based on the following assumed sizes for the column lengths:

Roof to second floor

152 × 152 UC 30

Second floor to first floor

203 × 203 UC 52

First floor to base

203 × 203 UC 52.

The eccentricities of the beam reactions and the column loads and moments from dead and imposed loads are shown in Figure 7.22.

3. Column design
   Roof to second floor
   Referring to Figure 7.22, the design load and moments at roof level are

Axial load $N_{Ed} = (1.35 \times 70.6) + (1.5 \times 17.1) = 121.0$ kN
Moment $M_y = (1.35 \times 6.07) + (1.5 \times 1.51) = 10.46$ kN-m

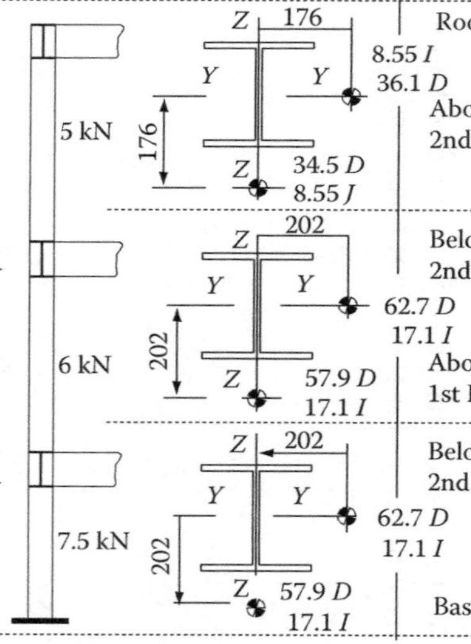

Column stack	Column sections	Position	Dead load	Imposed load	Dead M_x	Imposed M_x	Dead M_y	Imposed M_y
Roof	(176 section, 5 kN; Z–Y; 8.55 I, 36.1 D)	Roof	70.6	17.1	6.07	1.57	6.35	1.51
	(34.5 D, 8.55 I)	Above 2nd floor	76.6	17.1	3.34	0.99	3.52	0.99
2nd floor	(202 section, 6 kN; 62.7 D, 17.1 I)	Below 2nd floor	197.2	31.3	8.35	2.47	9.01	2.47
	(57.9 D, 17.1 I)	Above 1st Floor	203.2	31.3	5.84	1.79	6.33	1.73
1st floor	(202 section, 7.5 kN; 62.7 D, 17.1 I)	Below 2nd floor	323.8	85.5	5.84	1.79	6.33	1.73
	(57.9 D, 17.1 I)	Base	331.3	—	—	—	—	—

Loads are in kN. moments in kN-m

Figure 7.22 Loads and moments from actual and imposed loads.

$M_z = (1.35 \times 6.35) + (1.5 \times 1.51) = 10.84$ kN-m

Try 152 × 152 UC 30, the properties of which are

$A = 38.3$ cm²; $i_z = 3.82$ cm; $W_{el,y} = 221.2$ cm³
$W_{el,z} = 73.06$ cm³; $W_{pl,y} = 247.1$ cm³; $W_{pl,z} = 111.2$ cm³

The roof beam connections and column-section dimensions are shown in Figure 7.23a.
The yield strength from Table 3.1

$f_y = 275$ N/mm²

Flange, $c_f/t_f = 6.98 < 9.0\varepsilon$ – Class 1
Web, $c_w/t_w = 19.0 < 33\varepsilon$ – Class 1

The limiting proportions are from Table 5.2 of the code.

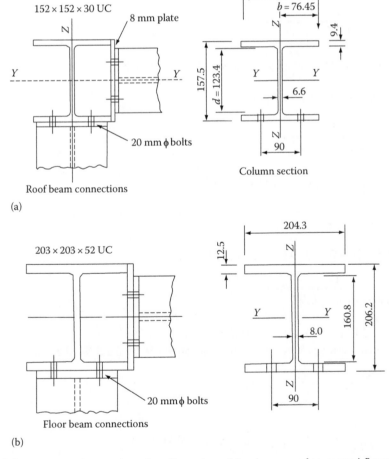

Figure 7.23 Column connections and section dimensions: (a) column–roof to second floor; (b) column–second floor to base.

Overall buckling check
The column is effectively held in position and partially restrained in direction at both ends. The axial load is applied at the centre of the column. The buckling lengths are:

Buckling lengths $L_{cr,y} = 0.85 \times 4000 = 3400$ mm
Buckling lengths $L_{cr,z} = 0.85 \times 4000 = 3400$ mm

Maximum slenderness ratio

$$\lambda_z = L_{cr,z}/i_z = 3400/38.2 = 89$$

From Table 6.2 of EN1993-1-1, for an S275 rolled H section, t_f less than 100 mm, $h/b \leq 1.2$, buckling about the major y–y axis, curve b, imperfection factor, $\alpha = 0.34$, buckling about the minor z–z axis, curve c, imperfection factor, $\alpha = 0.49$,

$$\lambda_1 = 93.9\varepsilon = 93.9 \times 0.92 = 86.4$$

$$\overline{\lambda_z} = \frac{\lambda_z}{\lambda_1} = \frac{89}{86.4} = 1.03$$

$$\Phi = 0.5\,[1 + 0.49(1.03 - 0.2) + 1.03^2] = 1.23$$

$$\chi_z = \frac{1}{1.23 + \sqrt{1.23^2 - 1.03^2}} = 0.53$$

Buckling resistance, $N_{b,z\,Rd} = 0.53 \times 38.3 \times 275/10 = 558.2$ kN

$$\overline{\lambda_{LT}} = 0.9\overline{\lambda_z} = 0.9 \times 1.03 = 0.93$$

$$\chi_{LT} = \frac{1}{\Phi_{LT} + \sqrt{\Phi_{LT}^2 - \overline{\lambda}_{LT}^2}}$$

$$\Phi_{LT} = 0.5\left[1 + \alpha_{LT}\left(\overline{\lambda}_{LT} - 0.2\right) + \overline{\lambda}_{LT}^2\right]$$

For $h/b \leq 2$, curve a, $\alpha_{LT} = 0.21$

$$\Phi_{LT} = 0.5 \times [1 + 0.21(0.93 - 0.2) + 0.93^2] = 1.00$$

$$\chi_{LT} = \frac{1}{1.0 + \sqrt{1.0^2 - 0.93^2}} = 0.73$$

$M_{b,Rd} = 0.73 \times 247.1 \times 275 \times 10^{-3} = 49.6$ kN-m
For simple construction, the interaction equation could be used for the buckling check:

$$\frac{N_{Ed}}{N_{b,z,Rd}} + \frac{M_{y,Ed}}{M_{b,Rd}} + 1.5\frac{M_{z,Ed}}{M_{z,Rd}} \leq 1.0$$

$$\frac{121}{558.2} + \frac{10.46}{49.6} + 1.5 \times \frac{10.84}{111.2 \times 275 \times 10^{-3}} = 0.96 \leq 1.0$$

The section is satisfactory.

Second floor to base
The same column section will be used from the second floor to the base. The lower column length between the first floor and the base will be designed.

Referring to Figure 7.22, the design load and moments just below the first-floor level are

$$N_{Ed} = (1.35 \times 323.8) + (1.5 \times 68.4) = 539.73 \text{ kN}$$

$$M_y = (1.35 \times 5.84) + (1.5 \times 1.73) = 10.48 \text{ kN-m}$$

$$M_z = (1.35 \times 6.33) + (1.5 \times 1.73) = 11.14 \text{ kN-m}$$

Try 203 × 203 UC 52, the properties of which are

$$A = 66.3 \text{ cm}^2; i_z = 5.18 \text{ cm}$$

$$W_{el,z} = 174 \text{ cm}^3; W_{pl,y} = 567 \text{ cm}^3$$

Overall buckling check
Axial load at centre of column

$$= 539.73 + (1.35 \times 3.75) = 544.79 \text{ kN}$$

$$\lambda_z = 0.85 \times 5000/51.8 = 82.0$$

$$\lambda_1 = 93.9\varepsilon = 93.9 \times 0.92 = 86.4$$

$$\overline{\lambda_z} = \frac{\lambda_z}{\lambda_1} = \frac{82.0}{86.4} = 0.95$$

From Table 6.2 of EN1993-1-1, for an S275 rolled H section, t_f less than 100 mm, $h/b \leq 1.2$, buckling about the major y–y axis, curve b, imperfection factor, $\alpha = 0.34$, buckling about the minor z–z axis, curve c, imperfection factor, $\alpha = 0.49$,

$$\Phi = 0.5 \left[1 + 0.49(0.95 - 0.2) + 0.95^2\right] = 1.14$$

$$\chi_z = \frac{1}{1.14 + \sqrt{1.14^2 - 0.95^2}} = 0.57$$

Buckling resistance, $N_{b,z\,Rd} = 0.57 \times 66.3 \times 275/10 = 1039 \text{ kN}$

$$\overline{\lambda_{LT}} = 0.9\overline{\lambda_z} = 0.9 \times 0.95 = 0.86$$

For $h/b \leq 2$, curve a, $\alpha_{LT} = 0.21$

$$\Phi_{LT} = 0.5 \times \left[1 + 0.21(0.86 - 0.2) + 0.86^2\right] = 0.94$$

$$\chi_{LT} = \frac{1}{0.94 + \sqrt{0.94^2 - 0.86^2}} = 0.76$$

$$M_{b,Rd} = 0.76 \times 567 \times 275 \times 10^{-3} = 118.5 \text{ kN-m}$$

For simple construction, the interaction equation could be used for the buckling check:

$$\frac{N_{Ed}}{N_{b,z,Rd}} + \frac{M_{y,Ed}}{M_{b,Rd}} + 1.5\frac{M_{z,Ed}}{M_{z,Rd}} \le 1.0$$

$$\frac{544.79}{1039} + \frac{10.48}{118.5} + 1.5 \times \frac{11.14}{174 \times 275 \times 10^{-3}} = 0.96 \le 1.0$$

The section is satisfactory.

7.7 SIDE COLUMN FOR A SINGLE-STOREY INDUSTRIAL BUILDING

7.7.1 Arrangement and loading

The cross section and side elevation of a single-storey industrial building are shown in Figure 7.24a and b. The columns are assumed to be fixed at the base and pinned at the top and act as partially propped cantilevers in resisting lateral loads. The top of the column is held in the longitudinal direction by the eaves member and bracing, as shown on the side elevation.
   The loading on the column is due to

1. Dead and imposed load from the roof and dead load from the walls and column
2. Wind loading on roof and walls
   The load on the roof consists of

   a. Dead load due to sheeting, insulation board, purlins and weight of truss and bracing. This is approximately 0.3–0.5 KN/m² on the slope length of the roof.
   b. Imposed load due to snow, erection and maintenance loads. This is taken as 0.75 KN/m² on plan area.

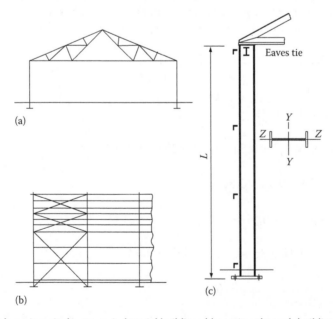

Figure 7.24 Side column in a single-storey industrial building: (a) section through building; (b) side elevation; (c) side column.

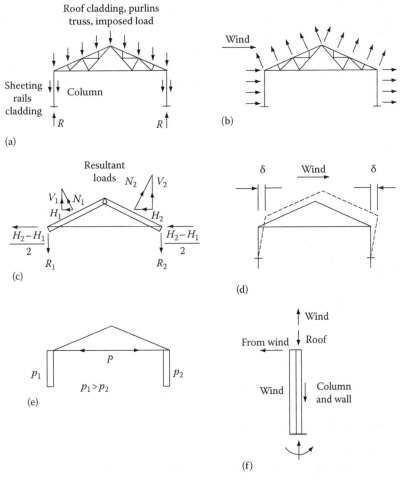

Figure 7.25 Loads on side column of an industrial building: (a) dead and imposed load; (b) wind load on roof and wall; (c) reaction from wind on roof; (d) frame deflection; (e) force in the bottom chord of the truss; (f) column loads.

The loading on the walls is due to sheeting, insulation board, sheeting rails and the weight of the column and bracing. The weight is approximately the same as for the roof.

The wind load depends on the location and dimensions of the building. The method of calculating the wind load is taken from BS EN1991-1-4. This is shown in the following example.

The breakdown and diagrams for the calculation of the loading and moments on the column are shown in Figure 7.25, and the following comments are made on these figures:

- The dead and imposed loads give an axial reaction R at the base of the column (see Figure 7.25a).
- The wind on the roof and walls is shown in Figure 7.25b. There may be a pressure or suction on the windward slope, depending on the angle of the slope. The reactions from wind on the roof only are shown in Figure 7.25c. The uplift results in vertical reactions R_1 and R_2. The net horizontal reaction is assumed to be divided equally between the two columns. This is $0.5(H_2 - H_1)$, where H_2 and H_1 are the horizontal components of the wind loads on the roof slopes.

- The wind on the walls causes the frame to deflect, as shown in Figure 7.25d. The top of each column moves by the same amount, δ. The wind P_1 and P_2 on each wall, taken as uniformly distributed, will have different values, and these result in a force P in the bottom chord of the truss, as shown in Figure 7.25e. The value of P may be found by equating deflections at the top of each column. For the case where P_1 is greater than P_2, there is a compression P in the bottom chord:

$$\frac{P_1 L^4}{8EI} - \frac{PL^3}{3EI} = \frac{P_2 L^4}{8EI} + \frac{PL^3}{3EI}$$

This gives

$$P = 3L \, (P_1 - P_2)/16$$

where

  I denotes the moment of inertia of the column about the y–y axis (same for each column)
  E is the Young's modulus
  L is the column height
  P_1 and P_2 are the distributed loads

The resultant loading on the column is shown in Figure 7.25f, where the horizontal point load at the top is

$$H = P + \left(\frac{H_2 - H_1}{2} \right)$$

The column moments are due entirely to wind load.

7.7.2 Column design procedure

1. Section classification
   Universal beams are often used for these columns where the axial load is small, but the moment due to wind is large. Referring to Figure 7.26a, the classification is checked as follows:

   1. Flanges are checked using Table 5.2 of the code where limits for c_f/t_f are given, where c_f is the flange outstand as shown in the figure and t_f is the flange thickness.
   2. Webs are in combined axial and flexural compression. The classification can be checked using Table 5.2 of the code. For example, from Table 5.2 for webs, generally Class 1 section has the limit:

   $$\frac{c}{t} \leq \frac{36\varepsilon}{\alpha}$$

   where
     c is the clear depth of web
     t is the thickness of web
     α is the stress ratio as defined in Table 5.2 of the code

2. Buckling length for axial compression
   Buckling lengths for cantilever columns connected by roof trusses are given in the following. The tops must be held in position longitudinally by eaves members connected to a braced bay.

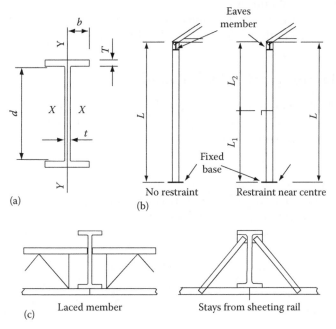

Figure 7.26 Side column design features: (a) column sector; (b) column conditions; (c) lateral support for column.

Two cases are shown in Figure 7.26b:
1. Column with no restraints

    y–y axis $L_{cr,y} = 1.5L$

    z–z axis $L_{cr,z} = 0.85L$

    If the base is not effectively fixed about the y–y axis, $L_{cr,y} = 1.0L$.
2. Column with restraints

    The restraint provides lateral support against buckling about the weak axis:

    y–y axis $L_{cr,y} = 1.5L$

    z–z axis $L_{cr,z} = 0.85L_1$ or L_2, whichever is the greater

    The restraint is often provided by a laced member or stays from a sheeting rail, as shown in Figure 7.26c.
3. Buckling length for calculating the buckling resistance moment

    The buckling lengths for the two cases shown in Figure 7.26b are
1. Column with no restraints

    The column is fixed at the base and restrained laterally and torsionally at the top. For normal loading, $L_{cr} = 0.85L$.
2. Column with restraints

    This is to be treated as a beam and the buckling length taken as
    $$L_{cr} = 0.85L_1 \text{ or } 1.0L_2 \text{ in the case shown}$$
4. Column design

    The column moment is due to wind and controls the design. The load combination is then dead plus imposed plus wind load. The load factor is taken from BS EN1990.

    The design procedure is shown in the example that follows.

5. Deflection at the column cap

The deflection at the column cap must not exceed the limit given in Table 8 of the code for a single-storey building. The limit is height/300.

7.7.3 Example: Design of a side column

A section through a single-storey building is shown in Figure 7.27. The frames are at 5 m centres and the length of the building is 30 m. The columns are pinned at the top and fixed at the base. The loading is as follows:

Roof

Dead load – measured on slope
Sheeting, insulation board, purlins and truss = 0.45 kN/m²
Imposed load – measured on plan = 0.75 kN/m²

Walls

Sheeting, insulation board, sheeting rails = 0.35 kN/m²

Column

Estimate = 3.0 kN

Wind load

The code of practice for wind loading is BS EN1994-1-1.

Determine the loads and moments on the side column and design the member using S275 steel. Note that the column is taken as not being supported laterally between the top and base.

1. Column loads and moments
    Dead and imposed load
        Roof

        Dead load = 10 × 5 × 0.45 × 10.77/10 = 24.23 kN
        Imposed load = 10 × 5 × 0.75 = 37.5 kN
        Walls = 6 × 5 × 0.35 = 10.5 kN
        Column = 3.0 kN
        Total load at base = 75.23 kN

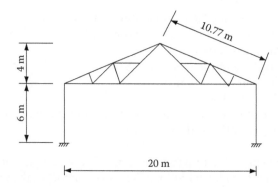

Figure 7.27 Section through building.

Wind load
Location: north-east England. The peak velocity wind pressure, $q_{p(z)} = 0.554$ kN/m². The wind pressure coefficients and wind loads for the building are shown in Figure 7.29b. The wind load normal to the walls and roof slope is given by

$$W = q_{p(z)} \, s \, L \, (C_{pe} - C_{pi})$$

where
    s is the frame centres
    L is the height of wall or length of roof slope
    $q_{p(z)}$ is the peak velocity wind pressure walls or roof slopes

The resultant normal loads on the roof and the horizontal and vertical resolved parts are shown in Figure 7.28c. The horizontal reaction is divided equally between each

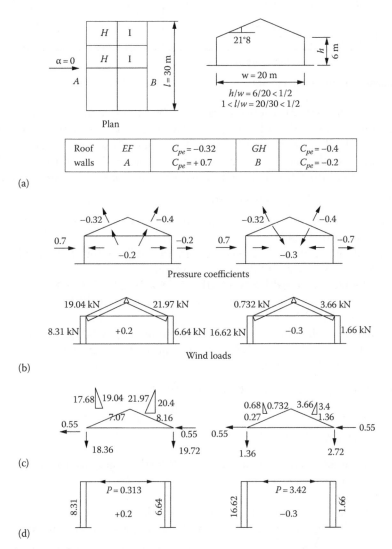

Figure 7.28 Wind-pressure coefficients and loads: (a) external pressure coefficients C_{pe}–wind angle $\alpha = 0$; (b) pressure coefficients and wind loads; (c) roof loads and reactions; (d) wind on walls.

Wind case	Internal pressure		Internal suction	
Column	Windward	Leeward	Windward	Leeward
Column	↓24.23	↓24.23	↓24.23	↓24.23
Imposed	↓39.5	↓37.5	↓37.5	↓37.5
Wind	↓18.36	↑19.72	↑1.36	↑2.72
	0.237	0.863	2.87	3.97
Wind wall column	8.31 ↓13.5	6.64 ↓13.5	16.6 ↓13.5	13.5 ↓ 1.66
Dead	↓37.73	↓37.73	↑37.73	↑37.73
Imposed	↓37.54	↓37.5	↑37.5	↑37.5
Wind	↓18.36	↑19.72	↓1.36	↓2.72
Wind moment	26.35	25.1	32.64	18.84

Loads are in kN, moments are in kN m

Figure 7.29 Summary of loads and moments.

support and the vertical reactions are found by taking moments about supports. The reactions at the top of the columns for the two wind-load cases are shown in the figure.

The wind loading on the walls requires the analysis set out in Section 7.7.1, where the column tops deflect by an equal amount and a force P is transmitted through the bottom chord of the truss. For the internal pressure case, see Figure 7.28d:

$P = 3(8.31 - 6.64)/16 = 0.313$ kN

The loads and moments on the columns are summarized in Figure 7.29.

2. Notional horizontal loads

To ensure stability, the structure is checked for notional horizontal load. The notional force from the roof loads is taken as the greater of

1% of the factored dead loads = $0.01 \times 1.35 \times 24.23 = 0.33$ kN or

0.5% of the factored dead load plus vertical imposed load = $0.005 \times (1.35 \times 24.23) + (1.5 \times 37.5)1 = 0.41$ kN

This load is applied at the top of each column and is taken to act simultaneously with 1.35 times the dead and 1.5 times the imposed vertical loads.

The design load at the base is

$P = (1.35 \times 37.73) + (1.5 \times 37.5) = 107.19$ kN

The moment is

$M = 0.41 \times 6 = 2.5$ kN-m

The design conditions for this case are less severe than those for the combination dead + imposed + wind loads.

3. Column design

The maximum design condition is for the wind-load case of internal pressure. For the windward column, the load combination is dead plus imposed plus wind loads:

Design load = $1.35 \times 37.73 + 1.5 \times 37.5 - 0.75 \times 1.36 = 106.2$ kN
Design moment = $1.5 \times 32.64 = 48.96$ kN-m

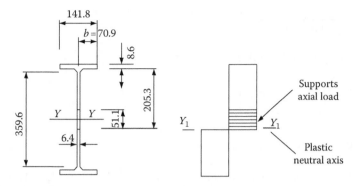

Figure 7.30 Column section.

Try 406 × 140 UB 39, the properties of which are

$A = 49.4$ cm^2; $W_{pl,y} = 721$ cm^3; $W_{el,y} = 627$ cm^3
$i_y = 15.9$ cm; $i_z = 2.87$ cm; $I_y = 12,500$ cm^4; $I_z = 410$ cm^4

Check the section classification using Table 5.2. The section dimensions are shown in Figure 7.30:

Yield strength $f_y = 275$ N/mm^2

Overall buckling check

Buckling lengths $L_{cr,y} = 1.5 \times 6000 = 9000$ mm
Buckling lengths $L_{cr,z} = 0.85 \times 6000 = 5100$ mm

Maximum slenderness ratio

$\lambda_y = L_{cr,y}/i_y = 9000/159 = 56.6$
$\lambda_z = L_{cr,z}/i_z = 5100/28.7 = 178$

From Table 6.2 of EN1993-1-1, for an S275 rolled H section, t_f less than 40 mm, $h/b > 1.2$, buckling about the major y–y axis, curve *a*, imperfection factor, $\alpha = 0.21$, buckling about the minor z–z axis, curve *b*, imperfection factor, $\alpha = 0.34$,

$\lambda_1 = 93.9\varepsilon = 93.9 \times 0.92 = 86.4$

$$\overline{\lambda_y} = \frac{\lambda_y}{\lambda_1} = \frac{56.6}{86.4} = 0.66$$

$\Phi = 0.5\,[1 + 0.21(0.66 - 0.2) + 0.66^2] = 0.77$

$$\chi_y = \frac{1}{0.77 + \sqrt{0.77^2 - 0.66^2}} = 0.86$$

Side column for a single-storey industrial building 191

$$\overline{\lambda_z} = \frac{\lambda_z}{\lambda_1} = \frac{178}{86.4} = 2.06$$

$\Phi = 0.5\,[1 + 0.34(2.06 - 0.2) + 2.06^2] = 2.94$

$$\chi_z = \frac{1}{2.94 + \sqrt{2.94^2 - 2.06^2}} = 0.2$$

Buckling lengths $L_{cr} = 0.85 \times 6000 = 5100$ mm

$$M_{cr} = C_1 \frac{\pi^2 EI_z}{L_{cr}^2} \left(\frac{I_w}{I_z} + \frac{L_{cr}^2 \, GI_T}{\pi^2 EI_z} \right)^{0.5}$$

For ratio of end moment, $\psi = 1.0$; $C_1 = 1.0$

$$M_{cr} = 1.0 \times \frac{\pi^2 \times 210,000 \times 4,100,000}{5,100^2} \left(\frac{155 \times 10^9}{4,100,000} + \frac{5,100^2 \times 81,000 \times 107,000}{\pi^2 \times 210,000 \times 4,100,000} \right)^{0.5}$$

$$M_{cr} = 82.86 \text{ kN-m}$$

$$\overline{\lambda_{LT}} = \sqrt{\frac{W_y f_y}{M_{cr}}} = \sqrt{\frac{721 \times 10^3 \times 275}{82.86 \times 10^6}} = 1.55$$

$$\chi_{LT} = \frac{1}{\Phi_{LT} + \sqrt{\Phi_{LT}^2 - \overline{\lambda}_{LT}^2}}$$

$$\Phi_{LT} = 0.5 \left[1 + \alpha_{LT} \left(\overline{\lambda}_{LT} - 0.2 \right) + \overline{\lambda}_{LT}^2 \right]$$

For $h/b > 2$, curve a, $\alpha_{LT} = 0.34$

$$\Phi_{LT} = 0.5 \times [1 + 0.34(1.55 - 0.2) + 1.55^2] = 1.93$$

$$\chi_{LT} = \frac{1}{1.93 + \sqrt{1.93^2 - 1.55^2}} = 0.33$$

Determination of interaction factors k_{ij} (Annex B in BS EN1993-1-1)
   Table B.3. Equivalent uniform moment factors, C_m
   For $\psi = 1.0$, C_{my}, $C_{mz} = 0.6 + 0.4(1.0) = 1.0 \geq 0.4$

$$k_{yy} = C_{my} \left(1 + \left(\overline{\lambda}_y - 0.2 \right) \frac{N_{Ed}}{\chi_y N_{Rk} / \gamma_{M1}} \right) \leq C_{my} \left(1 + 0.8 \frac{N_{Ed}}{\chi_y N_{Rk} / \gamma_{M1}} \right)$$

$$k_{yy} = 1.0 \left(1 + (0.66 - 0.2) \frac{106.2}{0.86 \times 4940 \times 10^{-3} \times 275 / 1.0} \right) = 1.04$$

$$k_{zy} = 0.6, \; k_{yy} = 0.63$$

$$\frac{N_{Ed}}{\chi_y N_{Rk} / \gamma_{M1}} + k_{yy} \frac{M_{y,Ed}}{\chi_{LT} M_{y,Rk} / \gamma_{M1}} + k_{yz} \frac{M_{z,Ed}}{M_{z,Rk} / \gamma_{M1}} \leq 1$$

$$\frac{106.2}{0.86 \times 1358 / 1.0} + 1.04 \frac{48.96}{0.33 \times 198.3 / 1.0} = 0.87$$

$$\frac{N_{Ed}}{\chi_z N_{Rk}/\gamma_{M1}} + k_{zy}\frac{M_{y,Ed}}{\chi_{LT}M_{y,Rk}/\gamma_{M1}} + k_{zz}\frac{M_{z,Ed}}{M_{z,Rk}/\gamma_{M1}} \le 1$$

$$\frac{106.2}{0.2\times1358/1.0} + 0.63\frac{48.96}{0.33\times198.3/1.0} = 0.86$$

The section is satisfactory.
  Deflection at column cap
  For the internal suction case

$$\delta = \frac{16.62\times10^3\times6000^4}{8\times210\times10^3\times12,500\times10^4} - \frac{2.87\times10^3\times6000^3}{3\times210\times10^3\times12,500\times10^4}$$

$$= 17.09 - 7.87 = 9.22 \text{ mm}$$

δ/height = 9.22/6000 = 1/651 < 1/300 NA to BS EN1993-1-1
  The column is satisfactory with respect to deflection.

7.8 CRANE COLUMNS

7.8.1 Types

Three common types of crane columns used in single-storey industrial buildings are shown in Figure 7.31. These are

1. A column of uniform section carrying the crane beam on a bracket
2. A laced crane column
3. A compound column fabricated from two universal beams or built up from plate

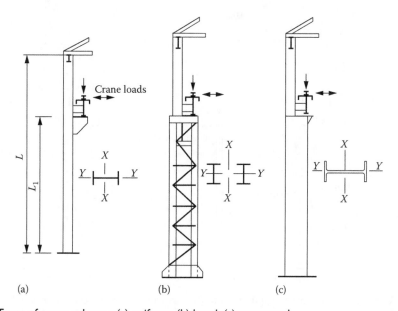

Figure 7.31 Types of crane columns: (a) uniform; (b) laced; (c) compound.

Only the design of a uniform column used for light cranes will be discussed here. Types (2) and (3) are used for heavy cranes.

7.8.2 Loading

A building frame carrying a crane is shown in Figure 7.32a. The hook load is placed as far as possible to the left to give the maximum load on the column. The building, crane and wind loads are shown in the figure in (b), (c) and (d), respectively.

7.8.3 Frame action and analysis

In order to determine the values of moments in the columns, the frame as a whole must be considered. Consider the frame shown in Figure 7.32a, where the columns are of uniform section pinned at the top and fixed at the base. The separate load cases are discussed.

1. Dead and imposed load
   The dead and imposed loads from the roof and walls are taken as acting axially on the column. The dead load from the crane girder causes moments as well as axial load in the column. (See the crane-wheel load case in the following.)
2. Vertical crane-wheel loads
   The vertical crane-wheel loads cause moments as well as axial load in each column. The moments applied to each column are unequal, so the frame sways (as shown in Figure 7.33a) and a force P is transmitted through the bottom chord.

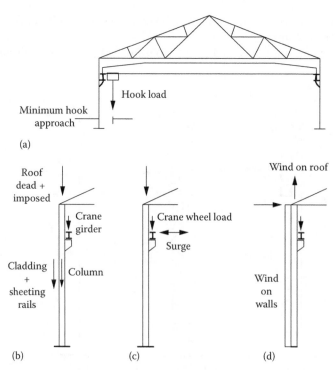

Figure 7.32 Loads on crane columns: (a) section; (b) dead and imposed load; (c) crane loads; (d) wind loads.

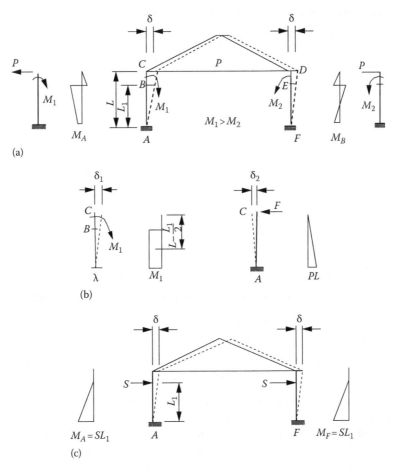

Figure 7.33 Vertical and horizontal crane-wheel loads and moments: (a) frame are column moments; (b) moments and load causing deflection in column *ABC*; (c) column moments due to crane surge.

Consider the column ABC in Figure 7.33a. The deflection at the top is calculated for the moment from the crane-wheel loads M_1 and force P separately using the moment area method. The separate moment diagrams are shown in Figure 7.33b.

The deflection due to M_1 is

$$\delta = M_1 L_1 (L - L_1/2)/EI$$

where
   L is the column height
   L_1 is the height to the crane rail
   EI is the column rigidity
   The deflection due to the load P at the column top is

$$\delta_2 = P L^3/3EI$$

The frame deflection is

$$\delta = \delta_1 - \delta_2$$

Equating deflections at the top of each column gives

$$\frac{M_1 L_1}{EI}\left(L - \frac{L_1}{2}\right) - \frac{PL^3}{3EI} = \frac{M_2 L_1}{EI}\left(L - \frac{L_1}{2}\right) + \frac{PL^3}{3EI}$$

where

$$P = 3L_1\left(L - \frac{L_1}{2}\right)\left(\frac{M_1 + M_2}{2L^3}\right)$$

The moments in the column can now be calculated.

If the self-weight of the crane girder applies a moment M to each column, then the force in the bottom chord is

$$P_1 = \frac{L_1 M}{L_3}\left(L - \frac{L_1}{2}\right)$$

3. Crane surge

In Figure 7.33c, the crane surge load S is the same each side and each column acts as a free cantilever. The loads and moments for this case are shown in the figure.

4. Wind loads

Wind loads on this type of frame were treated in Section 7.7.1.

5. Load combinations

The separate load combinations and load factors γ_f to be used in design are given in BS EN1990. The load cases and load factors are

(1) 1.35 dead + 1.5 imposed + 1.5 vertical crane load
(2) 1.35 dead + 1.5 imposed + 1.5 horizontal crane load
(3) 1.35 dead + 1.5 imposed + 1.5 (vertical and horizontal crane loads)
(4) 1.35 dead + 1.5 imposed + 1.5 (vertical and horizontal crane loads) + 0.75 wind

It may not be necessary to examine all cases. Note that in case (2) there is no impact allowance on the vertical crane-wheel loads.

7.8.4 Design procedure

1. Buckling lengths for axial compression

The buckling lengths for axial compression for a uniform column carrying the crane girder on a bracket are given in the following:

In Figure 7.31a, the buckling lengths are

y–y axis: $L_{cr,y} = 1.5L$
z–z axis: $L_{cr,z} = 0.85L$

The crane girder must be held in position longitudinally by bracing in the braced bays. If the base is not fixed in the z–z direction, $L_{cr,z} = 1.0L_1$.

2. Buckling length for calculating the buckling resistance moment

The crane girder forms an intermediate restraint to the cantilever column. In this case, the member is to be treated as a beam between restraints and the buckling length L_{cr}. A value of $L_{cr} = 0.85L_1$ may be used for this case.

3. Column design

The column is checked for overall buckling.

4. Deflection

The deflection limitation for columns in single-storey buildings applies. In the NA to BS EN1993-1-1, the limit for the column top is height/300. However, the code also states that in the case of crane surge and wind, only the greater effect of either needs to be considered in any load combination.

7.8.5 Example: Design of a crane column

1. Building Frame and Loading

The single-storey building frame shown in Figure 7.34a carries a 50 kN electric overhead crane. The frames are at 5 m centres and the length of the building is 30 m. The static crane-wheel loads are shown in Figure 7.34b. The crane beams are simply supported, spanning 5 m between columns, and the weight of a beam is approximately 4 kN. The arrangement of the column and crane beam with the end clearance and eccentricity is shown in Figure 7.34c.

Dead and imposed loads
The roof and wall loads are the same as for the building in Section 7.7.3. The loads are

Dead loads

Roof	= 24.23 kN
Walls	= 10.5 kN
Crane column + bracket	= 4.0 kN
Crane beam	= 4.0 kN
——————————————	———————
Dead load at column base	= 42.73 kN
——————————————	———————
Dead load at crane girder level	= 27.86 kN
Imposed load – Roof	= 37.5 kN

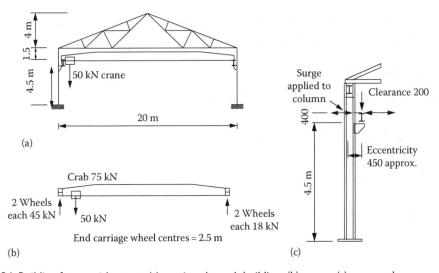

(a)

20 m

4 m
1.5
4.5 m

50 kN crane

(b)

Crab 75 kN

2 Wheels each 45 kN   50 kN   2 Wheels each 18 kN

End carriage wheel centres = 2.5 m

(c)

Surge applied to column   Clearance 200

400

4.5 m

Eccentricity 450 approx.

Figure 7.34 Building frame with crane: (a) section through building; (b) crane; (c) crane column.

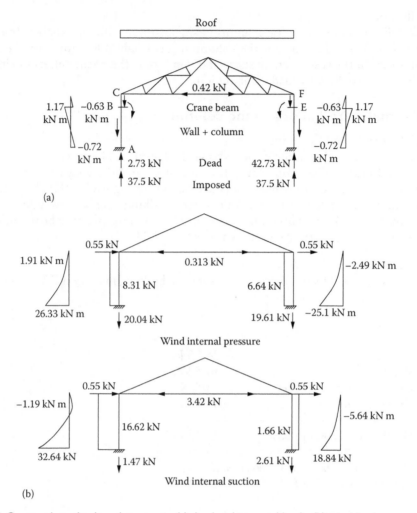

Figure 7.35 Crane column loads and moments: (a) dead and imposed loads; (b) wind loads.

The eccentric dead load of the crane beams causes small moments in the columns. In Figure 7.35a, the applied moment to each column

$$M = 4 \times 0.45 = 1.8 \text{ kN-m}$$

The load P_1 in the bottom chord of the truss (see preceding text)

$$P_1 = \frac{3 \times 4.5 \times 1.8}{6^3}\left(6 - \frac{4.5}{2}\right) = 0.42 \text{ kN}$$

Column moments

$$3M_{BC} = 0.42 \times 1.5 = -0.63 \text{ kN-m}$$

$$M_{BA} = 1.8 - 0.63 = 1.17 \text{ kN-m}$$

$$M_{AB} = 1.8 - (0.42 \times 6) = -0.72 \text{ kN-m}$$

Wind loads

The wind loads are the same as for the building in Section 7.7.3 and wind load and column moments are shown in Figure 7.35b.

Vertical crane-wheel loads

The crane-wheel loads, including impact, are

Maximum wheel loads = 45 + 25% = 56.25 kN

Light side-wheel loads = 18 + 25% = 22.5 kN

To determine the maximum column reaction, the wheel loads are placed equidistant about the column, as shown in Figure 7.36a. The column reaction and moment for the maximum wheel loads are

$R_1 = 2 \times 56.25 \times 3.75/5 = 84.375$ kN

$M_1 = 84.3 \times 0.45 = 37.87$ kN-m

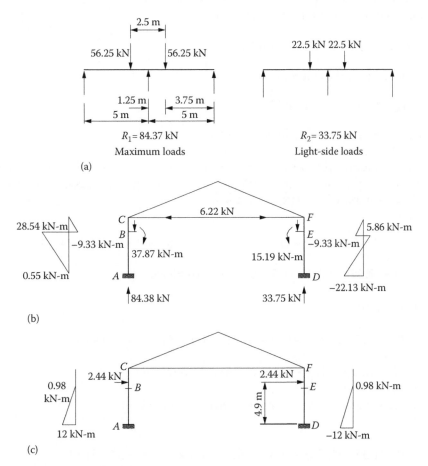

Figure 7.36 Crane column loads and moments: (a) column reactions from crane wheel loads; (b) vertical crane wheel loads; (c) horizontal crane wheel loads.

For the light side-wheel loads,

$$R_2 = 2 \times 22.5 \times 3.75/5 = 33.75 \text{ kN}$$

$$M_2 = 33.75 \times 0.45 = 15.19 \text{ kN}$$

The load in the bottom chord is (see above):
The moments for column ABC are

$$P_2 = \frac{3 \times 4.5}{2 \times 6^3}\left(6 - \frac{4.5}{2}\right)(37.87 + 15.19) = 6.22\,\text{kN}$$

$$M_{BC} = -6.22 \times 1.5 = -9.33 \text{ kN-m}$$

$$M_{BA} = 37.87 - 9.33 = 28.54 \text{ kN-m}$$

$$M_A = 37.87 - (6.22 \times 6) = 0.55 \text{ kN-m}$$

These moments and the moments for column DEF are shown in Figure 7.36b.
Crane surge loads
The horizontal surge load per wheel = 0.1(50 + 15)/4 = 1.63 kN.
The column reaction from surge loads

$$R_3 = 2 \times 1.63 \times 3.75/5 = 2.45 \text{ kN}$$

The moments at the column base are

$$M_A = 2.45 \times 4.9 = 12.0 \text{ kN-m}$$

The loads and moments are shown in Figure 7.36c.
2. Design load combinations
Consider column ABC with wind internal suction case, maximum crane-wheel loads and crane surge. The design loads and moments for three load combinations are
a. Dead + imposed + vertical crane loads
Base

$$P = (1.35 \times 42.73) + (1.5 \times 37.5) + (1.5 \times 84.38) = 240.51 \text{ kN}$$

$$M = (1.35 \times 0.72) + (1.5 \times 0.55) = 1.80 \text{ kN-m}$$

Crane girder

$$P = (1.35 \times 27.86) + (1.5 \times 37.5) + (1.5 \times 84.38) = 220.43 \text{ kN}$$

$$M = (1.35 \times 1.17) + (1.5 \times 28.54) = 44.39 \text{ kN-m}$$

b. Dead + imposed + vertical and horizontal crane loads
Base

$$P = (1.35 \times 42.73) + (1.5 \times 37.5) + (1.5 \times 84.38) = 240.51 \text{ kN}$$

$$M = -(1.5 \times 0.72) + 1.5(0.55 - 12) = -18.26 \text{ kN-m}$$

Crane girder

$$P = (1.35 \times 27.86) + (1.5 \times 37.5) + (1.5 \times 84.38) = 220.43 \text{ kN}$$

$$M = (1.5 \times 1.17) \times (28.54 + 0.98) = 51.81 \text{ kN-m}$$

c. Dead + imposed + vertical and horizontal crane loads+ wind
    Base

$$P = (1.35 \times 42.73) + (1.5 \times 37.5) + (1.5 \times 84.38) - 0.75 \times 1.47 = 239.40 \text{ kN}$$

$$M = (1.35 \times -0.72) + (1.5 \times 32.64) + (1.5 \times 12.0) + (0.75 \times 0.55) = 66.4 \text{ kN-m}$$

Crane girder

$$M = (1.35 \times 1.17) + (1.5 \times - 1.19) + (1.5 \times 28.54) + (0.75 \times 0.55) = 43.0 \text{ kN-m}$$

Note that design conditions arising from notional horizontal loads are not as severe as those in condition (3) earlier.

3. Column design
   Try 406 × 140 UB 46, the properties of which are

$$A = 59.0 \text{ cm}^2; \ W_{pl,y} = 888.4 \text{ cm}^3; \ W_{el,y} = 777.8 \text{ cm}^3$$

$$i_y = 16.29 \text{ cm}; \ i_z = 3.02 \text{ cm}; \ I_y = 15\ 647 \text{ cm}^4; \ I_z = 538 \text{ cm}^4$$

Case (3) is the most severe load combination. Compressive strength

$$\lambda_y = 1.5 \times 6000/162.9 = 55.25, \quad \lambda_z = 0.85 \times 4500/30.2 = 112.6$$

From Table 6.2 of EN1993-1-1, for an S275 rolled H section, t_f less than 40 mm, $h/b >$ 1.2, buckling about the major y–y axis, curve a, imperfection factor, $\alpha = 0.21$, buckling about the minor z–z axis, curve b, imperfection factor, $\alpha = 0.34$,

$$\lambda 1 = 93.9\varepsilon = 93.9 \times 0.92 = 86.4$$

$$\overline{\lambda_y} = \frac{\lambda_y}{\lambda_1} = \frac{55.25}{86.4} = 0.64$$

$$\Phi = 0.5 \ [1 + 0.21(0.64 - 0.2) + 0.64^2] = 0.75$$

$$\chi_y = \frac{1}{0.75 + \sqrt{0.75^2 - 0.64^2}} = 0.87$$

$$\overline{\lambda_z} = \frac{\lambda_z}{\lambda_1} = \frac{112.6}{86.4} = 1.30$$

$$\Phi = 0.5 \ [1 + 0.34(1.3 - 0.2) + 1.3^2] = 1.53$$

$$\chi_z = \frac{1}{1.53 + \sqrt{1.53^2 - 1.30^2}} = 0.43$$

Buckling moment resistance
Buckling lengths $L_{cr} = 0.85 \times 4500 = 3825$ mm

$$M_{cr} = C_1 \frac{\pi^2 E I_z}{L_{cr}^2} \left(\frac{I_w}{I_z} + \frac{L_{cr}^2 \, G I_T}{\pi^2 E I_z} \right)^{0.5}$$

For ratio of end moment, $\psi = 0$; $C_1 = 1.879$

$$M_{cr} = 1.879$$

$$\times \frac{\pi^2 \times 210{,}000 \times 5{,}380{,}000}{3{,}825^2} \left(\frac{207 \times 10^9}{5{,}380{,}000} + \frac{3{,}825^2 \times 81{,}000 \times 190{,}000}{\pi^2 \times 210{,}000 \times 5{,}380{,}000} \right)^{0.5}$$

$M_{cr} = 346.87$ kN-m

$$\overline{\lambda_{LT}} = \sqrt{\frac{W_y f_y}{M_{cr}}} = \sqrt{\frac{888.4 \times 10^3 \times 275}{346.87 \times 10^6}} = 0.84$$

$$\chi_{LT} = \frac{1}{\Phi_{LT} + \sqrt{\Phi_{LT}^2 - \overline{\lambda}_{LT}^2}}$$

$$\Phi_{LT} = 0.5 \left[1 + \alpha_{LT} \left(\overline{\lambda}_{LT} - 0.2 \right) + \overline{\lambda}_{LT}^2 \right]$$

For $h/b > 2$, curve a, $\alpha_{LT} = 0.34$

$$\Phi_{LT} = 0.5 \times [1 + 0.34(0.84 - 0.2) + 0.84^2] = 0.96$$

$$\chi_{LT} = \frac{1}{0.96 + \sqrt{0.96^2 - 0.84^2}} = 0.70$$

Determination of interaction factors k_{ij} (Annex B in BS EN1993-1-1)
   Table B.3 of BS EN1993-1-1. Equivalent uniform moment factors, C_m
   For $\psi = 0$, C_{my}, $C_{mz} = 0.6 + 0.4(0) = 0.6 \geq 0.4$

$$k_{yy} = C_{my} \left(1 + \left(\overline{\lambda}_y - 0.2 \right) \frac{N_{Ed}}{\chi_y N_{Rk}/\gamma_{M1}} \right) \leq C_{my} \left(1 + 0.8 \frac{N_{Ed}}{\chi_y N_{Rk}/\gamma_{M1}} \right)$$

$$k_{yy} = 0.6 \left(1 + (0.64 - 0.2) \frac{239.4}{0.87 \times 5900 \times 10^{-3} \times 275/1.0} \right) = 0.65$$

$$k_{zy} = 0.6, \quad k_{yy} = 0.39$$

$$\frac{N_{Ed}}{\chi_y N_{Rk}/\gamma_{M1}} + k_{yy} \frac{M_{y,Ed}}{\chi_{LT} M_{y,Rk}/\gamma_{M1}} + k_{yz} \frac{M_{z,Ed}}{M_{z,Rk}/\gamma_{M1}} \leq 1$$

$$\frac{239.4}{0.87 \times 1622/1.0} + 0.65 \frac{66.4}{0.7 \times 244.3/1.0} = 0.43$$

$$\frac{N_{Ed}}{\chi_z N_{Rk}/\gamma_{M1}} + k_{zy}\frac{M_{y,Ed}}{\chi_{LT}M_{y,Rk}/\gamma_{M1}} + k_{zz}\frac{M_{z,Ed}}{M_{z,Rk}/\gamma_{M1}} \leq 1$$

$$\frac{239.4}{0.43 \times 1622/1.0} + 0.39\frac{66.4}{0.7 \times 244.3/1.0} = 0.49$$

The section is satisfactory.

Deflection at column cap
The reader should refer to Figures 7.35 and 7.36.
Deflection due to crane surge δ_s

$$EI\,\delta_s = 12 \times 10^6 \times 4900 \times 4367/2 = 1.284 \times 10^{14}.$$

Deflection due to wind, δ_w

$$EI\,\delta_w = 16620 \times 6000^3/8 - 2.87 \times 10^3 \times 6000^3/3 = 2.421 \times 10^{14.}$$

Add deflection from crane-wheel loads to that caused by wind load:

$$EI\,\delta = 2.421 \times 10^{14} + 37.87 \times 10^6 \times 4500 \times 3750 - 6220 \times 6000 \times 10^3/3$$

$$= 4.334 \times 10^{14}$$

$$\delta = \frac{4.334 \times 10^4}{210 \times 10^3 \times 15{,}647 \times 10^4} = 13.19\,\text{mm}$$

δ/height = 13.19/6000 = 1/455 < 1/300.

The deflection controls the column size.

7.9 COLUMN BASES

7.9.1 Types and loads

Column bases transmit axial load, horizontal load and moment from the steel column to the concrete foundation. The main function of the base is to distribute the loads safely to the weaker material.

The main types of bases used are shown in Figure 7.37. These are

1. Slab base
2. Gusseted base
3. Pocket base

With respect to slab and gusseted bases, depending on the values of axial load and moment, there may be compression over the whole base or compression over part of the base and tension in the holding-down bolts. Bases subjected to moments about the major axis only are considered here. Horizontal loads are resisted by shear in the weld between column and base plates, shear in the holding-down bolts and friction and bond between the base and the concrete. The horizontal loads are generally small.

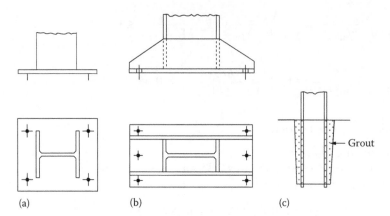

Figure 7.37 Column bases: (a) slab base; (b) gusseted base; (c) pocket base.

7.9.2 Design strengths

1. Base plates
   The yield strength of the plate f_y is given in Table 3.1 of BS EN1993-1-1.
2. Holding-down bolts
   The strengths of bolts are given in Table 3.1 of BS EN1993-1-8 (see Sections 10.2.2 and 10.2.3). The tensile stress area should be used in the design check for bolts in tension.
3. Concrete
   The column base is set on steel packing plates and grouted in. Concrete cube strengths vary from 25 to 40 N/mm². The bearing strength is given in BS EN1992-1-1 as f_{ck}/γ_m, where f_{ck} is the characteristic cylinder strength at 28 days. For design of pocket bases, the compressive strength of the structural concrete is taken from BS EN1992-1-1.

7.9.3 Axially loaded slab base

1. Code requirements and theory
   This type of base is used extensively with thick steel slabs being required for heavily loaded columns. The slab base is free from pockets where corrosion may start and maintenance is simpler than with gusseted bases.

   The design of slab bases with concentric loads is given in Section 6.2.5 of BS EN1993-1-8. This states that the bearing width, c, should not exceed

   $$c = t \left[\frac{f_y}{3 f_{jd} \gamma_{M0}} \right]^{0.5}$$

   where

   t is the thickness of the base plate
   f_y is the yield strength of the base plate
   f_{jd} is the design bearing strength of concrete $= 2/3 (f_{ck}/\gamma_c)$
   γ_c is the material safety factor for concrete $= 1.5$

   but not less than the flange thickness of the column supported, where c is the largest perpendicular distance from the edge of the effective portion of the base plate to the face of the column cross section and t_f is the flange thickness of the column.

2. Weld: column to slab

The code states in Section 4.13.3 that where the slab and column end are in tight contact, the load is transmitted in direct bearing. The surfaces in contact would be machined in this case. The weld only holds the base slab in position. Where the surfaces are not suitable to transmit the load in direct bearing, the weld must be designed to transmit the load.

7.9.4 Example: Axially loaded slab base

A column consisting of a 305 × 305 UC 198 carries an axial dead load of 1440 kN and an imposed load of 800 kN. Adopting a square slab, determine the size and thickness required. The cylinder strength of the concrete is 30 N/mm². Use Grade S275 steel:

Design load = (1.35 × 1440) + (1.5 × 800) = 3144 kN

The bearing strength of concrete = 2/3 × 30/1.5 = 13.4 N/mm²
Required area of base plate, $A_{req} = N_{Ed}/f_{jd}$
$$= 3144 \times 10^3/13.4 = 234{,}667 \text{ mm}^2$$
Provide 600 × 600 mm² square base plate:

The area provided, $A_{provide}$ = 360,000 mm² > A_{req}.
The arrangement of the column on the base plate is shown in Figure 7.38; from this, the effective area is

$$A_{eff} = (D + 2c)(B + 2c) - (D - 2c - 2T)(B - t)$$

$$= (DB + 2Bc + 2Dc + 4c^2) - (DB - 2Bc - 2BT - DT - Dt + 2tc + 2Tt)$$

$$= 4c^2 + (2D + 4B - 2t)c + (2BT + Dt - 2Tt)$$

where D and B are the depth and width of the universal steel column used, T and t are the flange and web thicknesses of the UC and c is the perpendicular spread distance as defined in Clause 6.2.5 of BS EN1993-1-8.

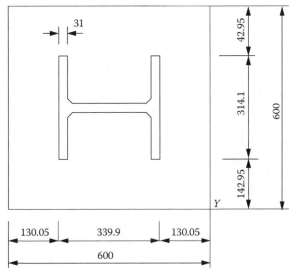

Figure 7.38 Axial loaded base: example.

The effective area is approximately equal to

$$A_{eff} = 4c^2 + c(\text{sectional perimeter}) + \text{sectional area} = A_{reg}.$$

Now, sectional perimeter = 1.87×10^3 mm sectional area = 25,200 mm²
Equating yields $4c^2 + (1.87 \times 10^3)c + (25.2 \times 10^3) = 234,667$
Solving the equation, $c = 93.38$.
   Check to ensure no overlapping occurs:

$$D + 2c = 339.9 + 2 \times 93.38 = 526 \text{ mm} < D_p = 600 \text{ mm (ok)}$$

$$B + 2c = 314.1 + 2 \times 93.38 = 501 \text{ mm} < B_p = 600 \text{ mm (ok)}$$

The thickness of the base plate is given by

$$t = \frac{c}{\left[f_y/3f_{jd}\gamma_{M0}\right]^{0.5}} = \frac{93.38}{\left[275/3 \times 13.4\right]^{0.5}} = 35.7 \text{ mm}$$

Assume that the thickness of plate is less than 40 mm. Design strength for S275 plate, $f_y = 275$ N/mm²:

   Provide 40 mm thick base plate is adequate.
   The column flange thickness is 31.4 mm. Make the base plate 40 mm thick.
   Use 6 mm fillet weld all round to hold the base plate in place. The surfaces are to be machined for direct bearing. The holding-down bolts are nominal, but four No. 24 mm diameter bolts would be provided.
   Base slab: 600 × 600 × 40 mm³ thick.

PROBLEMS

7.1   A Grade S275 steel column having 6.0 m effective length for both axes is to carry pure axial loads from the floor above. If a 254 × 254 UB 89 is available, check the ultimate load that can be imposed on the column. The self-weight of the column may be neglected.

7.2   A column has an effective length of 5.0 m and is required to carry an ultimate axial load of 250 kN, including allowance for self-weight. Design the column using the following sections:
   1. Universal column section
   2. Circular hollow section
   3. Rectangular hollow section

7.3   A column carrying a floor load is shown in Figure 7.39. The column can be considered as pinned at the top and the base. Near the mid-height, it is propped by a strut about the minor axis. The column section provided is a 457 × 152 UB 60 of Grade S275 steel. Neglecting its self-weight, what is the maximum ultimate load the column can carry from the floor above?

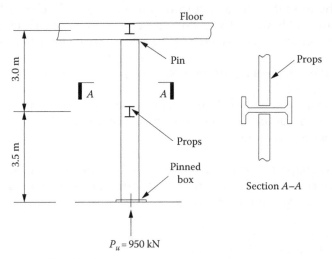

Figure 7.39 Column carrying floor load.

7.4   A Grade S275 steel 457 × 152 UB 60 used as a column is subjected to uniaxial bending about its major axis. The design data are as follows:

Ultimate axial compression = 1000 kN
Ultimate moment at top of column = +200 kN-m
Ultimate moment at bottom of column = –100 kN-m
Effective length of column = 7.0 m

Determine the adequacy of the steel section.

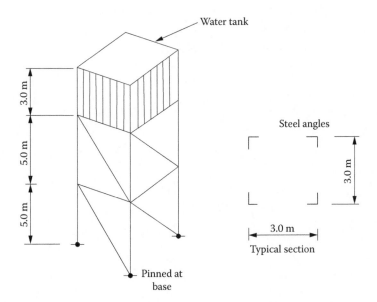

Figure 7.40 Support columns for water tower.

7.5   A column between floors of a multistorey building frame is subjected to biaxial bending at the top and bottom. The column member consists of a Grade S275 steel 305 × 305 UC158 section. Investigate its adequacy if the design load data are as follows:

Ultimate axial compression = 2300 kN
Ultimate moments
Top-about major axis = 300 kN-m, about minor axis = 50 kN-m
Bottom-about major axis = 150 kN-m, about minor axis = −80 kN-m
Buckling length of column = 6.0 m

7.6   A steel tower supports a water tank of size 3 × 3 × 3 m^3. The self-weight of the tank is 50 kN when empty. The arrangement of the structure is shown in Figure 7.40. Other design data are given in the following:

Unit weight of water = 9.81 kN/m^3
Design wind pressure = 1.0 kN/m^2.

Use Grade S275 steel angles for all members. Design the steel tower structure and prepare the steel drawings.

Chapter 8

Trusses and bracing

8.1 TRUSSES: TYPES, USES AND TRUSS MEMBERS

8.1.1 Types and uses of trusses

Trusses and lattice girders are framed elements resisting in-plane loading by axial forces in either tension or compression in the individual members. They are beam elements, but their use gives a large weight saving when compared with a universal beam designed for the same conditions.

The main uses for trusses and lattice girders in buildings are to support roofs and floors and carry wind loads. Pitched trusses are used for roofs, while parallel chord lattice girders carry flat roofs or floors.

Some typical trusses and lattice girders are shown in Figure 8.1a. Trusses in buildings are used for spans of about 10–50 m. The spacing is usually about 5–8 m. The panel length may be made to suit the sheeting or decking used and the purlin spacing adopted. Purlins need not be located at the nodes, but this introduces bending into the chord. The panel spacing is usually 1.5–4 m.

8.1.2 Truss members

Truss and lattice girder members are shown in Figure 8.1b. The most common members used are single and double angles, tees, channels and structural hollow sections. I and H and compound and built-up members are used in heavy trusses.

8.2 LOADS ON TRUSSES

The main types of loads on trusses are dead, imposed and wind loads. These are shown in Figure 8.1a.

8.2.1 Dead loads

The dead load is due to sheeting or decking, insulation, felt, ceiling if provided, weight of purlins and self-weight. This load may range from 0.3 to 1.0 kN/m². Typical values are used in the worked examples here.

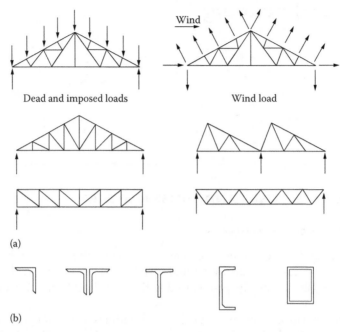

Figure 8.1 Roof trusses and lattice girders: (a) typical trusses and lattice girders; (b) truss members.

8.2.2 Imposed loads

The imposed load on roofs is taken from BS EN1991-1-3. This loading may be summarised as follows:

1. Where there is only access to the roof for maintenance and repair – 0.75 kN/m²
2. Where there is access in addition to that in (1) – 1.5 kN/m²

8.2.3 Wind loads

Readers should refer wind loading to the new BS EN1993-1-4. The wind load depends on the building dimensions and roof slope, among other factors. The wind blowing over the roof causes suction or pressure on the windward slope and suction on the leeward one (see Figure 8.1a). The loads act normal to the roof surface.

Wind loads are important in the design of light roofs where the suction can cause reversal of load in truss members. For example, a light angle member is satisfactory when used as a tie but buckles readily when required to act as a strut. In the case of flat roofs with heavy decking, the wind uplift will not be greater than the dead load, and it need not be considered in the design.

8.2.4 Application of loads

The loading is applied to the truss through the purlins. The value depends on the roof area supported by the purlin. The purlin load may be at a node point, as shown in Figure 8.1a, or between nodes, as discussed in Section 8.3.3 as follows. The weight of the truss is included in the purlin point loads.

8.3 ANALYSIS OF TRUSSES

8.3.1 Statically determinate trusses

Trusses may be simply supported or continuous, statically determinate or redundant and pin jointed or rigid jointed. However, the most commonly used truss or lattice girder is single span, simply supported and statically determinate. The joints are assumed to be pinned, though, as will be seen in actual construction, continuous members are used for the chords. A pin-jointed truss is statically determinate when

$$m = 2j - 3$$

where
  m denotes the number of members
  j denotes the number of joints

8.3.2 Load applied at the nodes of the truss

When the purlins are located at the node points, the following manual methods of analysis are used:

1. Force diagram – this is the quickest method for pitched-roof trusses.
2. Joint resolution – this is the best method for a parallel chord lattice girder.
3. Method of sections – this method is useful where it is necessary to find the force in only a few members.

The force diagram method is used for the analysis of the truss in Section 8.6. An example of use of the method of sections would be for a light lattice girder where only the force in the maximum loaded member would be found. The member is designed for this force and made uniform throughout.

   A matrix analysis program can be used for truss analysis. In Chapter 11, the roof truss for an industrial building is analysed using a computer program.

8.3.3 Loads not applied at the nodes of the truss

The case where the purlins are not located at the nodes of the truss is shown in Figure 8.2a. In this case, the analysis is made in two parts:

1. The total load is distributed to the nodes as shown in Figure 8.2b. The truss is analysed to give the axial loads in the members.
2. The top chord is now analysed as a continuous beam loaded with the normal component of the purlin loads as shown in Figure 8.2c. The continuous beam is taken as fixed at the ridge and simply supported at the eaves. The beam supports are the internal truss members. The beam is analysed by moment distribution. The top chord is then designed for axial load and moment.

8.3.4 Eccentricity at connections

Member meeting at a joint should be arranged so that their centroidal axes meet at a point. For bolted connections, the bolt gauge lines can be used in place of the centroidal axes.

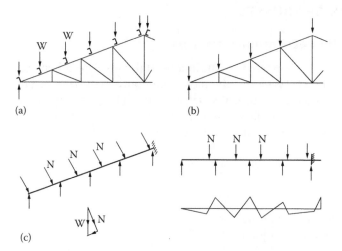

Figure 8.2 Loads applies between nodes of truss: (a) loads applied between nodes; (b) primary analysis – loads at nodes; (c) secondary analysis of top chord as a continuous beam.

If the joint is constructed with eccentricity, then the members and fasteners must be designed to resist the moment that arises. The moment at the joint is divided between the members in proportion to their stiffness.

8.3.5 Rigid-jointed trusses

Moments arising from rigid joints are important in trusses with short, thick members. Secondary stresses from these moments will not be significant if the slenderness of chord members in the plane of the truss is greater than 50 and that of most web members is greater than 100. Rigid-jointed trusses may be analysed using a matrix stiffness analysis program.

8.3.6 Deflection of trusses

The deflection of a pin-jointed truss can be calculated using the strain energy method. The deflection at a given node is

$$\delta = \frac{\sum NuL}{AE}$$

where
   N is the force in a truss member due to the applied loads
   u is the force in a truss member due to unit load applied at the node and in the direction of the required deflection
   L is the length of a truss member
   A is the area of a truss member
   E is the Young's modulus

A computer analysis gives the deflection as part of the output.

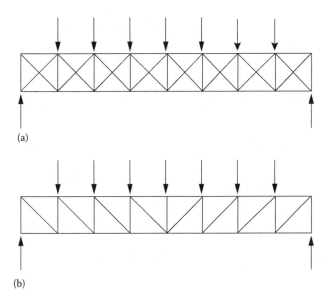

Figure 8.3 Cross-braced lattice girder: (a) cross-braced redundant truss; (b) analysis with tension diagonals.

8.3.7 Redundant and cross-braced trusses

A cross-braced wind girder is shown in Figure 8.3. To analyse this as a redundant truss, it would be necessary to use the computer analysis program.

However, it is usual to neglect the compression diagonals and assume that the panel shear is taken by the tension diagonals, as shown in Figure 8.3b. This idealisation is used in design of cross bracing (see Section 8.7).

8.4 DESIGN OF TRUSS MEMBERS

8.4.1 Design conditions

The member loads from the separate load cases – dead, imposed and wind – must be combined and factored to give the critical conditions for design. Critical conditions often arise through reversal of load due to wind, as discussed in the succeeding text. Moments must be taken into account if loads are applied between the truss nodes.

8.4.2 Struts

1. *Maximum slenderness ratios*
   For lightly loaded members, these limits often control the size of members. The maximum ratios are
   a.   Members resisting other than wind load – 180
   b.   Members resisting self-weight and wind load – 250
   c.   Members normally acting as ties but subject to reversal of stress due to wind – 350
   Deflection due to self-weight should be checked for members whose slenderness exceeds 180. If the deflection exceeds the ratio length/1000, the effect of bending should be taken into account in design.

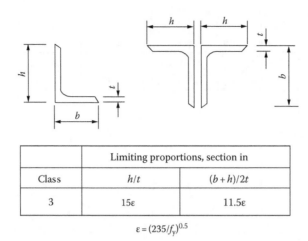

	Limiting proportions, section in	
Class	h/t	$(b+h)/2t$
3	15ε	11.5ε

$$\varepsilon = (235/f_y)^{0.5}$$

Figure 8.4 Limiting proportions for single- and double-angle struts.

2. *Limiting proportions of angle struts*

To prevent local buckling, limiting width/thickness ratios for single angles and double angles with components separated are given in Table 5.2 of BS EN1993-1-1. These are shown in Figure 8.4.

3. *Effective lengths for compression chords*

The compression chord of a truss or lattice girder is usually a continuous member over a number of panels or, in many cases, its entire length. The chord is supported in its plane by the internal truss members and by purlins at right angles to the plane as shown in Figure 8.5.

The length of chord members is as follows:

a. In the plane of the truss-panel length – L_1

b. Out of the plane of the truss–purlin spacing – L_2

The rules can be used to determine the buckling lengths. The slenderness ratios for single- and double-angle chords are shown in Figure 8.5. Note that truss joints reduce the in-plane value of the buckling length.

4. *Effective lengths of discontinuous internal truss members*

Discontinuous internal truss members, a single angle or double angle connected to a gusset at each end, are shown in Figure 8.5. The buckling lengths for the cases where the connections contain at least two fasteners or the equivalent in welding. The non-dimensional slenderness is shown in Figure 8.5. The length L_3 is the distance between truss nodes.

5. *Design procedure*

The end eccentricity for discontinuous struts may be ignored and the design made for an axially loaded member.

For single angles or double angles with members separated, the buckling resistance for Class 1, 2 and 3 sections is given by

$$N_{b,Rd} = \frac{\chi A f_y}{\gamma_{M1}}$$

where A denotes gross area.

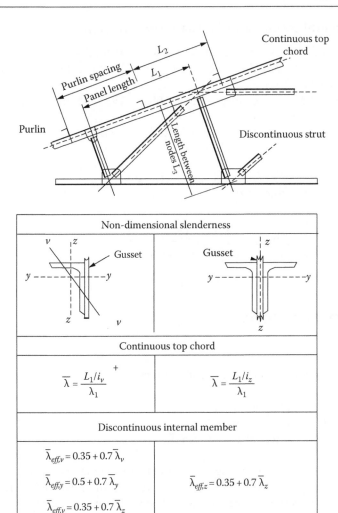

Figure 8.5 Slenderness ratios for truss members.

From Table 6.2 in the code, the strut curve *b* is selected to obtain the buckling resistance, $N_{b,Rd}$.

If the section is Class 4, the effective area method is used to determine the member compressive strength (refer to Section 7.4.9 for the method and example).

8.4.3 Ties

The effective area is used in the design of discontinuous angle ties. This was set out in Section 6.4 earlier. Tension chords are continuous throughout all or the greater part of their length. Checks will be required at end connections and splices.

8.4.4 Members subject to reversal of load

In light roofs, the uplift from wind can be greater than the dead load. This causes a reversal of load in all members. The bottom chord is the most seriously attached member and must be supported laterally by a lower chord bracing system, as shown in Figure 1.2. It must be checked for tension due to dead and imposed loads and compression due to wind load.

8.4.5 Chords subjected to axial load and moment

Angle top chords of trusses may be subjected to axial load and moment, as discussed in Section 8.3.3 earlier. The buckling capacity for axial load is calculated in accordance with Section 8.4.2 (3) earlier.

The reader should refer to Clause 6.3.3 in BS EN1993-1-1 for the design of combined bending and axial compression.

The interaction expressions for combined axial and bending are

$$\frac{N_{Ed}}{\chi_y N_{Rk}/\gamma_{M1}} + k_{yy}\frac{M_{y,Ed}}{\chi_{LT}M_{y,Rk}/\gamma_{M1}} + k_{yz}\frac{M_{z,Ed}}{M_{z,Rk}/\gamma_{M1}} \leq 1$$

$$\frac{N_{Ed}}{\chi_z N_{Rk}/\gamma_{M1}} + k_{zy}\frac{M_{y,Ed}}{\chi_{LT}M_{y,Rk}/\gamma_{M1}} + k_{zz}\frac{M_{z,Ed}}{M_{z,Rk}/\gamma_{M1}} \leq 1$$

8.5 TRUSS CONNECTIONS

8.5.1 Types

The following types of connections are used in trusses:

1. Column cap and end connections
2. Internal joints in welded construction
3. Bolted site joints – internal and external

The internal joints may be made using a gusset plate or the members may be welded directly together. Some typical connections using gussets are shown in Figure 8.6. In these joints, all welding is carried out in the fabrication shop. The site joints are bolted.

8.5.2 Joint design

Joint design consists of designing the bolts, welds and gusset plate.

1. *Bolted joints*: The load in the member is assumed to be divided equally between the bolts. The bolts are designed for direct shear and the eccentricity between the bolt gauge line and the centroidal axis is neglected (see Figure 8.7a). The bolts and gusset plate are checked for bearing.
2. *Welded joints*: In Figure 8.7a, the weld groups can be balanced as shown. That is, the centroid of the weld group is arranged to coincide with the centroidal axis of the angle in the plane of the gusset. The weld is designed for direct shear.

   If the angle is welded all round, the weld is loaded eccentrically, as shown in Figure 8.7b. However, the eccentricity is generally not considered in practical design because much more weld is provided than is needed to carry the load.
3. *Gusset plate*: The gusset plate transfers loads between members. The thickness is usually selected from experience, but it should be at least equal to that of the members to be connected.

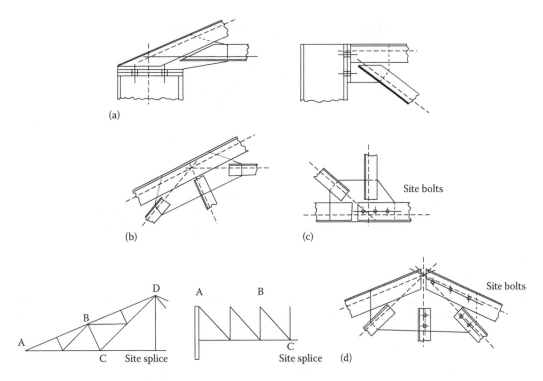

Figure 8.6 Truss and lattice girder connections: (a) end connections – A; (b) internal connection – B; (c) site splice – C; (d) ridge joint – D.

The actual stress conditions in the gusset are complex. The direct stress in the plate can be checked at the end of the member assuming that the load is dispersed at 30° as shown in Figure 8.7c. The direct load on the width of dispersal *b* should not exceed the design strength of the gusset plate. In joints where members are close together, it may not be possible to disperse the load. In this case, a width of gusset equal to the member width is taken for the check.

8.6 DESIGN OF A ROOF TRUSS FOR AN INDUSTRIAL BUILDING

8.6.1 Specification

A section through an industrial building is shown in Figure 8.8a. The frames are at 5 m centres and the length of the building is 45 m. The purlin spacing on the roof is shown in Figure 8.8b. The loading on the roof is as follows:

1. Dead load – measured on the slope length

Sheeting and insulation board	= 0.25 kN/m²
Purlins	= 0.1 kN/m²
Truss	= 0.1 kN/m²
Total dead load	= 0.45 kN/m²

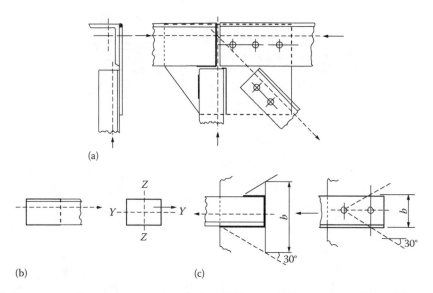

(a)

(b)   (c)

Figure 8.7 Truss connections and gusset plate design: (a) shop welded – site bolted joint; (b) eccentrically
loaded weld group; (c) effective width of gusset plate.

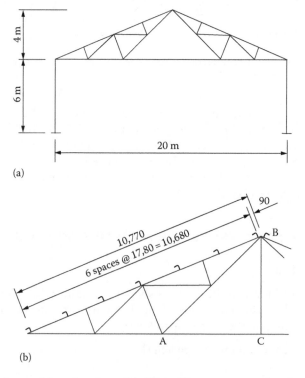

(a)

(b)

Figure 8.8 Pitched-roof truss: (a) section through building; (b) arrangement of purlins.

2. Imposed load measured on plan = 0.75 kN/m²
   Imposed load measured on slope = 0.75 × 10/10.77 = 0.7 kN/m²

   Design the roof truss using angle members and gusseted joints. The truss is to be
   fabricated in two parts for transport to site. Bolted site joints are to be provided at *A*,
   B and *C* as shown in Figure 8.8b.

8.6.2 Truss loads

1. *Dead and imposed loads*

   Because of symmetry, only one-half of the truss is considered.

   Dead loads:
   > End panel points = 1/8 × 0.45 × 10.77 × 5 = 3.03 kN
   > Internal panel points = 2 × 3.03 = 6.06 kN

   Imposed loads:
   > End panel point = 3.03 × 0.7/0.45 = 4.71 kN
   > Internal panel points = 2 × 4.71 = 9.42 kN

   The dead loads are shown in Figure 8.10.

2. *Wind loads*

   > Basic velocity pressure, $q_b = 0.5 \times \rho \times v_b^2 = 0.5 \times 1.226 \times 22^2 = 296$ N/m²
   > Peak velocity pressure $q_p(z) = c_e(z) \times q_b = 2.3 \times 296 = 0.68$ kN/m²

   The external pressure coefficients C_{pe} from the internal pressure coefficients C_{pi} are shown in Figure 8.9. The values used are where there is only a negligible probability of a dominant opening occurring during a severe storm. C_{pi} is taken as the more onerous of the values +0.2 or −0.3.

   For the design of the roof truss, the condition of maximum uplift is the only one that need be investigated. A truss is selected from Section HI of the roof shown in Figure 8.9a, where C_{pe} is a maximum and C_{pi} is taken as +0.2, the case of internal pressure. The wind load normal to the roof is

   $$0.68\,(C_{pe} - C_{pi})$$

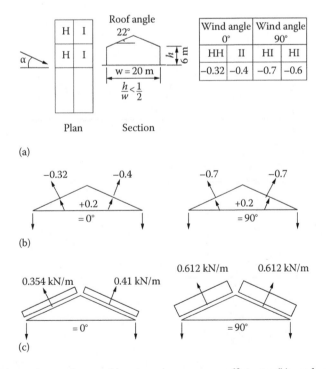

Figure 8.9 Wind loads on the roof truss: (a) external pressure coefficients; (b) roof pressure coefficients; (c) roof wind loads.

The wind loads on the roof are shown in Figure 8.9c for the two cases of wind transverse and longitudinal to the building.

The wind loads at the panel points normal to the top chord for the case of wind longitudinal to the building are

End panel points = 1/8 × 0.612 × 10.77 × 5 = 4.12 kN
Internal panel points = 8.24 kN

The wind loads are shown in Figure 8.11.

8.6.3 Truss analysis

1. *Primary forces its truss members*
Because of symmetry of loading in each case, only one half of the truss is considered. The truss is analysed by the force diagram method and the analyses are shown in Figures 8.10 and 8.11. Note that members 4–5 and 5–6 must be replaced by the fictitious member 6–X to locate point 6 on the force diagram. Then point 6 is used to find points 4 and 5.

The dead-load case is analysed and the forces due to the imposed loads are found by proportion. The case for maximum wind uplift is analysed. The forces in the members of the truss are tabulated for dead, imposed and wind load in Table 8.1.

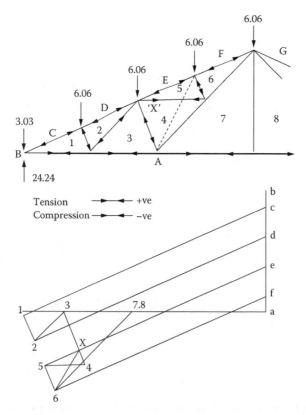

Figure 8.10 Dead-load analysis.

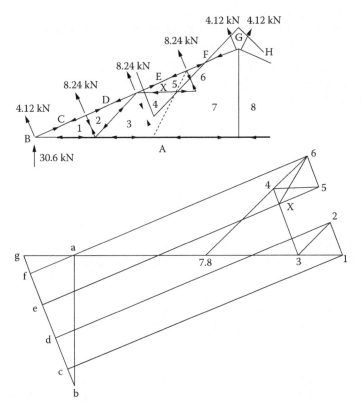

Figure 8.11 Wind-load analysis.

Table 8.1 Forces in members of roof truss (kN)

Member	Dead load	Imposed load	Wind load
Top chord			
C-1	−57.1	−88.8	72.1
D-2	−54.9	−85.4	72.1
E-5	−52.6	−81.8	74.9
F-6	−50.3	−78.2	74.9
Bottom chord			
A-1	53.9	82.4	−66.9
A-3	45.5	−70.8	−55.9
A-7	30.3	47.1	−32.1
Struts			
1–2	−5.6	−8.7	8.2
3–4	−11.3	−17.6	17.6
5–6	−5.6	−8.7	8.2
Ties			
2–3	7.6	11.8	−11.1
4–5	7.6	11.8	−14.1
4–7	15.2	23.6	−23.7
6–7	22.7	35.3	−34.8
7–8	0.0	0.0	0.0

2. *Moments in the top chord*

The top chord is analysed as a continuous beam for moments caused by the normal component of the purlin load from dead load:

Purlin load = $1.78 \times 5 \times 0.35 = 3.12$ kN
Normal component = $3.12 \times 10/10.77 = 2.89$ kN
End purlin $L = 2.89 \times 0.98/1.78 = 1.59$ kN

The top-chord loading is shown in Figure 8.12a. The fixed end moments are

Span AB

$M_A = 0$
$M_{BA} = 2.89 \times 1.78(2.69^2 - 1.78^2)/2 \times 2.69^2 = 1.44$ kN-m

Span BC

$M_{BC} = 2.89[(0.87 \times 1.82^3) + (2.65 \times 0.04^2)]/2.69^2 = 1.15$ kN-m
$M_{CB} = 2.89[(1.82 \times 0.87^2) + (0.04 \times 2.65^2)]/2.69^2 = 0.66$ kN-m

Span CD

$M_{CD} = 2.89 \times 1.74 \times 0.95^2/2.69^2 = 0.63$ kN-m
$M_{DC} = 2.89 \times 0.95 \times 1.74^2/2.69^2 = 1.15$ kN-m

Span DE

$M_{DE} = 2.89 \times 0.82 \times 1.87^2/2.69^2 + 1.59 \times 2.6 \times 0.09^2/2.69^2 = 1.15$ kN-m
$M_{ED} = 2.89 \times 1.87 \times 0.82^2/2.69^2 + 1.59 \times 2.6 \times 0.09^2/2.69^2 = 0.51$ kN-m

The distribution factors are

Joint B $BA:BC = 0.75:1 = 0.43:0.57$
Joints C and $D = 0.5:0.5$

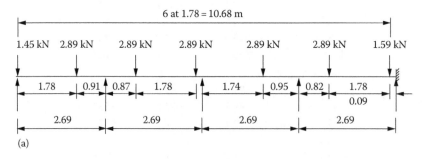

(a)

	0.43	0.57		0.5	0.5		0.5	0.5	
0.0	1.44	−1.15		0.66	−0.63		1.15	−1.15	0.63
	−0.12	−0.17		−0.02	−0.02		0.0	0.0	
	0.0	−0.01		−0.09	0.0		0.0	0.0	0.0
	0.0	0.0		0.05	0.05		0.0	0.0	
0.0	1.32	−1.32		0.6	−0.6		1.15	−1.15	0.63

(b)

Figure 8.12 Top-chord analysis: (a) top chord dead loads; (b) moment distribution, top chord analysis.

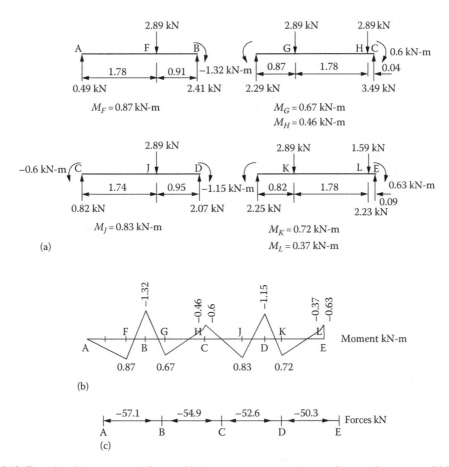

Figure 8.13 Top-chord moments and forces: (a) separate spans, reactions and internal moments; (b) bending moment diagram; (c) axial forces.

The moment distribution is shown in Figure 8.12b and the reactions and internal moments for the separate spans are shown in Figure 8.13a. The bending moment diagram for the complete top chord is shown in Figure 8.13b and the axial loads from the force diagram in Figure 8.13c.

8.6.4 Design of truss members

Part of the member design depends on the joint arrangement. The joint detail and design are included with the member design. Full calculations for joint design are not given in all cases.

1. *Top chord*
   The top chord is to be a continuous member with a site joint at the ridge.

   Member C-1 at eaves
   The maximum design conditions are at B (Figure 8.13).

   Dead + imposed load:

   Compression $F = -(1.35 \times 57.1) - (1.5 \times 88.8) = -210.3$ kN
   Moment $M_B = -(1.35 \times 1.32) - (1.5 \times 1.32 \times 0.7/0.35) = -5.74$ kN-m

Dead + wind load:

Tension $F = -57.1 + (1.5 \times 72.1) = 51$ kN

Try two no. $100 \times 100 \times 10$ RSA angles with 10 mm thick gusset plate. The gross section is shown in Figure 8.14a. The properties are

$A = 19.2 \times 2 = 38.4$ cm^2; $i_y = 3.05$ cm, $i_z = 4.51$ cm
$W_{el,y} = 24.8 \times 2 = 49.6$ cm^3

Design strength $f_y = 275$ N/mm^2.
Check the section classification using Figure 8.4:

$$\frac{b}{t} = \frac{100}{10} = 10 < 15\varepsilon$$

$$\frac{b+h}{2t} \times \frac{200}{20} = 10 < 11.5\varepsilon$$

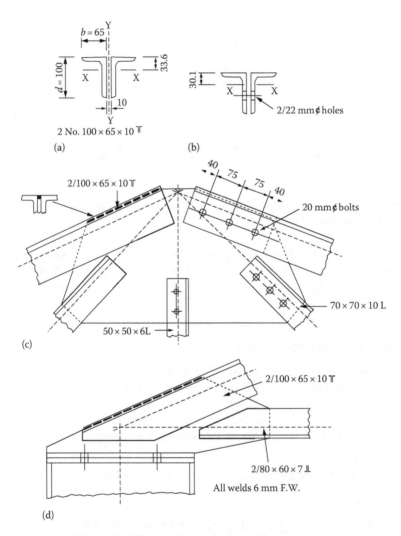

Figure 8.14 Top-chord design details: (a) top chord; (b) net section at site joint; (c) ridge joint; (d) eaves joint.

The section is Class 3.

The slenderness is the maximum of

$$\frac{L_{cr}}{i_y} = 0.7 \times \frac{2690}{30.5} = 61.8$$

$$\frac{L_{cr}}{i_z} = \frac{1780}{45.1} = 39.5$$

From Table 6.2 of EN1993-1-1, for a S275 rolled RSA section, curve b, $\alpha = 0.34$

$$\lambda_1 = 93.9\varepsilon = 93.9 \times 0.92 = 86.4$$

$$\bar{\lambda} = \frac{\lambda}{\lambda_1} = \frac{61.8}{86.4} = 0.72$$

$$\Phi = 0.5 \, [1 + 0.34(0.72 - 0.2) + 0.72^2] = 0.85$$

$$\chi = \frac{1}{0.85 + \sqrt{0.85^2 - 0.72^2}} = 0.77$$

Buckling resistance, $N_{b,Rd} = 0.77 \times 38.4 \times 275/10 = 813$ kN

$$\overline{\lambda_{LT}} = 0.9\overline{\lambda_z} = 0.41$$

From Table 6.3 and 6.4, curve d, $\alpha_{LT} = 0.76$

$$\Phi_{LT} = 0.5 \times [1 + 0.76(0.41 - 0.2) + 0.41^2] = 0.66$$

$$\chi_{LT} = \frac{1}{0.66 + \sqrt{0.66^2 - 0.41^2}} = 0.85$$

Interaction expression

$$\frac{222}{813} + \frac{5.74}{0.85 \times 49.6 \times 275 \times 10^{-3}} = 0.77 \leq 1.0$$

Provide two no. 100 × 100 × 10 angles.

The member will be satisfactory for the case of dead plus wind load.

Member F-6 at ridge

The ridge joint is shown in Figure 8.14c. A bolted site joint is provided in the chord on one side. The design conditions are

Compression $F = (1.35 \times 50.3) + (1.5 \times 78.2) = 185.2$ kN
Moment $M_E = -(1.35 \times 0.57) - (1.5 \times 0.57 \times 0.7/0.35) = 2.48$ kN-m

The net section is shown in Figure 8.14b:

Net area = 34.0 cm²; $W_{el,y}$ = 40.69 cm³ (minimum)
$M_{b,Rd}$ = 0.85 × 275 × 40.69 × 10⁻³ = 9.51 kN-m

Interaction expression for local capacity is

$$\frac{185.2}{275 \times 34 \times 10^{-1}} + \frac{2.48}{9.51} = 0.46 \leq 1.0$$

The section is satisfactory.

Eaves joint (see Figure 8.14d)
The member is connected to both sides of the gusset, so the gross section is effective in resisting load (see Section 4.6.3.3 of the code).

Compression F = 210.3 kN
Length of 6 mm fillet, strength 0.981 kN/mm, required
= 210.3/0.981 = 214 mm

This may be balanced around the member as shown in Figure 8.14d. More weld has been provided than needed.

The bearing capacity of the gusset is checked at the end of the member on a width of 100 mm. No dispersal of the load is considered because of the compact arrangement of the joint:

Bearing capacity = 275 × 100 × 10/10³ = 275 kN
The gusset is satisfactory.

Ridge joint (see Figure 8.14c)
Try three no. 20 mm diameter Class 4.6 bolts at the centres shown in Figure 8.14c. From the blue book, the double shear value is 94.1 kN. The bolts resist

Direct shear = 185.2 kN, moment = 2.48 kN
Direct shear per bolt = 185.2/3 = 61.7 kN
Shear due to moment = 2.48/0.15 = 16.5 kN
Resultant shear = $(61.7^2 + 16.5^2)^{0.5}$ = 63.8 kN

The bolts are satisfactory.

The shop-welded joint must be designed for moment and shear. The weld is shown in Figure 8.14c. The gusset will be satisfactory.

2. *Bottom chord*
The bottom chord is to have two site joints at P and R, as shown in Figure 8.15.

Member A-1
The design conditions are

Dead + imposed load:

Tension F = (1.35 × 53) + (1.5 × 82.4) = 195.15 kN

Dead + wind load:

Compression F = 53 − (1.5 × 66.9) = −47.35 kN

Try two no. 80 × 60 × 7 RSA angles. The section is shown in Figure 8.15b. The properties are

A = 18.76 cm²; i_y = 2.50 cm; i_z = 2.66 cm

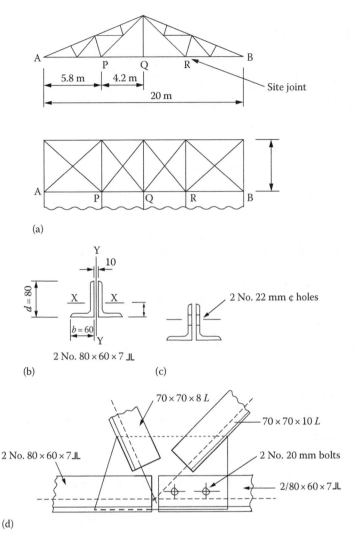

Figure 8.15 Bottom-chord design details: (a) lower chard bracing; (b) bottom chord; (c) net section; (d) site joint at *P*.

The section is Class 3.

When the bottom chord is in compression due to uplift from wind, lateral supports will be provided at *P*, *Q* and *R* by the lower chord bracing shown in Figure 8.15a. The effective length for buckling about the *z–z* axis is 5800 mm:

$L_E/i_z = 5800/26.6 = 218$, $\lambda_1 = 86.4$
Buckling resistance, $N_{b,Rd} = 0.21 \times 18.76 \times 275/10 = 108.3$ kN

At end *A*, the angles are connected to both sides of the gusset:

$N_{t,Rd} = 275 \times 18.76/10 = 515.9$ kN

Provide two no. 80 × 60 × 7 angles. The wind load controls the design.

The connection to the gusset is shown in Figure 8.15d. The length of 6 mm weld required = 206.04/0.9 = 228.9 mm.

The weld is placed as shown in Figure 8.15d. The gusset is satisfactory.

Member A-7

Dead + imposed load:

Tension $F = (1.35 \times 30.3) + (1.5 \times 47.1) = 111.6$ kN

Dead + wind load:

Compression $= 30.3 - (1.4 \times 32.1) = -17.85$ kN

Member A-7 is connected to the gusset by two no. 20 mm diameter bolts in double shear as shown in Figure 8.15d.

Net area $= 18.76 - 2 \times 22 \times 7/10^2 = 15.68$ cm^2

$$N_{t,Rd} = \frac{0.9 f_u A_{net}}{\gamma_{M2}} = 0.9 \times 430 \times 15.68 \times 10^{-1} / 1.25 = 485.4 \text{ kN}$$

The member will also be satisfactory when acting in compression due to wind. The shear capacity of the joint $= 2 \times 94.1 = 188.2$ kN > 111.6 kN.

3. *Internal members*

Members 4–7 and 6–7

Design for the maximum load in members 6–7

Dead + imposed load:

Tension $F = (1.35 \times 22.7) + (1.5 \times 35.3) = 83.6$ kN

Dead + wind load:

Compression load $= 22.7 - (1.5 \times 34.8) = -29.5$ kN

Try 70 × 70 × 10 angle. The member lengths and section are shown in Figure 8.16. The properties are

$A = 13.1$ cm^2; $i_y = 2.09$ cm; $i_v = 1.36$ cm

The slenderness values are calculated as follows (see Figure 8.5).

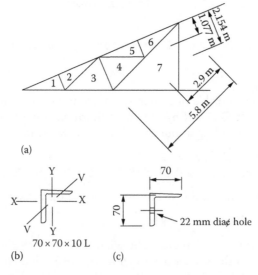

(a)

(b) 70 × 70 × 10 L

(c)

Figure 8.16 Design for members 4–7, 6–7 and 3–4: (a) internal member lengths; (b) section; (c) net section at site joint.

Members 6–7 buckling about the v–v axis:

$$\overline{\lambda}_v = \frac{2900/13.6}{86.4} = 2.47$$

Members 6–7 and 4–7 buckling laterally:

$$\overline{\lambda}_z = \frac{5800/20.9}{86.4} = 3.21$$

Buckling resistance, $N_{b,Rd} = 0.12 \times 13.1 \times 275/10 = 43.2$ kN
This is satisfactory.

The end of member 6–7 is connected at the ridge by bolts, as shown in Figure 8.14c. The net section is shown in Figure 8.16c:

Net area of connected leg = $10(65 - 22) = 430$ mm^2
Area of unconnected leg = $10 \times 65 = 650$ mm^2
Net area $A_{net} = 430 + 650 = 1080$ mm^2

$$N_{t,Rd} = \frac{0.6 \times 1080 \times 430 \times 10^{-3}}{1.25} = 222.9 \text{ kN} > 83.6 \text{ kN}$$

This is satisfactory.

Joints for members 4–7 and 6–7

Use 20 mm bolts in clearance holes.
Single shear value = 47 kN (Table 4.2).
Number of bolts required = 83.6/47 = 1.8. Use three bolts.

The bolts are shown in Figure 8.14c and the welded connection is also shown on the figure. The connection to the site joint at P is shown in Figure 8.15d.

Members 3–4
The design loads are dead + imposed load:

Compression $F = -(1.35 \times 11.3) - (1.5 \times 17.6) = -41.65$ kN.

Dead + wind load:

Tension $F = -11.3 + (1.5 \times 17.6) = 15.1$ kN.
Use $70 \times 70 \times 8$ angle.

Other internal members
All other members are to be $50 \times 50 \times 6$ angles. The design for these members is not given.
4. *Truss arrangement*
A drawing of the truss is shown in Figure 8.17 and details for the main joints are shown in Figures 8.14 and 8.15.

8.7 BRACING

8.7.1 General considerations

Bracing is required to resist horizontal loading in buildings designed to the simple design method. The bracing also generally stabilises the building and ensures that the framing is square. It consists of the diagonal members between columns and trusses and is usually placed in the end bays. The bracing carries the load by forming lattice girders with the building members.

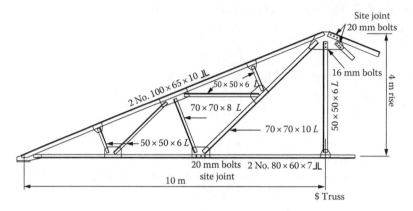

Figure 8.17 Arrangement drawing of truss.

8.7.2 Bracing for single-storey industrial buildings

The bracing for a single-storey building is shown in Figure 8.18a. The internal frames resist the transverse wind load by bending in the cantilever columns. However, the gable frame can be braced to resist this load, as shown. The wind blowing longitudinally causes pressure and suction forces on the windward and leeward gables and wind drag on the roof and walls. These forces are resisted by the roof and wall bracing shown.

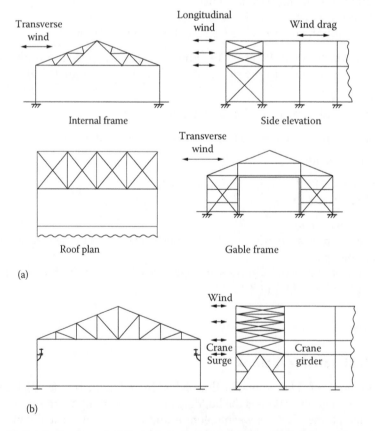

Figure 8.18 Bracing for single-storey building: (a) single-storey building; (b) building with a crane.

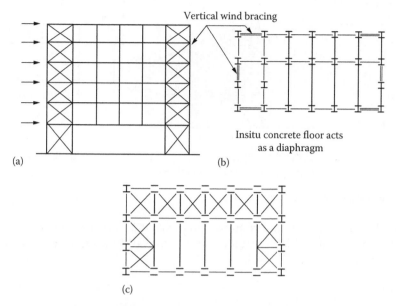

Figure 8.19 Bracing for multistorey building: (a) floor elevation; (b) floor plan; (c) floor bracing.

If the building contains a crane, an additional load due to the longitudinal crane surge has to be taken on the wall bracing. A bracing system for this case is shown in Figure 8.18b.

8.7.3 Bracing for a multistorey building

The bracing for a multistorey building is shown in Figure 8.19. Vertical bracing is required on all elevations to stabilise the building. The wind loads arc is applied at floor level. The floor slabs transmit loads on the internal columns to the vertical lattice girders in the end bays. If the building frame and cladding are erected before the floors are constructed, floor bracing must be provided as shown in Figure 8.19c. Floor bracing is also required if precast slabs not effectively tied together are used.

8.7.4 Design of bracing members

The bracing can be single diagonal members or cross members. The loading is generally due to wind or crane surge and is reversible. Single-bracing members must be designed to carry loads in tension and compression. With cross bracing, only the members in tension are assumed to be effective and those in compression are ignored.

The bracing members are the diagonal web members of the lattice girder formed with the main building members, column, building truss chords, purlins, eaves and ridge members, floor beams and roof joists. The forces in the bracing members are found by analysing the lattice girder. The members are designed as ties or struts, as set out in Section 8.4 earlier. Bracing members are often very lightly loaded and minimum-size sections are chosen for practical reasons.

8.7.5 Example: Bracing for a single-storey building

The gable frame and bracing in the end bay of a single-storey industrial building are shown in Figure 8.20. The end bay at the other end of the building is also braced. The length of the

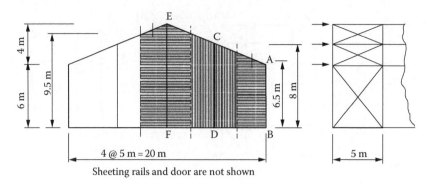

Figure 8.20 Gable frame and bracing in end bay.

building is 50 m and the truss and column frames are at 5 m centres. Other building dimensions are shown in Figure 8.20. Design the roof and wall bracing to resist the longitudinal wind loading using Grade S275 steel.

1. *Wind load*

Basic wind speed, v_b = 22 m/s
Basic velocity pressure, $q_b = 0.5 \times \rho \times v_b^2 = 0.5 \times 1.226 \times 22^2 = 296$ N/m²
Peak velocity pressure $q_p(z) = c_e(z) \times q_b = 2.3 \times 296 = 0.68$ kN/m²

The pressure coefficients on the end walls from are set out in Figure 8.21a.

C_{pe} = 0.85. This is not affected by the internal pressure: wind pressure = 0.85 × 0.68 = 0.578 kN/m².

The method for calculating the frictional drag is given in BSEN1991-1-4.

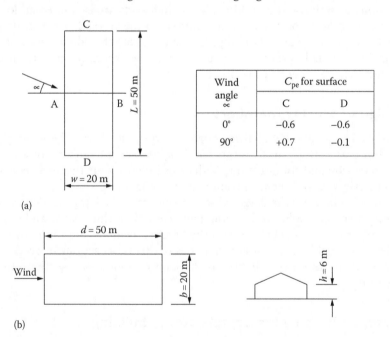

Wind angle $\propto$	C_{pe} for surface	
	C	D
0°	−0.6	−0.6
90°	+0.7	−0.1

(a)

(b)

Figure 8.21 Data for calculating wind loads: (a) external pressure coefficients – end walls; (b) dimensions for calculating frictional drag.

Cladding – corrugated plastic-coated steel sheet.

Factor $C_f = 0.02$ for surfaces with corrugations across the wind direction (refer to Figure 8.21b).

$d/h = 50/6 = 8.33 > 4$

$d/b = 50/20 = 2.5 < 4$

The ratio d/h is greater than 4, so the drag must be evaluated.

For $h < b$, the frictional drag is given by

F roof $= C_f qbd(d - 4h)$
$= 0.02 \times 0.68 \times 20(50 - 24) = 7.1$ kN
F walls $= C_f q2h(d - 4h)$
$= 0.02 \times 0.68 \times 12(50 - 24) = 4.24$ kN

The total load is the sum of the wind load on the gable ends and the frictional drag. The load is divided equally between the bracings at each end of the building.

2. *Loads on bracing* (see Figure 9.20)

Point E

Load from the wind on the end gable column EF

$= 9.5 \times 5 \times 0.5 \times 0.578 = 13.73$ kN

Reaction at E, top of the column $= 6.86$ kN
Load at E from wind drag on the roof

$= 0.5 \times 7.1 \times 0.25 = 0.9$ kN

Total load at $E = 7.76$ kN

Point C

Load from the wind on the end gable column CD

$= 8 \times 5 \times 0.5 \times 0.578 = 11.56$ kN

Reaction at C, top of the column $= 5.78$ kN
Load C from wind drag on the roof $= 0.9$ kN
Total load at $C = 6.68$ kN

Point A

Load from the wind on building column AB

$= 2.5 \times 6.5 \times 0.5 \times 0.578 = 4.69$ kN

Reaction at A, top of the column $= 2.35$ kN
Load at A from wind drag on roof and wall

$= (0.125 \times 7.1 \times 0.5) + (0.5 \times 4.24 \times 0.25) = 0.97$ kN

Total load at $A = 3.32$ kN

3. *Roof bracing*

The loading on the lattice girder formed by the bracing and roof members and the forces in the bracing members are shown in Figure 8.22a. Note that the members of the cross bracing in compression have not been shown. Forces are transmitted through the purlins in this case. The maximum loaded member is AH:

Design load $= 1.5 \times 15.5 = 23.3$ kN.
Try $50 \times 50 \times 6$ angle with two no. 16 mm diameter Class 4.6 bolts in the end connections.
Bolt capacity $= 2 \times 30.1 = 60.2$ kN.

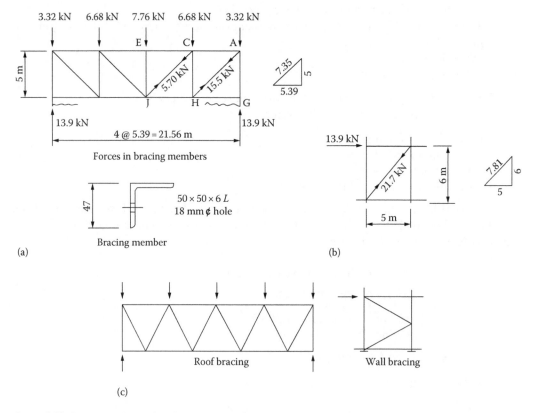

Figure 8.22 Bracing and member forces: (a) roof bracing; (b) wall bracing; (c) single bracing system.

Referring to Figure 8.22a

Net area $A_e = (47 - 18)6 + 282 = 456$ mm²

$$N_{t,Rd} = \frac{0.6 \times 456 \times 430 \times 10^{-3}}{1.25} = 94.1 \text{ kN}$$

Make all the members the same section.

4. *Wall bracing*

The load on the wall bracing and the force in the bracing member are shown in Figure 8.22b.

Design load = 1.5 × 21.7 = 32.5 kN. Provide 50 × 50 × 6 angle.

5. *Further considerations*

Design for load on one gable:

In a long building, the bracing should be designed for the maximum load at one end. This is the external pressure and internal suction on the gable plus one half of the frictional drag on the roof and walls. In this case, if the design is made on this basis,

roof bracing (design load in AH = 25.1 kN) and wall bracing (design load = 39.2 kN), the 50 × 50 × 6 angles will be satisfactory

Single-bracing system:
In the aforementioned analysis, cross bracing is provided and the purlins form part of the bracing lattice girder. This arrangement is satisfactory when angle purlins are used. However, if cold rolled purlins are used, the bracing system should be independent of the purlins. A suitable system is shown in Figure 8.22c, where the roof-bracing members support the gable columns. Circular hollow sections are often used for the bracing members.

8.7.6 Example: Bracing for a multistorey building

The framing plans for an office building are shown in Figure 8.23. The floors and roof are cast in in situ reinforced concrete slabs that transmit the wind load from the internal columns to the end bracing. Design the wind bracing using Grade S275 steel.

1. *Wind loads*
   The data for the wind loading are
   The site in the country with the closest distance to the sea upwind 2 km, at the first floor with the height of 5 m above ground, takes

   Basic wind velocity, v_b = 26 m/s

In Table 4 of BS EN1991-1-4, the value of $c_e(z)$ for height difference is as follows:

First floor H = 5 m $c_e(z)$ = 1.62
Second floor H = 10 m $c_e(z)$ = 1.78
Roof H = 15 m $c_e(z)$ = 1.85

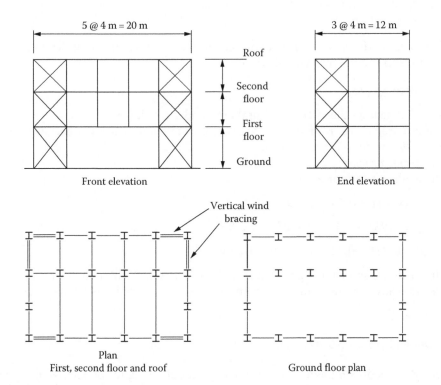

Figure 8.23 Framing plans for a multi-storey building.

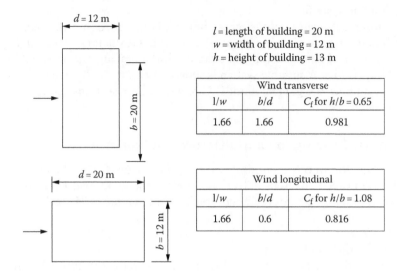

l = length of building = 20 m
w = width of building = 12 m
h = height of building = 13 m

Wind transverse		
l/w	b/d	C_f for $h/b = 0.65$
1.66	1.66	0.981

Wind longitudinal		
l/w	b/d	C_f for $h/b = 1.08$
1.66	0.6	0.816

Figure 8.24 Wind loads: force coefficients.

The wind speed design and dynamic pressures are

First floor v_b = 26.0 m/s
$q_b = 0.5 \times 1.226 \times 26.0/10^3 = 0.42$ kN/m^2
Second floor $v_b = 26.0 \times (1.78/1.62) = 28.5$ m/s
$q_b = 0.49$ kN/m^2
Roof level $v_b = 26.0 \times (1.85/1.62) = 29.6$ m/s
$q_b = 0.54$ kN/m^2

The force coefficients C_f for wind on the building as a whole are shown in Figure 8.24 for transverse and longitudinal wind:

$$\text{Force} = C_f q A_e$$

where A_e denotes the effective frontal area under consideration.
    Wind drag on the roof and walls need not be taken into account because neither the ratio d/h nor d/h is greater than 4.

2. *Transverse bracing*
    The loads at the floor levels are

$$P = 0.981 \times 0.54 \times 2 \times 10 = 10.6 \text{ kN}$$

$$Q = 0.981 \times 0.49 \times 4 \times 10 = 19.2 \text{ kN}$$

$$R = 0.981 \times 0.42 \times 4.5 \times 10 = 18.5 \text{ kN}$$

The loads are shown in Figure 8.25a and the forces in the bracing members in tension are also shown in the figure. The member sizes are selected (see Figure 8.25c).

Member QT, RU

Design load = $1.5 \times 42.1 = 63.15$ kN.

Provide 50 × 50 × 6 angle. The tension capacity allowing for one no. 18 mm diameter hole was calculated as 94.1 kN in Section 8.7.5 (2) earlier.

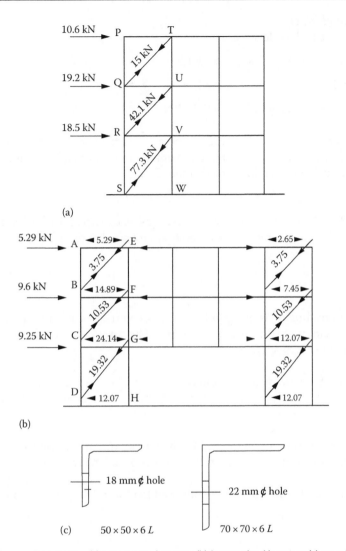

Figure 8.25 Bracing member design: (a) transverse bracing; (b) longitudinal bracing; (c) member sizes selected.

Using 16 mm diameter Grade 4.6 bolts, capacity 30 kN in single shear,

Member QT – provide two bolts at each end.
Member RU – provide three bolts at each end.

Member S V

Design load = 1.5 × 77.3 = 116 kN.
Try 70 × 70 × 6 angle with 20 mm diameter bolts with capacity in single shear of 47 kN per bolt.
No. of bolts required at each end = 3. For the angle,
Net area = 538.7 mm²
Tension capacity = 111.2 kN
Note that a 60 × 60 × 6 angle with 16 mm bolts could be used, but five bolts would be required in the end connection.

3. *Longitudinal bracing*

The loads at the floor levels are

$$A = 0.816 \times 0.54 \times 2 \times 6 = 5.29 \text{ kN}$$

$$B = 0.816 \times 0.49 \times 4 \times 6 = 9.6 \text{ kN}$$

$$C = 0.816 \times 0.42 \times 4.5 \times 6 = 9.25 \text{ kN}$$

The loads are shown in Figure 8.25b and are divided between the bracings at each end of the building. Those in the bracing members in tension are shown in Figure 8.25b. The maximum design load for member DG is

$$= 1.5 \times 19.32 = 29 \text{ kN}$$

Provide $50 \times 50 \times 6$ angles for all members. Tension capacity 94.1 kN with one 18 mm diameter hole. Two no. 16 mm diameter bolts are required at the ends of all bracing member.

PROBLEMS

8.1   A flat roof building of 18 m span has 1.5 m deep trusses at 4 m centres. The trusses carry purlins at 1.5 m centres. The total dead load is 0.7 kN/m² and the imposed load is 0.75 kN/m²:

1.   Analyse the truss by joint resolution.
2.   Design the truss using angle sections with welded internal joints and bolted field splices.

8.2   A roof truss is shown in Figure 8.26. The trusses are at 6 m centres, the length of the building is 36 m and the height to the eaves is 5 m. The roof loading is

Dead load = 0.4 kN/m² (on slope); imposed load = 0.75 kN/m² (on plan)

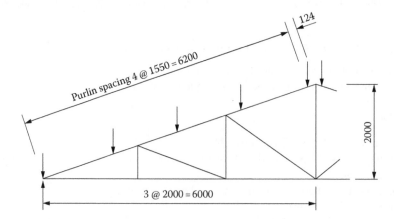

Figure 8.26 Roof truss arrangement.

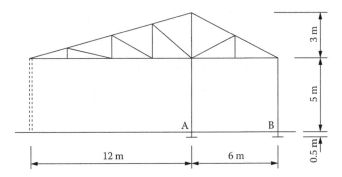

Figure 8.27 Building frame with cantilever truss.

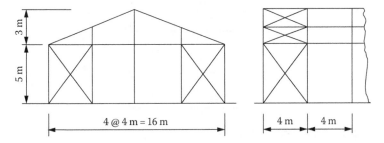

Figure 8.28 Bracing for a factory building.

The wind load is to be estimated using BS EN1991-1-4. The building is located on the outskirts of a city and the basic wind velocity is 26 m/s.
1. Analyse the truss for the roof loads.
2. Analyse the top chord for the loading due to the purlin spacing shown. The dead load from the roof and purlins is 0.32 kN/m².
3. Design the truss.

8.3 A section through a building is shown in Figure 8.27. The roof trusses are supported on columns at A and B and cantilever out to the front of the building. The front has roller doors running on tracks on the floor. The frames are at 6 m centres and the length of the building is 48 m. The roof load is

Dead load = 0.45 kN/m² (on slope); imposed load = 0.75 kN/m² (on plan)

The wind loads are to be in accordance with BS EN1991-1-4. The basic wind velocity is 26 m/s and the location is in the suburbs of a city. The structure should be analysed for wind load for the two conditions of doors opened and closed. Analyse and design the truss.

8.4 The end framing and bracing for a single-storey building are shown in Figure 8.28. The location of the building is on an industrial estate on the outskirts of a city in the north-east of England. The length of the building is 32 m. The wind loads are to be in accordance with BS EN1991-1-4. Design the bracing.

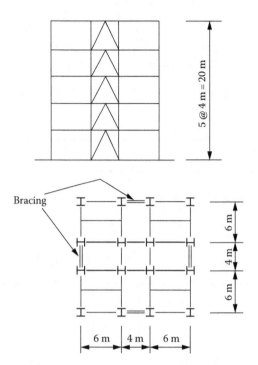

Figure 8.29 Framing for a square-tower building.

8.5   The framing for a square-tower building is shown in Figure 8.29. The bracing is similar on all four faces. The building is located in a city centre in an area where the basic wind velocity is 30 m/s. Design the bracing.

Chapter 9

Portal frames

9.1 DESIGN AND CONSTRUCTION

9.1.1 Portal type and structural action

The single-storey clear-span building is in constant demand for warehouses, factories and many other purposes. The clear internal appearance makes it much more appealing than a trussed roof building and it also requires less maintenance and heating. The portal may be of three-pinned, pinned-base or fixed-base construction as shown in Figure 9.1a. The pinned-base portal is the most common type adopted because of the greater economy in foundation design over the fixed-base type.

In plane, the portal resists the following loads by rigid frame action (see Figure 9.1b):

1. Dead and imposed loads acting vertically
2. Wind causing horizontal loads on the walls and generally uplift loads on the roof slopes

In the longitudinal direction, the building is of simple design and diagonal bracing is provided in the end bays to provide stability and resist wind load on the gable ends and wind friction on sides and roof (see Figure 9.1d).

9.1.2 Construction

The main features in modern portal construction shown in Figure 9.1c are

1. *Columns*: uniform universal beam section.
2. *Rafters*: universal beams with haunches usually of sections 30%–40% lighter than the columns.
3. *Eaves and ridge connections*: site-bolted joints using Grade 8.8 bolts where the haunched ends of the rafters provide the necessary lever arm for design. Local joint stiffening is required.
4. *Base*: nominally pinned with two or four holding down bolts.
5. *Purlins and sheeting rails*: cold-formed sections spaced at not greater than 1.75–2 m centres.
6. *Stays from purlins and rails*: these provide lateral support to the inside flange of portal frame members.
7. *Gable frame*: a braced frame at the gable ends of the buildings.
8. *Bracing*: provided in the end bay in roof and walls.
9. *Eaves and ridge ties:* may be provided in larger-span portals, though now replaced by stays from purlins or sheeting rails.

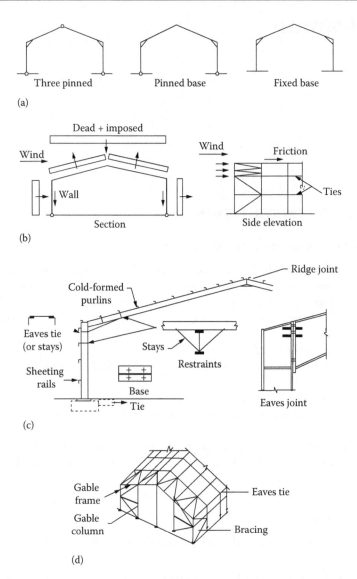

Figure 9.1 Portal frames: (a) portal types; (b) loading; (c) portal construction; (d) gable frame and bracing.

9.1.3 Foundations

The pinned-base portal is generally adopted because it is difficult to ensure fixity without piling and it is more economical to construct. It is also advantageous to provide a tie through the ground slab to resist horizontal thrust due to dead and imposed load as shown in Figure 9.1c.

9.1.4 Design outline

Either elastic or plastic design may be used. Plastic design gives the more economical solution and is almost universally adopted. The design process for the portal consists of

1. Analysis – elastic or plastic
2. Design of members taking into account of flexural and lateral torsional buckling with provision of restraints to limit out-of-plane buckling

3. Sway stability check in the plane of the portal
4. Joint design with provision of stiffeners to ensure all parts are capable of transmitting design actions
5. Serviceability check for deflection at eaves and apex

Procedures for elastic and plastic design are discussed in the succeeding text.

9.2 ELASTIC DESIGN

9.2.1 Design procedures

The procedures are summarised as follows:

1. Analysis is made using factored loads.
2. The stability of the frame should be checked using Section 5 of the code.
3. The capacity and the buckling resistance of members are checked using Section 6 of the code.

9.2.2 Portal analysis

The most convenient manual method of analysis is to use formulae from the *Steel Designers Manual* (see further reading at the end of this chapter). A general load case can be broken down into separate cases for which solutions are given, and then these results are recombined. Computer analysis is the most convenient method to use, particularly for wind loads and load combinations. The output gives design actions and deflections.

The critical load combination for design is 1.35× dead load +1.5× imposed load. The wind loads mainly cause uplift and the moments are generally in the opposite direction to those caused by the dead and imposed loads. The bending moment diagram for the dead and imposed load case is given in Figure 9.2. This shows the inside flange of the column and rafter near the eaves to be in compression and hence the need for lateral restraints in those areas.

As noted in Section 9.1.2, portals are constructed with haunched joints at the eaves. The primary purpose of the haunch is to provide the lever arm to enable the bolted site joints to be made. It is customary to neglect the haunch in elastic analysis. If the haunch is large in comparison with the rafter length, more moment is attracted to the eaves and a more accurate solution is obtained if it is taken into consideration. This can be done by dividing the haunch into a number of parts and using the properties of the centre of each part in a computer analysis.

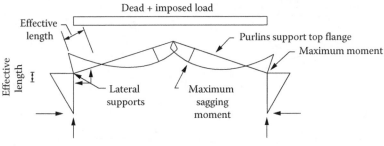

Bending moment diagram

Figure 9.2 Bending moment diagram.

9.2.3 Buckling lengths

Member design depends on the buckling lengths of members in the plane and normal to the plane of the portal. Effective lengths control the design strengths for axial load and bending.

9.2.3.1 In-plane buckling lengths

The method for estimating the in-plane effective lengths for portal members was developed by Fraser and is reproduced with his kind permission. Computer programs for matrix stability analysis are also available for determining buckling loads and effective lengths. Only the pinned-base portal with symmetrical uniform loads (i.e. the critical load case dead + imposed load on the roof) is considered. Reference should be made to Fraser for unsymmetrical load cases (see further reading at the end of this chapter).

If the roof pitch is less than 10°, the frame may be treated as a rectangular portal and the buckling length of the column can be found using the buckling length equations. For pinned-base portals with a roof slope greater than 10°, Fraser gives the following equation for obtaining the column buckling length that is a close fit to results from matrix stability analyses:

$$k_c = \frac{L_{cr}}{L_c} = 2 + 0.45 G_R$$

where

$$G_R = \frac{L_r I_c}{L_c I_r}$$

L_{cr} is the buckling length of column
L_c and L_r are the actual length of column and rafter, respectively
I_c and I_r are the second moment of area of column and rafter

Similar results can be obtained using the limited frame sway chart if, as Fraser suggests, the pitched-roof portal is converted to an equivalent rectangular frame. The beam length is made equal to the total rafter length as shown in Figure 9.3.

Column top

$$k_1 = \frac{I_c / L_c}{(I_c / L_c) + (I_r / L_r)}$$

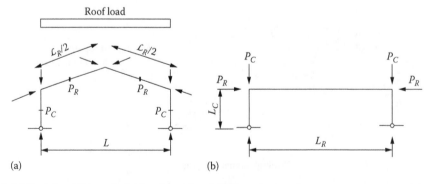

Figure 9.3 (a) Portal and (b) equivalent rectangular portal.

Column base

$k_2 = 1.0$ for pinned base
$k_2 = 0.5$ for fixed base

The effective length of the rafter can be obtained from the effective length of the column because the entire frame collapses as a unit at the critical load.

Column

$$P_{CC} = \frac{\pi E I_c}{K_c^2 L_c^2}$$

Rafter

$$P_{RC} = \frac{\pi E I_r}{K_r^2 (L_c/2)^2}$$

Hence,

$$K_r = K_c \frac{2L_c}{L_R} \sqrt{\frac{P_C I_r}{P_R I_c}}$$

where
E is the Young's modulus, $K_r = L_{cr,r}/L_r$
$L_{cr,r}$ is the buckling length of rafter
L_R is the actual length of rafter
P_{CC}, P_{RC} is the critical loads for column and rafter = load factor × design loads
P_C, P_R is the average design axial load

9.2.3.2 Out-of-plane effective length

The purlins and sheeting rails restrain the outside flange of the portal members. It is essential to provide lateral support to the side flange, i.e. a torsional restraint to both flanges at critical points to prevent out-of-plane buckling. For the pinned-base portal with the bending moment diagram for dead and imposed load shown in Figure 9.2, supports to the inside flange are required at

1. Eaves (may be stays from sheeting rails or a tie)
2. Within the top half of the column (more than one support may be necessary)
3. Near the point of contraflexure in the rafter (more than one support may be necessary)

The lateral supports and effective lengths about the minor axis for flexural buckling and lateral torsional buckling are shown in Figure 9.2.

9.2.4 Column design

The column is a uniform member subjected to axial load and moment with moment predominant. The design procedure for the critical load case of dead + imposed load is as follows:

9.2.4.1 Buckling resistance

Estimate non-dimensional slenderness $\overline{\lambda}_y$ and $\overline{\lambda}_z$

where

$L_{cr,y}$ and $L_{cr,z}$ are the buckling lengths for y–y and z–z axis
i_y and i_z are radii of gyration for y–y and z–z axis, respectively

Calculate the reduction factor, χ.
Determine the buckling resistance, $N_{b,Rd} = (\chi A f_y)/\gamma_{M1}$.

9.2.4.2 Moment capacity

For Class 1 and 2 section,

$$M_{c,Rd}\frac{W_{pl}f_y}{\gamma_{M0}}$$

where

f_y is the design strength
$W_{pl,y}$ is the plastic modulus

9.2.4.3 Buckling resistance moment

The bending moment diagram and buckling length are shown in Figure 9.2.
Calculate the non-dimensional slenderness, $\overline{\lambda}_{LT}$, and reduction factor, χ_{LT}.

$$M_{b,Rd} = \chi_{LT}\frac{W_y f_y}{\gamma_{M1}}$$

9.2.4.4 Interaction expressions

$$\frac{N_{Ed}}{\chi_y N_{Rk}/\gamma_{M1}} + k_{yy}\frac{M_{y,Ed}}{\chi_{LT}M_{y,Rk}/\gamma_{M1}} + k_{yz}\frac{M_{z,Ed}}{M_{z,Rk}/\gamma_{M1}} \leq 1$$

$$\frac{N_{Ed}}{\chi_z N_{Rk}/\gamma_{M1}} + k_{zy}\frac{M_{y,Ed}}{\chi_{LT}M_{y,Rk}/\gamma_{M1}} + k_{zz}\frac{M_{z,Ed}}{M_{z,Rk}/\gamma_{M1}} \leq 1$$

where

N_{Ed}, $M_{y,Ed}$ and $M_{z,Ed}$ are the design values of the compression force and the maximum moments about the y–y and z–z axis along the member
χ_y and χ_z are the reduction factors due to flexural buckling from Cl. 6.3.1 from BS EN1993-1-1
χ_{LT} is the reduction factor due to lateral torsional buckling from Cl. 6.3.2 from BS EN1993-1-1
k_{yy}, k_{yz}, k_{zy}, k_{zz} are the interaction factors in accordance to Annex A or B in BS EN1993-1-1

9.2.5 Rafter design

The rafter is a member haunched at both ends with the moment distribution shown in Figure 9.2. The portion near the eaves has compression on the inside. Beyond the point of contraflexure, the inside flange is in tension and is stable. The top flange is fully restrained by the purlins.

If the haunch length is small, it may conservatively be neglected and the design made for the maximum moment at the eaves. A torsional restraint is provided at the first or second purlin point away from the eaves, and the design is made in the same way as for the column. Rafter design taking the haunch into account is considered under plastic design in Section 9.3.

9.2.6 Example: Elastic design of a portal frame

9.2.6.1 Specification

The pinned-base portal for an industrial building is shown in Figure 9.4a. The portals are at 5 m centres and the length of the building is 40 m. The building load are

Roof dead load measured on slope:

Sheeting	$= 0.10$ kN/m²
Insulation	$= 0.15$ kN/m²
Purlins (P145/170, 3.97 kg/m at 1.5 m c/c)	$= 0.03$ kN/m²
Rafter 457 × 191 UB 67	$= 0.13$ kN/m²
Total dead load	$= 0.41$ kN/m²
Imposed snow load on plan	$= 0.60$ kN/m²
Imposed services load	$= 0.15$ kN/m²
Total imposed load	$= 0.75$ kN/m²

Wind load – BS EN1991-1-4 – location: North of England, outskirts of the city, 50 m above sea level, 100 m to sea

Carry out the following work: estimate the building loads.
Analyse the portal using elastic theory with a uniform section throughout.
Design the section for the portal using Grade S275 steel.

9.2.6.2 Loading

Roof – Dead = $0.41 \times 5 \times 10.77/10 = 2.21$ kN/m

Imposed = $0.75 \times 5 = 3.75$ kN/m

Design = $(1.35 \times 2.21) + (1.5 \times 3.75) = 8.61$ kN/m

Notional horizontal load at each column top:

$= 0.5 \times 0.005 \times 20 \times 8.61 = 0.43$ kN

Wind – basic wind speed

$V_{b,0} = 23$ m/s

Basic velocity pressure, $q_b = 0.5 \times \rho \times v_b^2 = 0.5 \times 1.226 \times 23^2 = 324$ N/m²

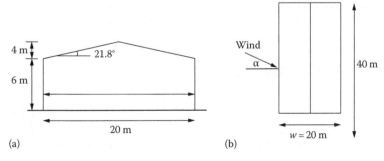

(a)   (b)

Figure 9.4 Pinned-base portal: (a) portal; (b) plan.

Table 9.1 Wind loads

Element	Eff. height (m)	Factor $C_{e(z)}$	Basic velocity pressure, q_b	Peak velocity pressure, $q_p(z)$	External pressure coefficient, $\alpha = 0°$		C_{pe}, $\alpha = 90°$
					Windward	Leeward	Side
Roof	10	2.40	0.324	0.78	−0.25	−0.45	−0.6
Walls	6	2.05	0.324	0.67	+0.65	−0.15	−0.8

The $c_{e(z)}$ factor, the basic velocity pressure q_b, peak velocity pressure $q_b(z)$ and external pressure coefficients C_{pe} for the portal are shown in Table 9.1. The internal pressure coefficients C_{pi} from the wind code are +0.2 or −0.3 for the case where there is a negligible probability of a dominant opening occurring during a severe storm.

Wind load, $w = q_{p(z)}(C_{pe} - C_{pi})C_a$

The diagrams for the characteristic loads are shown in Figure 9.5 as follows.

9.2.6.3 Analysis

The manual analysis for the design dead + imposed load case is set out here. Bending moment diagrams are given for the separate load cases in Figure 9.6.

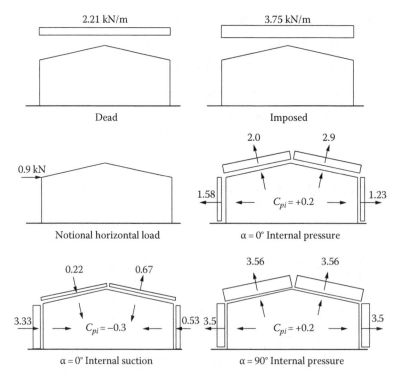

Figure 9.5 Load diagrams.

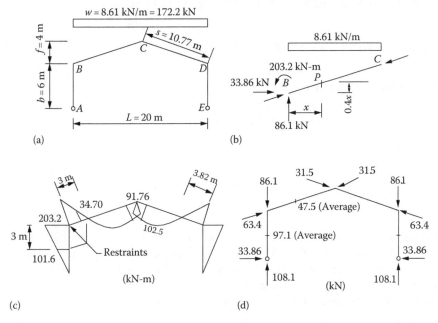

Figure 9.6 Forces and moment diagrams: (a) frame and loads; (b) rafter loads; (c) bending moment diagram; (d) member axial loads.

Frame constants (refer to Figure 9.6a)

$$K = \frac{b}{s} = 0.557$$

$$\phi = \frac{f}{b} = 0.667$$

$$m = 1 + \phi = 1.667$$

$$B = 2(K + 1) + m = 4.781$$

$$C = 1 + 2m = 4.334$$

$$N = B + mC = 12.006$$

Moments and reactions

$$M_B = wL^2(3 + 5m)/16N = -203.2 \text{ kN-m}$$

$$M_C = \left(\frac{wL^2}{8}\right) + mM_B = 91.76 \text{ kN-m}$$

$$H = \frac{203.2}{6} = 33.86 \text{ kN}$$

Referring to Figure 9.6b, the moment at any point P in the rafter is

$$M_p = 86.1x - 203.2 - 33.86 \times 0.4x - \frac{8.61x^2}{2}$$

Put $M_p = 0$ and solve to give $x = 3.55$ m.
Put $dM_p/dx = 0$ and solve to give $x = 8.43$ m
and maximum sagging moment = 102.5 kN-m.

Thrust at B:

$$= 86.1 \times \frac{4}{10.77} + 33.86 \times \frac{10}{10.77} = 63.4 \text{ kN}$$

The bending moment diagram and member axial loads are shown in Figure 9.6c and d, respectively. Other values required for design are given in the appropriate figures. The maximum design moments for the separate load cases shown in Figure 9.5 and the moment diagram in Figure 9.7 are given for comparison:

Maximum negative moment at the eaves are as follows:

1. 1.35 × dead + 1.5 × imposed

$$M_B = -[1.35 \times 52.1 + 1.5 \times 88.5]$$

$$= -203.09 \text{ kN-m.}$$

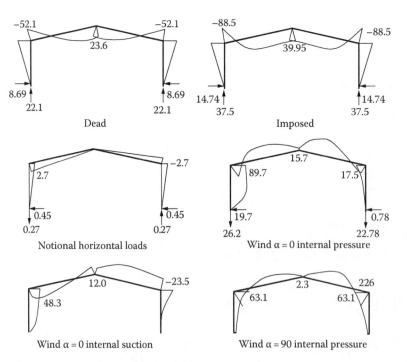

Figure 9.7 Bending moments (kN-m) and reactions (kN).

2. $1.35 \times$ dead + $1.05 \times$ imposed $+0.75 \times$ wind internal suction + notional horizontal load

$$M_D = -[1.35 \times 52.1 + 1.05 \times 88.5 + 0.75 \times 23.5 + 2.7]$$

$$= -183.59 \text{ kN-m}$$

Maximum reversed moment at the eaves are as follows:

– $1.0 \times$ dead + $1.5 \times$ wind internal pressure

$$M_B = -52.16 + 1.5 \times 89.7 = 82.39 \text{ kN-m}$$

9.2.6.4 Column design

Trial section
Try 457×152 UB 60, Grade 275, uniform throughout:

$$A = 75.9 \text{ cm}^2, \quad i_y = 18.3 \text{ cm}, \quad i_z = 3.23 \text{ cm}, \quad w_{pl,y} = 1280 \text{ cm}^3$$

Compression resistance – applied load, $N_{Ed} = 86.1$ kN
In-plane slenderness
Fraser's formula

$$\frac{L_{cr}}{L} = 2 + 0.45\left(\frac{2 \times 10.77}{6}\right) = 3.62$$

Slenderness ratio, $\lambda_y = L_{cr,y}/i_y = 3.62 \times 6000/183 = 118.7$
Out-of-plane slenderness ratio with restraint at the mid-height of the column as shown in Figure 9.6c:
Slenderness ratio, $\lambda_z = L_{cr,z}/i_z = 3000/32.3 = 92.9$
From Table 6.2 of EN1993-1-1, for an S275 rolled H section, t_f less than 40 mm, $h/b > 1.2$, buckling about the major y–y axis, curve a, imperfection factor, $\alpha = 0.21$, buckling about the minor z–z axis, curve b, imperfection factor, $\alpha = 0.34$

$$\lambda_1 = 93.9\varepsilon = 93.9 \times 0.92 = 86.4$$

$$\overline{\lambda_y} = \frac{\lambda_y}{\lambda_1} = \frac{118.7}{86.4} = 1.37$$

$$\Phi = 0.5\,[1 + 0.21(1.37 - 0.2) + 1.37^2] = 1.56$$

$$\chi_y = \frac{1}{1.56 + \sqrt{1.56^2 - 1.37^2}} = 0.43$$

$$\overline{\lambda_z} = \frac{\lambda_z}{\lambda_1} = \frac{92.9}{86.4} = 1.08$$

$$\Phi = 0.5 \ [1 + 0.34(1.08 - 0.2) + 1.08^2] = 1.23$$

$$\chi_z = \frac{1}{1.23 + \sqrt{1.23^2 - 1.08^2}} = 0.55$$

Moment capacity $M_{c,Rd} = 275 \times 1280 \times 10^{-3} = 352$ kN-m.
For buckling moment resistance,
  Applied moment $M = 203.2$ kN-m
  The bending moment diagram is shown in Figure 9.6c.
  Buckling lengths $L_{cr} = 3000$ mm:

$$M_{cr} = C_1 \frac{\pi^2 EI_z}{L_{cr}^2} \left(\frac{I_w}{I_z} + \frac{L_{cr}^2 \ GI_T}{\pi^2 EI_z} \right)^{0.5}$$

For ratio of end moment, $\psi = 101.6/203.2 = 0.5$; $C_1 = 1.323$:

$$M_{cr} = 1.323 \times \frac{\pi^2 \times 210,000 \times 7,950,000}{3,000^2} \left(\frac{387 \times 10^9}{7,950,000} + \frac{3,000^2 \times 81,000 \times 338,000}{\pi^2 \times 210,000 \times 7,950,000} \right)^{0.5}$$

$M_{cr} = 611$ kN-m

$$\overline{\lambda_{LT}} = \sqrt{\frac{W_y f_y}{M_{cr}}} = \sqrt{\frac{1280 \times 10^3 \times 275}{611 \times 10^6}} = 0.76$$

$$\chi_{LT} = \frac{1}{\Phi_{LT} + \sqrt{\Phi_{LT}^2 - \overline{\lambda}_{LT}^2}}$$

$$\Phi_{LT} = 0.5 \left[1 + \alpha_{LT} \left(\overline{\lambda}_{LT} - 0.2 \right) + \overline{\lambda}_{LT}^2 \right]$$

For $h/b > 2$, curve b, $\alpha_{LT} = 0.34$:

$$\Phi_{LT} = 0.5 \times [1 + 0.34(0.76 - 0.2) + 0.76^2] = 0.89$$

$$\chi_{LT} = \frac{1}{0.89 + \sqrt{0.89^2 - 0.76^2}} = 0.74$$

Determination of interaction factors k_{ij} (Annex B in BS EN1993-1-1)
  Table B.3 of BS EN1993-1-1. Equivalent uniform moment factors, C_m
  For $\psi = 0.5$, C_{my}, $C_{mz} = 0.6 + 0.4(0.5) = 0.8 \geq 0.4$

$$k_{yy} = C_{my} \left(1 + \left(\overline{\lambda}_y - 0.2 \right) \frac{N_{Ed}}{\chi_y N_{Rk}/\gamma_{M1}} \right) \leq C_{my} \left(1 + 0.8 \frac{N_{Ed}}{\chi_y N_{Rk}/\gamma_{M1}} \right)$$

$$k_{yy} = 0.8\left(1 + 0.8\frac{86.1}{0.43 \times 7590 \times 10^{-3} \times 275/1.0}\right) = 0.86$$

$$k_{zy} = 0.6\ k_{yy} = 0.52$$

Interaction expressions

$$\frac{N_{Ed}}{\chi_y N_{Rk}/\gamma_{M1}} + k_{yy}\frac{M_{y,Ed}}{\chi_{LT}M_{y,Rk}/\gamma_{M1}} \leq 1$$

$$\frac{86.1}{0.43 \times 2087.3/1.0} + 0.86\frac{203.2}{0.74 \times 352/1.0} = 0.77$$

$$\frac{N_{Ed}}{\chi_z N_{Rk}/\gamma_{M1}} + k_{zy}\frac{M_{y,Ed}}{\chi_{LT}M_{y,Rk}/\gamma_{M1}} \leq 1$$

$$\frac{86.1}{0.55 \times 2087.3/1.0} + 0.52\frac{203.2}{0.74 \times 352/1.0} = 0.48$$

The section is satisfactory.

9.2.6.5 Rafter design check

Compression force, $N_{Ed} = 63.4$ kN

The average compressive forces for the rafter and column are shown in Figure 9.6d. In-plane slenderness

$$\frac{L_{cr}}{L_{R/2}} = \frac{3.62 \times 2 \times 6}{21.54}\sqrt{\frac{97.1}{47.5}} = 2.88$$

$$\frac{L_{cr}}{i_y} = 2.88 \times \frac{10,770}{183} = 167$$

Out-of-plane slenderness

$$L_{cr,z} = 3000 \text{ mm (Figure 9.6c)}$$

Slenderness ratio, $\lambda_z = L_{cr,z}/i_z = 3000/32.3 = 92.9$

From Table 6.2 of EN1993-1-1, for an S275 rolled H section, t_f less than 40 mm, $h/b > 1.2$, buckling about the major y–y axis, curve a, imperfection factor, $\alpha = 0.21$, buckling about the minor z–z axis, curve b, imperfection factor, $\alpha = 0.34$

$$\lambda_1 = 93.9\varepsilon = 93.9 \times 0.92 = 86.4$$

$$\overline{\lambda_y} = \frac{\lambda_y}{\lambda_1} = \frac{167}{86.4} = 1.93$$

$\Phi = 0.5 \, [1 + 0.21(1.93 - 0.2) + 1.93^2] = 2.54$

$$\chi_y = \frac{1}{2.54 + \sqrt{2.54^2 - 1.93^2}} = 0.24$$

$$\overline{\lambda_z} = \frac{\lambda_z}{\lambda_1} = \frac{92.9}{86.4} = 1.08$$

$\Phi = 0.5 \, [1 + 0.34(1.08 - 0.2) + 1.08^2] = 1.23$

$$\chi_z = \frac{1}{1.23 + \sqrt{1.23^2 - 1.08^2}} = 0.55$$

Moment capacity $M_{c,Rd} = 275 \times 1280 \times 10^{-3} = 352$ kN-m
Buckling moment resistance:

Applied moment $M = 203.2$ kN-m
The bending moment diagram is shown in Figure 9.6c.
Buckling lengths $L_{cr} = 3000$ mm

$$M_{cr} = C_1 \frac{\pi^2 E I_z}{L_{cr}^2} \left(\frac{I_w}{I_z} + \frac{L_{cr}^2 \, G I_T}{\pi^2 E I_z} \right)^{0.5}$$

For ratio of end moment, $\psi = 34.7/203.2 = 0.2$; $C_1 = 1.563$

$$M_{cr} = 1.563 \times \frac{\pi^2 \times 210,000 \times 7,950,000}{3,000^2} \left(\frac{387 \times 10^9}{7,950,000} + \frac{3,000^2 \times 81,000 \times 338,000}{\pi^2 \times 210,000 \times 7,950,000} \right)^{0.5}$$

$M_{cr} = 722$ kN-m

$$\overline{\lambda_{LT}} = \sqrt{\frac{W_y f_y}{M_{cr}}} = \sqrt{\frac{1280 \times 10^3 \times 275}{722 \times 10^6}} = 0.70$$

$$\chi_{LT} = \frac{1}{\Phi_{LT} + \sqrt{\Phi_{LT}^2 - \overline{\lambda}_{LT}^2}}$$

$$\Phi_{LT} = 0.5 \left[1 + \alpha_{LT} \left(\overline{\lambda}_{LT} - 0.2 \right) + \overline{\lambda}_{LT}^2 \right]$$

For $h/b > 2$, curve b, $\alpha_{LT} = 0.34$

$\Phi_{LT} = 0.5 \times [1 + 0.34(0.70 - 0.2) + 0.70^2] = 0.83$

$$\chi_{LT} = \frac{1}{0.83 + \sqrt{0.83^2 - 0.70^2}} = 0.78$$

Determination of interaction factors k_{ij} (Annex B in BS EN1993-1-1)

Table B.3 of BS EN1993-1-1. Equivalent uniform moment factors, C_m
For $\psi = 0.2$, C_{my}, $C_{mz} = 0.6 + 0.4(0.2) = 0.68 \geq 0.4$

$$k_{yy} = C_{my}\left(1 + \left(\overline{\lambda}_y - 0.2\right)\frac{N_{Ed}}{\chi_y N_{Rk}/\gamma_{M1}}\right) \leq C_{my}\left(1 + 0.8\frac{N_{Ed}}{\chi_y N_{Rk}/\gamma_{M1}}\right)$$

$$k_{yy} = 0.68\left(1 + 0.8\frac{63.4}{0.24 \times 7590 \times 10^{-3} \times 275/1.0}\right) = 0.69$$

$$k_{zy} = 0.6 \quad k_{yy} = 0.53$$

Interaction expressions

$$\frac{N_{Ed}}{\chi_y N_{Rk}/\gamma_{M1}} + k_{yy}\frac{M_{y,Ed}}{\chi_{LT} M_{y,Rk}/\gamma_{M1}} \leq 1$$

$$\frac{63.4}{0.24 \times 2087.3/1.0} + 0.69\frac{203.2}{0.78 \times 352/1.0} = 0.64$$

$$\frac{N_{Ed}}{\chi_z N_{Rk}/\gamma_{M1}} + k_{zy}\frac{M_{y,Ed}}{\chi_{LT} M_{y,Rk}/\gamma_{M1}} \leq 1$$

$$\frac{63.4}{0.55 \times 2087.3/1.0} + 0.53\frac{203.2}{0.74 \times 352/1.0} = 0.47$$

The section is satisfactory.

9.3 PLASTIC DESIGN

9.3.1 Design provisions

Plastic design may be used in the design of structures and elements provided that the following main conditions are met:

1. The loading is predominantly static.
2. Structural steels with stress–strain diagrams as shown in Figure 3.1 are used. The plastic plateau permits hinge formation and rotation necessary for moment redistribution at plastic moments to occur.
3. Member sections are plastic where hinges occur. Members not containing hinges are to be compact.
4. Torsional restraints are required at hinges and within specified distances from the hinges.

Provisions regarding plastic design of portals are not given in EN1993-1-1. However, other important provisions deal with overall sway stability and column and rafter stability are given in Annex BB.2.

9.3.2 Plastic analysis: Uniform frame members

Plastic analysis is set out in books such as that by Horne and only plastic analysis for the pinned-base portal is discussed here. The plastic hinge (the formation of which is shown in Figure 4.8) is the central concept. This rotates to redistribute moments from the elastic to the plastic moment distribution.

Referring to Figure 9.8a, as the load is increased, hinges form first at the points of maximum elastic moments at the eaves. Rotation occurs at the eaves' hinges with thereafter acting like a simply supported beam, taking more load until two further hinges form near the ridge, when the rafter collapses. The plastic bending moment diagram and collapse mechanism are shown in Figure 9.8. In general, the number of hinges required to convert the portal to a mechanism is one more than the statical indeterminacy. With unsymmetrical loads such as dead b wind load, two hinges only form to cause collapse.

For the location of the hinges to be correct, the plastic moment at the hinges must not be exceeded at any point in the structure. That is why, in Figure 9.8b, two plastic hinges form at each side of the ridge and not one only at the ridge at collapse. The critical mechanism is the one that gives the lowest value for the collapse load. The collapse mechanism that occurs depends on the form of loading.

Plastic analysis for the pinned-base portal is carried out in the following stages:

1. The frame is released to a statically determinate state by inserting rollers at one support.
2. The free bending moment diagram is drawn.
3. The reactant bending moment diagram due to the redundant horizontal reaction is drawn.
4. The free and reactant moment diagrams are combined to give the plastic bending-moment diagram with sufficient hinges to cause the frame or part of it (e.g. the rafter) to collapse. As mentioned earlier, the plastic moment must not be exceeded.

The process of plastic analysis for the pinned-base portal is shown for the case of dead and imposed load on the roof in Figure 9.9. The frame is taken to be uniform throughout and the bending moment diagrams are drawn on the opened-out frame. The case of dead + imposed + wind load is treated in the design example.

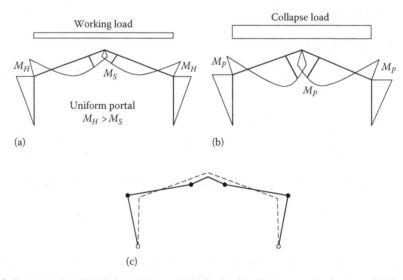

Figure 9.8 Collapse mechanism of pinned-base portal frame: (a) elastic moment diagram; (b) plastic moment diagram; (c) collapse mechanism.

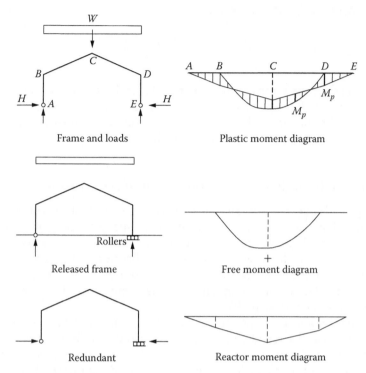

Figure 9.9 Plastic moment diagram.

The exact location of the hinge near the ridge must be found by successive trials or mathematically if the loading is taken to be uniformly distributed. Referring to Figure 9.10, for a uniform frame, hinge X is located by

$$g = h + \frac{2fx}{L}$$

The free moment at X in the released frame is

$$M_y = \frac{wLx}{2} - \frac{wx^2}{2}$$

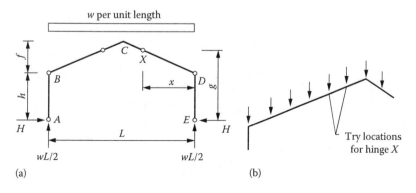

(a)

(b)

Figure 9.10 Plastic hinge locations: (a) frame and hinge locations; (b) loads approved at purlin points.

The plastic moments at B and X are equal, i.e.

$$M_p = Hb = M_x - Hg$$

Put $dH/dx = 0$ and solve for x and calculate H and M_p. Symbols used are shown in Figure 9.10.

If the load is taken to be applied at the purlin points as shown in Figure 9.10b, the hinge will occur at a purlin location. The purlins may be checked in turn to see which location gives the maximum value of the plastic moment M_p.

9.3.3　Plastic analysis: Haunches and non-uniform frame

Haunches are provided at the eaves and ridge primarily to give a sufficient lever arm to form the bolted joints. The haunch at the eaves causes the hinge to form in the column at the bottom of the haunch. This reduces the value of the plastic moment when compared with the analyses for a hinge at the eaves intersection. The haunch at the ridge will not affect the analyses because the hinge on the rafter forms away from the ridge. The haunch at the eaves is cut from the same UB as the rafter. The depth is about twice the rafter depths and the length is often made equal to span/10.

It is also more economical to use a lighter section for the rafter than for the column. The non-uniform frame can be readily analysed as discussed as follows. It is also essential to ensure that the haunched section of the rafter at the eaves remains elastic. That is, the maximum stress at the end of the haunch must not exceed the design strength, f_y:

$$f_y = \frac{F}{A} + \frac{M}{Z}$$

where
  F is the axial force
  M is the moment
  A is the area
  Z is the elastic modulus

The analysis of a frame with haunched rafter and lighter rafter than column section is demonstrated with reference to Figure 9.11a.

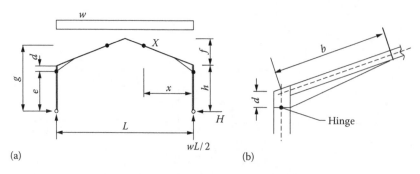

Figure 9.11　Non-uniform portal frame: (a) non-uniform portal; (b) haunch.

Frame dimensions are shown in Figure 9.11.

Hinge in column: $M_p = He$

Hinge in rafter: $qM_p = M_x - Hg = qHe$

where
M_p is the plastic moment of resistance of the column
qM_p is the plastic moment of resistance of the rafter
q is the normally 0.6–0.7,
M_x is the free moment at X in the released frame and $g = h + 2fx/L$

Put $dH/dx = 0$ and solve to give x and so obtain H and M_p.

9.3.4 Section design

At hinge locations, design is made for axial load and plastic moment (Figure 9.12).
For Class 1 and 2 sections subjected to axial and bending,

$$M_{N,y,Rd} = M_{pl,y,Rd} \frac{(1-n)}{(1-0.5a)} \quad \text{but} \quad M_{N,y,Rd} \leq M_{pl,y,Rd}$$

For $n \leq a$, $M_{N,z,Rd} = M_{pl,z,Rd}$

For $n \leq a$, $M_{N,z,Rd} = M_{pl,z,Rd}\left[1 - \left(\frac{n-a}{1-a}\right)^2\right]$

where $n = N_{Ed}/N_{pl,Rd}$

$$a = \frac{A - 2bt_f}{A} \quad \text{but} \quad a \leq 0.5$$

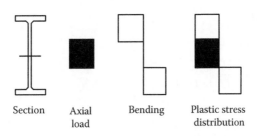

Section   Axial   Bending   Plastic stress
           load   distribution

Figure 9.12 Stress diagram.

9.4 IN-PLANE STABILITY

The in-plane stability of a portal frame should be checked under each load combination. Three procedures for checking the overall sway stability of a portal can be used:

1. Sway-check method
2. Amplified moments method
3. Second-order analysis

These ensure that the elastic buckling load is not reached and that the effects of additional moments due to deflection of the portal are taken into account.

9.4.1 Sway-check method

The sway-check method may be used to verify the in-plane stability of portal frames in which each bay satisfies the following conditions:

1. Span, L, does not exceed five times the mean height h of the columns.
2. Height, h_r, of the apex above the tops of the columns does not exceed 0.25 times the span, L.

9.4.1.1 Limiting horizontal deflection at eaves

For gravity loads (load combination 1), the horizontal deflection calculated by linear elastic analysis at the top of the columns due to the notional horizontal loads without any allowance for the stiffening effects of cladding should not exceed $h/1000$, where h is the column height. The notional loads are 0.5% of the factored roof dead and imposed loads applied at the column tops.

9.4.1.2 Limiting span/depth ratio of the rafter

The $h/1000$ sway criterion for gravity loads may be assumed to be satisfied if

$$\frac{L_b}{h} \le \frac{51.5L}{\Omega h_c}\left(\frac{\rho}{4+\rho L_r/L}\right)\left(\frac{235}{f_y}\right)$$

in which

$$L_b = L - \left(\frac{2h_h}{h_s+h_h}\right)L_h$$

For single bay, $\rho = \left(\frac{2I_c}{I_r}\right)\left(\frac{L}{H_c}\right)$

For multi-bay, $\rho = \left(\frac{I_c}{I_r}\right)\left(\frac{L}{H_c}\right)$

and $\Omega = \dfrac{W_r}{W_o}$

where

> h_r is the cross-sectional depth of the rafter
> h_h is the additional depth of the haunch
> h_s is the depth of the rafter, allowing for its slope
> H_c is the mean column height
> I_c is the in-plane second moment of area of the column (taken as zero if the column is not rigidly connected to the rafter, or if the rafter is supported on a valley beam)
> I_r is the in-plane second moment of area of the rafter
> L is the span of the bay
> L_b is the effective span of the bay
> L_h is the length of a haunch
> L_r is the total developed length of the rafters
> f_{yr} is the design strength of the rafters in N/mm^2
> W_o is the value of W_r for plastic failure of the rafters as a fixed-ended beam of span L
> W_r is the total factored vertical load on the rafters of the bay

9.4.1.3 Horizontal load

For load combinations that include wind loads or other significant horizontal loads, allowance may be made for the stiffening effects of cladding in calculating the notional horizontal deflections. Provided that the $h/1000$ sway criterion is satisfied for gravity loads, then for load cases involving horizontal loads, the required load factor α_r for frame stability should be determined using

$$\alpha_r = \frac{\alpha_{sc}}{\alpha_{sc} - 1}$$

in which α_{sc} is the smallest value, considering every column, determined from

$$\alpha_{sc} = \frac{h_i}{200\delta_i}$$

using the notional horizontal deflection δ_i for the relevant load case.

If $\alpha_{sc} < 5$, second-order analysis should be used.

α_{sc} may be approximate using

$$\alpha_{sc} = \frac{234 h_r L}{\Omega H_c L_b} \left(\frac{\rho}{4 + \rho L_r / L} \right) \left(\frac{235}{f_y} \right)$$

9.4.2 Amplified moments method

For each load case, the in-plane stability of a portal frame may be checked using the lowest elastic critical load factor α_{cr} for that load case. This should be determined taking account of the effects of all the members on the in-plane elastic stability of the frame as a whole.

In this method, the required load factor α_r for frame stability should be determined from the following:

$$\alpha_{cr} \geq 10: \alpha_r = 1.0$$

$$10 \geq \alpha_{cr} \geq 5: \alpha_r = \frac{0.9\alpha_{cr}}{\alpha_{cr} - 1}$$

If $\alpha_{cr} < 5$, the amplified moments method should not be used.

For pinned-base portal,

$$\alpha_{cr} = \frac{3EI_r}{L_s[1 + (1.2/(I_cL_r/I_rH_c))N_cH_c + 0.3N_rL_s]}$$

For fixed-base portal,

$$\alpha_{cr} = \frac{5E(10 + (I_cL_r)/(I_rH_c))}{5N_rL_s^2/I_r + 2((I_cL_r)/(I_rH_c))N_cH_c^2/I_c}$$

where
   N_c is the axial compression in column from elastic analysis
   N_r the axial compression in rafter from elastic analysis
   L_s is the rafter length along the slope (eaves to apex)

9.4.3 Second-order analysis

The in-plane stability of a portal frame may be checked using either elastic or elastic–plastic second-order analysis. When these methods are used, the required load factor α_r for frame stability should be taken as 1.0. Further information on second-order analysis can be found on *Plastic Design of Single-Storey Pitched-Roof Portal Frames to Eurocode 3 by SCI*.

9.5 RESTRAINTS AND MEMBER STABILITY

9.5.1 Restraints

Restraints are required to ensure that

1. Plastic hinges can form in the deep I sections used.
2. Overall flexural buckling of the column and rafter about the minor axis does not occur.
3. There is no lateral torsional buckling of an unrestrained compression flange on the inside of members.

The code requirements regarding restraints and member satiability are set out as follows. A restraint should be capable of resisting 2.5% of the compressive force in the member or part being restrained.

9.5.2 Column stability

The column contains a plastic hinge near the top at the bottom of the haunch. Below the hinge, it is subjected to axial load and moment with the inside flange in compression. The code states in Section 6.3.5 that torsional restraints (i.e. restraints to both flanges) must be provided at or within member depth $h/2$ from a plastic hinge.

In a member containing a plastic hinge, the maximum distance from the restraint at the hinge to the adjacent restraint depends on whether or not restraint to the tension flange is taken into account. The following procedures apply:

9.5.2.1 Restraint to tension flange not taken into account

This is the conservative method where the distance from the hinge restraint to the next restraint is given by the following:

For column and three-flange haunches,

$$L_m = \frac{38i_z}{\sqrt{1/57.4(N_{Ed}/A) + 1/756C_1^2\left(W_{pl,y}^2/AI_t\right)(f_y/235)^2}}$$

For two-flange haunches,

$$L_m = 0.85\frac{38i_z}{\sqrt{1/57.4(N_{Ed}/A) + 1/756C_1^2\left(W_{pl,y}^2/AI_t\right)(f_y/235)^2}}$$

where

N_{Ed} is the compressive force [N] in the member

i_z is the radius of gyration about the minor axis in the segment

C_1 is taken from literature or conservatively taken as 1.0

When this method is used, no further checks are required. The locations of restraints at hinge H and at G, L_m below H are shown in Figure 9.13a. It may be necessary to introduce a further restraint at F below G, in which case column lengths GF and FA would be checked for buckling resistance for axial load and moment. The buckling length for the x–x axis may be estimated for the portal as set out in Section 9.2.3.

The buckling length for the y–y axis is taken as HA in Figure 9.13, the distance between the plastic hinge and the base. The buckling lengths for the z–z axis are GF and FA. Note that compliance with the sway stability check ensures that the portal can safely resist in-plane buckling and additional moments due to frame deflections.

The code specifies that the buckling resistance moment $M_{b,Rd}$ calculated using a buckling length L_m equal to the spacing of the tension flange restraints must exceed the equivalent uniform moment for that column length. The restraints are shown in Figure 9.13b. The column length AF is checked as set out earlier.

9.5.3 Rafter stability near the ridge

The tension flange at the hinge in the rafter near the ridge is on the inside, and no restraints are provided. In the portal, two hinges form last near the ridge. A purlin is required at or near the hinge and purlins should be placed at a distance not exceeding L_m on each side of the hinge. The purlin arrangement is shown in Figure 9.13c. In wide-span portals (say, 30 m or over), restraints should be provided to the inside flange.

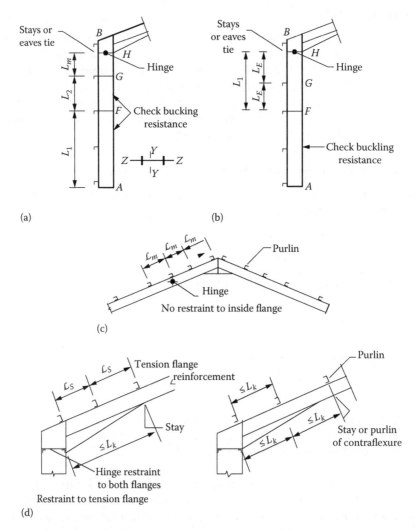

Figure 9.13 Column and rafter restraints: (a) no restraint to tension flange; (b) restraint to tension flange; (c) restraints near ridge; (d) column and rafter restraints.

9.5.4 Rafter stability at haunch

The requirements for rafter stability are set out in Section 6.3.5 and Annex BB.3 of the code. The haunch and the rafter are in compression on the inside from the eaves for a distance to the second or third purlin up from the eaves where the point of contraflexure is usually located:

9.5.4.1 Stable length of segment

In the uniform beam segments with I or H cross sections with $h/t_f \leq 40\varepsilon$ under linear moment and without significant axial compression, the stable length may be taken from

$$L_{stable} = 35\varepsilon\, i_z \quad \text{for } 0.635 \leq \psi \leq 1$$

$$L_{stable} = (60 - 40\psi)\, \varepsilon\, i_z \quad \text{for } -1 \leq \psi \leq 0.635$$

$$\psi = M_{Ed,min}/M_{pl,Rd} = \text{ratio of end moments in the segment}$$

9.5.4.2 Restraint to tension flange not taken into account

The maximum distance between restraints to the compression flange must not exceed L_m as set out in column stability earlier or that to satisfy the overall buckling expression given in Section 6.3 of the code. For the tapered member, the minimum value of i_z and the maximum value of h/t_f are used. No restraint is to be assumed at the point of contraflexure.

9.5.4.3 Stable length between torsional restraints

A method for determining spacing of lateral restraints is given in Clause BB.3.1.2 of BS EN1993-1-1. Lateral torsional buckling effect may be ignored where the length L of the segment of a member between the restrained section at a plastic hinge location and the adjacent torsional restraint subject to a constant moment is not greater than L_k

where

$$L_k = \frac{(5.4 + (600 f_y/E))(h/t_f) i_z}{\sqrt{5.4 (f_y/E)(h/t_f)^2 - 1}}$$

9.5.4.4 Restraint to tension flange taken into account

When the tension flange is restrained at intervals by the purlins, possible restraint locations for the compression flange are shown in Figure 9.13d. The code requirements are

1. The distance between restraints to the tension flange, L_E, must not exceed L_m, or alternatively, the interaction expression for overall buckling given in Section 6.3 of the code based on a buckling length L_E must be satisfied.
2. The distance between restraints to the compression flange must not exceed L_s specified in Annex BB 3.1.2. in BS EN1993-1-1.

A more rigorous method that takes into account the lateral restraints along the tension flange is given in Annex BB of BS EN1993-1-1.

The purlins provide restraint to the top flange, and they must be connected by two bolts and have a depth not less than one-quarter of the rafter depth. A vertical lateral restraint should not be automatically assumed at the point of contraflexure without provision of stays.

9.6 SERVICEABILITY CHECK FOR EAVES DEFLECTION

The national annex of BS EN1993-1-1 specifies that the deflection at the column top in a single-storey building is not to exceed height/300 unless such deflection does not damage the cladding. The deflections to be considered are due to the unfactored imposed and wind loads. If necessary, an allowance can be made for dead-load deflections in the fabrication.

A formula for horizontal deflection at the eaves due to uniform vertical load on the roof is derived. Deflections for wind load should be taken from a computer analysis.

Referring to Figure 9.14, because of symmetry of the frame and loading, the slope at the ridge does not change. The slope θ at the base is equal to the area of the M/EI diagram between the ridge and the base given by

$$\theta = \frac{1}{EI_r} \left[\frac{wL^2 \cdot L_s}{24} + H \cdot L_s \left(h + \frac{f}{2} \right) - \frac{VL \cdot L_s}{4} \right] + \frac{Hh^2}{2EI_c}$$

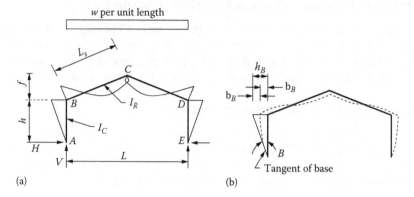

Figure 9.14 Deflection for portal frame: (a) loading and moment diagram; (b) deflected structure.

The deflection at the eaves is then

$$\delta_B = h\theta - \frac{Hh^3}{6EI_c}$$

where
 H is the horizontal reaction at the base
 V is the vertical reaction at base
 w is the characteristic imposed load on the roof
 I_c and I_r are the second moments of area of the column and rafter, respectively

Frame dimensions are shown in Figure 9.14.

9.7 DESIGN OF JOINTS

9.7.1 Eaves joint

The eaves joint arrangement is shown in Figure 9.15a. The steps in the joint design check are

9.7.1.1 Joint forces

Take moment about X:

$$\frac{Vd}{2} - M_p + T_a = 0$$

Bolt tension

$$T = \frac{(M_p - (Vd/2))}{a}$$

Compression

$$C = T + H$$

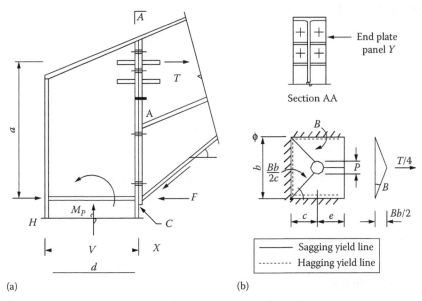

(a)   (b)

Figure 9.15 Eaves joint: (a) joint arrangement; (b) yield line pattern panel Y.

Haunch flange force

$$F = C \sec \phi$$

9.7.1.2 Bolt design

Tension

$$F_t = \frac{T}{4}$$

Shear

$$F_v = \frac{V}{8}$$

For a given bolt size with capacities F_t in tension and F_v in shear, the interaction expression is

$$\frac{F_{v,Ed}}{F_{v,Rd}} + \frac{F_{t,Ed}}{1.4F_{t,Rd}} \leq 1.0$$

9.7.1.3 Column flange check and end-plate design

Adopt a yield line analysis (see Horne and Morris in the further reading at the end of this chapter). The yield line pattern is shown in Figure 9.15b for one panel of the end plate or column flange. The hole diameter is v.

Work done by the load

$$= \frac{T}{4} \cdot \frac{b\theta}{2}$$

Work done in the yield lines

$$4m\theta(c + e) - mv\theta(1 + \cos\phi) + \frac{mb}{c}\left[b - \frac{v}{2}\sin\phi\right]$$

Equate the expressions and solve for m. The plate thickness required

$$t = \left(\frac{4m}{f_y}\right)^{0.5}$$

9.7.1.4 Haunch flange

Flange force $F < p_y \times$ flange area. The haunch section is checked for axial load and moment in the haunch stability check.

9.7.1.5 Column stiffener

Design the stiffener for force C (see Section 5.3.7).

9.7.1.6 Column web

Check for shear V (see Section 4.6.2).

9.7.1.7 Column and rafter webs

These webs are checked for tension T. The small stiffeners distribute the load.

9.7.1.8 Welds

The fillet welds from the end plate to rafter and on the various stiffeners must be designed.
The main check calculations are shown in the example in Section 9.8.

9.7.2 Ridge joint

The ridge joint is shown in Figure 9.16. The bolt forces can be found by taking moments about Z:

$$T = \frac{(M - Hh)}{8}$$

where H is the horizontal reaction in the portal and M the ridge moment.
Joint dimensions are shown in Figure 9.16. Other checks such as for end-plate thickness and weld sizes are made in the same way as for the eaves joint.

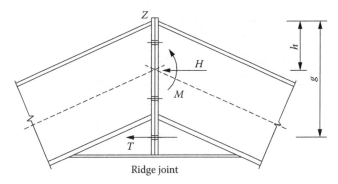

Figure 9.16 Ridge joint.

9.8 DESIGN EXAMPLE OF A PORTAL FRAME

9.8.1 Specification

Redesign the pinned-base portal specified in Section 9.2.6 using plastic design. The portal shown in Figure 9.17a has haunches at the eaves 1.5 m long and 0.45 m deep. The rafter moment capacity is to be approximately 75% of that of the column.

9.8.2 Analysis: Dead and imposed load

The frame dimensions, loading and plastic hinge locations are shown in Figure 9.17a. The plastic moments in the column at the bottom of the haunch and in the rafter at x from the eaves are given by the following expressions:

Column: $M_p = 5.55H$
Rafter: $0.75M_p = 90.9x - 9.09x^2/2 - H(6 + 0.4x)$
Reduce to give

$$H = \frac{90.9x - 4.55x^2}{10.16 + 0.4x}$$

Put $dh/dx = 0$ and collect terms to give

$$x^2 + 50.7x - 507.4 = 0$$

Solve $x = 8.56$ m

$H = 32.7$ kN

$M_p = 181.5$ kN-m – column

$0.75M_p = 136.1$ kN-m – rafter

The plastic bending moment diagram with moments needed for design is shown in Figure 9.17b and the coexistent thrusts in Figure 9.18c.

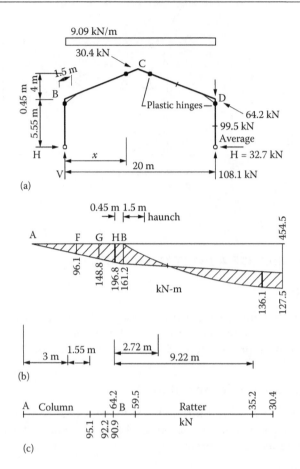

Figure 9.17 Analysis: dead + imposed load. (a) Frame and loads; (b) plastic bending moment diagram; (c) rusts in column and rafter.

9.8.3 Analysis: Dead + imposed + wind load

The frame is analysed for the load case:

[dead + imposed + wind internal suction]

The roof wind loads acting normally are resolved vertically and horizontally – and added to the dead and imposed load. The frame and hinge locations are shown in Figure 9.18a and the released frame, loads and reactions in Figure 9.18b.

The free moment in the rafter at distance x from the eaves point can be expressed by

$$M_x = 66.5x - 57.24 - 3.52x^2$$

The equations for the plastic moments at the hinges are

Columns: $M_p = 5.55H + 2.59$
Rafter: $0.75M_p = 66.5x - 57.24 - 3.52x^2 - H(6 + 0.4x)$
This gives

$$H = \frac{66.5x - 59.18 - 3.52x}{10.16 + 0.4x}$$

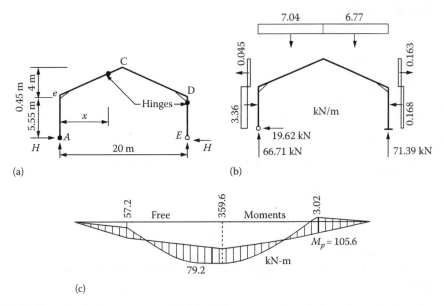

Figure 9.18 Analysis: dead + imposed + wind load. (a) Frame; (b) released frame; (c) plastic bending moment diagram.

Put $dh/dx = 0$ and solve to give $x = 8.39$ m

$H = 18.6$ kN, $M_p = 105.1$ kN

The plastic bending moment diagram is shown in Figure 9.18c. The moments are less than for the dead + imposed load case.

9.8.4 Column design

The design actions at the column hinge are

$M_p = 181.8$ kN-m, $N_{c,Ed} = 92.2$ kN

$W_{pl,y} = 181.8 \times 10^3/275 = 661.1$ cm³

Try 406 × 140 UB 46

$W_{pl,y} = 888$ cm³, $A = 59$ cm², $i_y = 163$ cm, $i_z = 3.02$ cm

$I_y = 15,600$ cm⁴

$n = 92.2 \times 10/(59 \times 275) = 0.056 < a$

$$a = \frac{A - 2bt_f}{A} = \frac{5900 - 2 \times 142.2 \times 11.2}{5900} = 0.46$$

Reduced plastic moment

$$M_{N,yRd} = M_{pl,y,Rd} \frac{(1-n)}{(1-0.5a)} \leq M_{pl,y,Rd}$$

$$M_{N,yRd} = 888 \times 275 \times 10^{-3} = 244.2 \text{ kN-m}$$

The section is satisfactory.

9.8.5 Rafter section

The design actions at the rafter hinge are

$M_p = 136.1$ kN-m

$N_{Ed} = 35.2$ kN

$W_{pl,y} = 136.1 \times 10^3/275 = 494.9$ cm^3

Try 356 × 127 UB 39

$W_{pl,y} = 654$ cm^3, $A = 49.4$ cm^2, $i_x = 14.3$ cm, $i_y = 2.69$ cm

$W_{el,y} = 572$ cm^3, $i_y = 10,100$ cm^4

$n = 35.5 \times 10/(49.4 \times 275) = 0.026$

The section is satisfactory.

Check that the rafter section at the end of the haunch remains elastic under factored loads. The actions are

$M = 96.9$ kN-m, $N_{Ed} = 59.5$ kN

$$\frac{N_{Ed}}{A} + \frac{M}{W_{el,y}} = \frac{59.5 \times 10}{49.4} + \frac{96.9 \times 10^3}{572} = 181.4 \text{ N/mm}^2$$

This is less than $f_y = 275$ N/mm^2. The section remains elastic.

9.8.6 Column restraints and stability

A stray is provided to restrain the hinge section at the eaves. The distance to the adjacent restraint using the conservative method is

$$L_m = \frac{38 \times 30.2}{\sqrt{\frac{1}{57.4}\left(\frac{92.2 \times 10^3}{59 \times 10^2}\right) + \frac{1}{756 \times 1.26}\left(\frac{(888 \times 10^3)^2}{59 \times 10^2 \times 19 \times 10^4}\right)\left(\frac{275}{235}\right)^2}} = 1013.7 \text{ mm}$$

The arrangement for the column restraints is shown in Figure 9.19a. The column is checked between the second and the third restraints G and F over a length of 1.55 m. The moments and thrusts are shown in Figure 9.17.

$M_F = 98.1$ kN-m

$M_G = 148.8$ kN-m

$F_G = 95.1$ kN

The effective lengths and slenderness ratios are

$$k_c = \left[2 + \frac{0.45 \times 21.54 \times 15,600}{6 \times 10,100}\right] = 4.49$$

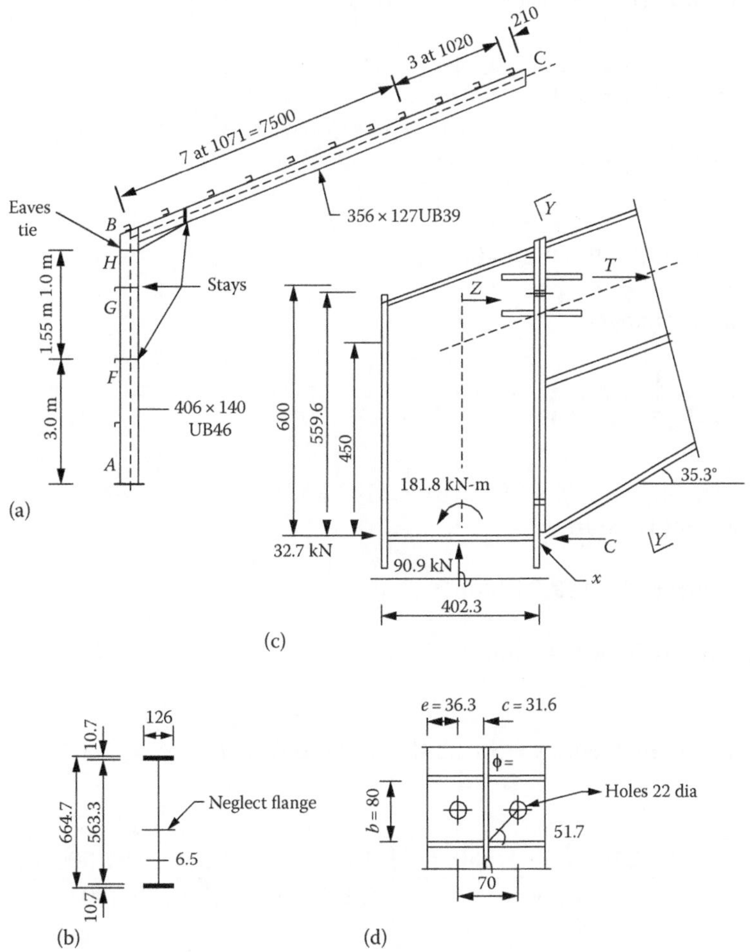

Figure 9.19 Restraints and eaves joint: (a) portal members and restraints; (b) rafter section YY; (c) eaves joint; (d) part of end plate view Z.

$$\frac{L_{cr,y}}{i_y} = 4.49 \times \frac{6000}{163} = 165.3$$

$$\frac{L_{cr,z}}{i_z} = \frac{1550}{30.2} = 51.3$$

From Table 6.2 of EN1993-1-1, for an S275 rolled H section, t_f less than 40 mm, $h/b > 1.2$, buckling about the major y–y axis, curve a, imperfection factor, $\alpha = 0.21$, buckling about the minor z–z axis, curve b, imperfection factor, $\alpha = 0.34$

$$\lambda_1 = 93.9\varepsilon = 93.9 \times 0.92 = 86.4$$

$$\overline{\lambda_y} = \frac{\lambda_y}{\lambda_1} = \frac{165.3}{86.4} = 1.91$$

$$\Phi = 0.5 \; [1 + 0.21(1.91 - 0.2) + 1.91^2] = 2.50$$

$$\chi_y = \frac{1}{2.50 + \sqrt{2.50^2 - 1.91^2}} = 0.24$$

$$\overline{\lambda_z} = \frac{\lambda_z}{\lambda_1} = \frac{51.3}{86.4} = 0.59$$

$$\Phi = 0.5 \; [1 + 0.34(0.59 - 0.2) + 0.59^2] = 0.74$$

Member buckling capacity

$$\chi_z = \frac{1}{0.74 + \sqrt{0.74^2 - 0.59^2}} = 0.84$$

Moment capacity $M_{c,Rd} = 275 \times 880 \times 10^{-3} = 242$ kN-m
  Buckling moment resistance

Applied moment $M = 148.8$ kN-m

Buckling lengths $L_{cr} = 1550$ mm

For ratio of end moment, $\psi = 98.1/148.8 = 0.66; \; C_1 = 1.21$

$$M_{cr} = 1.21 \times \frac{\pi^2 \times 210,000 \times 5,380,000}{1,550^2} \left(\frac{155 \times 10^9}{5,380,000} + \frac{1,550^2 \times 81,000 \times 190,000}{\pi^2 \times 210,000 \times 5,380,000} \right)^{0.5}$$

$M_{cr} = 1006.6$ kN-m

$$\overline{\lambda_{LT}} = \sqrt{\frac{W_y f_y}{M_{cr}}} = \sqrt{\frac{880 \times 10^3 \times 275}{1006 \times 10^6}} = 0.49$$

For $h/b > 2$, curve b, $\alpha_{LT} = 0.34$

$$\Phi_{LT} = 0.5 \times [1 + 0.34(0.49 - 0.2) + 0.49^2] = 0.67$$

$$\chi_{LT} = \frac{1}{0.67 + \sqrt{0.67^2 - 0.49^2}} = 0.89$$

Table B.3 of BS EN1993-1-1. Equivalent uniform moment factors, C_m
  For $\psi = 0.66$, C_{my}, $C_{mz} = 0.6 + 0.4(0.5) = 0.86 \geq 0.4$

$$k_{yy} = C_{my} \left(1 + (\overline{\lambda_y} - 0.2) \frac{N_{Ed}}{\chi_y N_{Rk}/\gamma_{M1}} \right) \leq C_{my} \left(1 + 0.8 \frac{N_{Ed}}{\chi_y N_{Rk}/\gamma_{M1}} \right)$$

$$k_{yy} = 0.86 \left(1 + 0.8 \frac{95.1}{0.24 \times 5900 \times 10^{-3} \times 275/1.0} \right) = 1.03$$

$$k_{zy} = 0.6 \quad k_{yy} = 0.62$$

Interaction expressions

$$\frac{N_{Ed}}{\chi_y N_{Rk}/\gamma_{M1}} + k_{yy} \frac{M_{y,Ed}}{\chi_{LT} M_{y,Rk}/\gamma_{M1}} \leq 1$$

$$\frac{95.1}{0.24 \times 1662.5/1.0} + 1.03 \frac{148.8}{0.89 \times 242/1.0} = 0.95$$

$$\frac{N_{Ed}}{\chi_z N_{Rk}/\gamma_{M1}} + k_{zy} \frac{M_{y,Ed}}{\chi_{LT} M_{y,Rk}/\gamma_{M1}} \leq 1$$

$$\frac{95.1}{0.55 \times 1662.5/1.0} + 0.62 \frac{148.8}{0.89 \times 242/1.0} = 0.53$$

Satisfactory.

9.8.7 Rafter restraints and stability

The rafter section at the eaves is shown in Figure 9.19b. The section properties are calculated to give

$$A = 64.9 \text{ cm}^2, \quad W_{pl,y} = 1353 \text{ cm}^3, \quad i_z = 2.35 \text{ cm}$$

At the large end of the haunch, at the eaves,

$F = 64.2$ kN from Figure 9.17

$M_y = 196.2$ kN-m from Figure 9.17

$$L_m = \frac{38 \times 23.5}{\sqrt{\frac{1}{57.4} \left(\frac{64.2 \times 10^3}{64.9 \times 10^2} \right) + \frac{1}{756 \times 1} \left(\frac{\left(1353 \times 10^3\right)^2}{64.9 \times 10^2 \times 15 \times 10^4} \right) \left(\frac{275}{235} \right)^2}} = 472 \text{ mm}$$

At the small end of the haunch, consider as a rolled section 356×127 UB 39 for which $A = 49.4 \text{ cm}^2$, $W_{pl,y} = 654 \text{ cm}^3$. The actions at the end of the haunch are

$F = 59.5$ kN from Figure 9.17

$M_x = 96.9$ kN-m from Figure 9.17

$$L_m = \frac{38 \times 26.9}{\sqrt{\frac{1}{57.4} \left(\frac{59.5 \times 10^3}{49.4 \times 10^2} \right) + \frac{1}{756 \times 1.563} \left(\frac{\left(654 \times 10^3\right)^2}{49.4 \times 10^2 \times 15 \times 10^4} \right) \left(\frac{275}{235} \right)^2}} = 1090 \text{ mm}$$

An additional purlin must be located at, say, 450 mm from the eaves with stays to the compression flange.

At the hinge location near the ridge, the purlins will be spaced at 1020 mm as shown in Figure 9.19a.

9.8.8 Sway stability

Using sway-check method in Section 9.4.1, $L/h < 5$; $h_r < 0.25L$; use limiting span/depth of rafter ratio. The haunch depth 604.7 mm (Figure 9.19b) is less than twice the rafter depth, the effective span $L_b = 20$ m:

$$\rho = \left(\frac{2 \times 15,600}{10,100}\right)\left(\frac{20}{6}\right) = 10.29$$

$$M_p = 654 \times 275 \times 10^{-3} = 179.9 \text{ kN-m} - \text{rafter}$$

$$= W_o L/16$$

$$W_o = 143.9 \text{ kN}$$

$$\Omega = 9.09 \times 20/143.9 = 1.26$$

$$\frac{L}{D} = \frac{20,000}{352.8} = 56.7 \le \frac{51.5 \times 20}{1.26 \times 6}\left(\frac{10.29}{4 + 10.29 \times 21.54/20}\right)\left(\frac{235}{275}\right) = 79.4$$

Satisfactory.

9.8.9 Serviceability: Deflection at eaves

An elastic analysis for the imposed load on the roof of 3.75 kN/m gives $H = 15.26$ kN and $V = 37.5$ kN (see Sections 9.2.2 and 9.2.6). From Section 9.6,

$$\theta = \frac{1}{210 \times 10^3 \times 10,100 \times 10^4} \times \left[\begin{array}{c} \dfrac{3.75 \times 20,000^2 \times 10,770}{24} + 15,260 \\ \times 10,770\left(6,000 + \dfrac{4,000}{2}\right) - \dfrac{37,500 \times 20,000 \times 10,770}{4} \end{array}\right]$$

$$+ \frac{15,260 \times 6,000^2}{2 \times 210 \times 10^3 \times 15,600 \times 10^4} = 6.695 \times 10^{-3} \text{ rad}$$

Deflection at eaves

$$S_B = 6,000 \times 6.695 \times 10^{-3} - \frac{15,260 \times 6,000^3}{6 \times 210 \times 10^3 \times 15,600 \times 10^4} = 23.4 \text{ mm} = \text{height}/256$$

This exceeds $h/300$; however, the metal sheeting will be able to accommodate this deflection.

9.8.10 Design of joints

The arrangement for the eaves joint is shown in Figure 9.19c. Selected check calculations only are given.

1. *Joint Forces*

   Take moment about X:

$$T = \frac{[181.8 - 90.9 \times 0.402/2]}{0.6} = 272.5 \text{ kN}$$

2. *Bolts*

$$F_t = \frac{272}{4} = 68.13 \text{ kN}$$

$$F_v = \frac{90.0}{6} = 15.15 \text{ kN}$$

   Try 20 mm Grade 8.8 bolt. From the *SCI blue book*,

$$F_{t,Rd} = 141 \text{ kN}, \quad F_{v,Rd} = 94.1 \text{ kN}$$

$$\frac{15.15}{94.1} + \frac{68.13}{1.4 \times 141} = 0.5 \leq 1.0$$

   The bolts are satisfactory.

3. *Column flange*

   See Figure 9.19d and Section 9.3.8. The yield line analysis gives

$$\frac{272.5 \times 10^3 \times 80}{8} = 4m \times 67.9 - m \times 22(1 + 0.62) + \frac{m \times 80}{31.6}\left(80 - \frac{22 \times 078}{2}\right)$$

$$= 416.8m$$

   $m = 6537.9$ N mm/mm

$$t = \left(\frac{4 \times 6537.9}{275}\right)^{0.5} = 9.75 \text{ mm}$$

   The flange thickness 11.2 mm is adequate. The rafter end plate can be made 12 mm thick.

4. *Column web shear*

$$F_v = 0.6 \times 275 \times 402.3 \times 6.9 \times 10^3 = 458.8 kN > 272.8 \text{ kN}$$

   Satisfactory.

   A similar design procedure is carried out for the ridge joint.

9.8.11 Comments

9.8.11.1 Wind uplift

The rafters on portals with low roof angles and light roof dead loads require checking for reverse bending due to wind uplift. Restraints to inside flanges near the ridge are needed to stabilise the rafter. Joints must be also checked for reverse moments.

The portal frames designed earlier has a relatively high roof angle and heavy roof dead load. Checks for load, $-1.4 \times$ wind $+ 1.0 \times$ dead, and for wind angles $0°$ and $90°$ show that the frame remains in the elastic range and the rafter is stable without adding further restraints (see Figure 9.13).

9.8.11.2 Eaves joint design

In designing the joint, the effect of axial load is usually ignored, and the bolts are sized to resist moment only. In the aforementioned case, $M = 180.8$ kN-m and the bolt tension $T = 75.3$ kN < 110 kN. The shear force is taken by the bottom two bolts.

FURTHER READING FOR PORTAL DESIGN

Davies, J. M., In-plane stability in portal frames, *The Structural Engineer*, 68(8), 141–147, 1990.

Davies, J. M., The stability of multi-bay portal frames, *The Structural Engineer* 69(12), 223–229, 1991.

Fraser, D. J., Effective lengths in gable frames, sway not prevented, *Civil Engineering Transaction, Institution of Engineers, Australia* CE22(3), 1980.

Fraser, D. J., Stability of pitched roof frames, *Civil Engineering Transaction, Institution of Engineers, Australia* CE28(1), 1986.

Horne, M. R. and Merchant, W., *The Stability of Frames*, Pergamon Press, Oxford, U.K., 1965.

Horne, M. R., *Plastic Theory of Structures*, Nelson, London, U.K., 1971.

Horne, M. R. and Morris, L. J., *Plastic Design of Low Rise Frames*, Collins, London, U.K., 1981.

The Steel Construction Institute, *Plastic Design of Single-Storey Pitched-Roof Portal Frames to Eurocode 3*, The Steel Construction Institute, Ascot, U.K., 1995.

Davison, B. and Owens, G. W., *Steel Designers Manual*, Wiley-Blackwell, U.K., 2012.

The Steel Construction Institute, *Steel Building Design: Design Data*, The Steel Construction Institute, Ascot, U.K., 2009.

Chapter 10

Connections

10.1 TYPES OF CONNECTIONS

Connections are needed to join

1. Members together in trusses and lattice girders
2. Plates together to form built-up members
3. Beams to beams, trusses, bracing, etc. to columns in structural frames
4. Columns to foundations

Some typical connections are shown in Figure 10.1. Only simple connections are considered in this chapter and end connections for beams and column bases are treated in Chapters 4 and 7, respectively.

Connections may be made by

- Bolting ordinary or non-preloaded bolts in standard clearance or oversize holes
- Preloaded bolt
- Welding – fillet and butt welds

The categories of bolted connections are given in Cl. 3.4 in BS EN1993-1-8.

10.2 NON-PRELOADED BOLTS

10.2.1 Bolts, nuts and washers

The ISO metric 'black' hexagon head non-preloaded bolt shown in Figure 10.2 with nut and washer is the most commonly used structural fastener in the industry. The bolts, in the three common bolt classes given as follows, are specified in BS EN 1993-1-8. The specification for ISO metric 'precision' hexagon bolts and nuts, which are manufactured to tighter dimensional tolerances, is given in BS EN ISO4016.

Bolt class	Yield strength (N/mm²)	Ultimate tensile strength (N/mm²)
4.6	240	400
8.8	640	800
10.9	900	1000

The main diameters used are
10, 12, 16, 20, (22), 24, (27) and 30 mm
The sizes shown in brackets are not preferred.

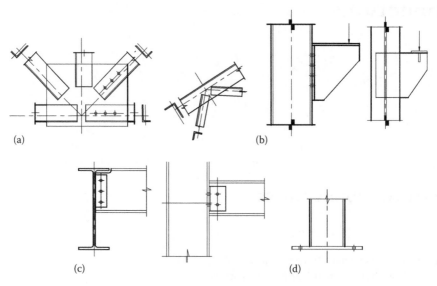

Figure 10.1 Typical connections: (a) internal truss joints; (b) brackets; (c) beam connection; (d) column base.

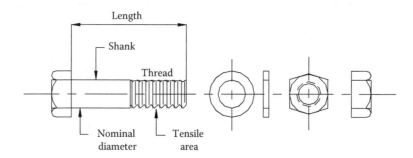

Figure 10.2 Hexagon head bolt, nut and washer.

10.2.2 Direct shear joints

Bolts may be arranged to act in single or double shear, as shown in Figure 10.3. Provisions governing spacing, edge and end distances are set out in Table 3.3 of BS EN1993-1-8. The principal provisions in normal conditions are

1. The minimum spacing is 2.2 times the bolt hole diameter, d_0.
2. The maximum spacing in unstiffened plates in the direction of stress is $14t$, where t is the thickness of the thinner plate connected.
3. The minimum edge and end distance as shown in Figure 10.3 from a rolled machine-flame cut or plane edge is $1.2d_0$, where d_0 is the hole diameter.
4. The maximum edge distance is $8t$ or 125 mm.

The standard dimensions of holes for non-preloaded bolts are now specified in Cl 3.6.1. It depends on the diameter of the bolt and the type of bolt hole. In addition to the usual

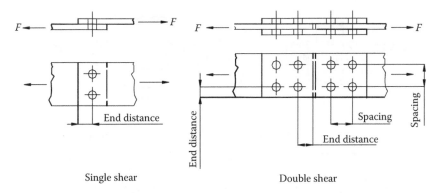

Figure 10.3 Bolts in single and double shear.

standard clearance, oversize, short and long slotted holes and kidney-shaped slotted hole are now permitted in the revised code. As in usual practice, larger-diameter hole is required as the bolt diameter increases. For example, the diameter of a standard clearance hole will be 22 mm for a 20 mm diameter bolt and 33 mm for a 30 mm diameter bolt.

A shear joint can fail in the following four ways:

1. By shear on the bolt shank
2. By bearing on the member or bolt
3. By tension in the member
4. By shear at the end of the member

The typical failure modes in a shear joint are shown in Figure 10.4a. These failures noted earlier can be prevented by taking the following measures:

1. For modes 1 and 2, provide sufficient bolts of suitable diameter
2. For mode 3, design tension members for the effective area (see Chapter 6)
3. Provide sufficient end distance for mode 4

In addition, a new failure mode, block shear, has been observed in a shear joint involving a group of bolts as shown in Figure 10.4b, and a check against this failure mode is required in the BS EN1993-1-8.

Also, the code now recognizes the beneficial effect of strain hardening and permits the effect of bolt holes on the plate shear capacity to be ignored. The design of bolted subjected to shear and/or tension is set out in Table 3.4 of BS EN1993-1-8. The basic provisions are the following:

1. Effective area resisting shear A
   When the shear plane occurs in the threaded portion of the bolt,

   $$A = A_s$$

   where A_s is the nominal tensile stress area of the bolt.
   When the shear plane occurs in the non-threaded portion,

   $$A = \text{bolt shank area based on the nominal diameter}$$

   For a more conservative design, the tensile stress area A_s may be used throughout.

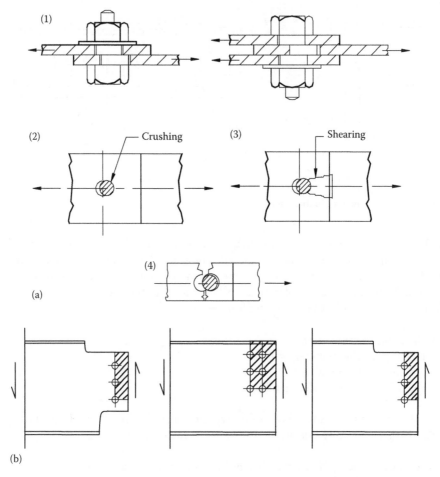

Figure 10.4 (a) Failure modes of a bolted joint: (1) single and double shear; (2) bearing on plate and bolt; (3) end shear failure; (4) tension failure of plate. (b) Block shear failure.

2. Shear resistance per shear plane

$$F_{v,Rd} = \frac{\alpha_v f_{ub} A}{\gamma_{M2}}$$

where f_{ub} is the ultimate tensile strength for bolts given in Table 3.1 in BS EN1993-1-8.
   Where the shear plane passed through the threaded portion of the bolt (A is the tensile stress area of the bolt A),

For Classes 4.6, 5.6 and 8.8,

   $\alpha_v = 0.6$

For Classes 4.8, 5.8, 6.8 and 10.9,

   $\alpha_v = 0.5$

where the shear plane passed through the unthreaded portion of the bolt (A is the gross cross-sectional area of the bolt)

3. Block shear

Block shear failure should also be checked to prevent shear failure through a group of bolt holes at a free edge. The combined block shear capacity for both the shear and the tension edges or faces in a shear joint (see Figure 10.4b) is given by

$$V_{eff,1,Rd} = \frac{f_u A_{nt}}{\gamma_{M2}} + \frac{1}{\sqrt{3}}\left(\frac{f_y A_{nv}}{\gamma_{M0}}\right)$$

where

A_{nt} is the net area subject to tension
A_{nv} is the net area subject to shear

4. Bearing resistance

$$F_{b,Rd} = \frac{k_1 \alpha_b f_u dt}{\gamma_{M2}}$$

where α_b is the smallest of α_d, f_{ub}/f_u or 1.0; in the direction of load transfer,

For end bolts, $\alpha_d = \dfrac{e_1}{3d_0}$

For inner bolts, $\alpha_d = \dfrac{p_1}{3d_0} - \dfrac{1}{4}$

Perpendicular to the direction of load transfer,

For edge bolts, k_1 is the smallest of $2.8\dfrac{e_1}{3d_0} - 1.7 \text{ or } 2.5$

For inner bolts, k_1 is the smallest of $1.4\dfrac{p_2}{3d_0} - 1.7 \text{ or } 2.5$

10.2.3 Direct-tension joints

1. Tension resistance

$$F_{t,Rd} = \frac{k_2 f_{ub} A_s}{\gamma_{M2}}$$

where
$k_2 = 0.9$
$k_2 = 0.63$ for countersunk bolt

The prying force Q adds directly to the tension in the bolt. Referring to Figure 10.5,

Total bolt tension, $F = \dfrac{W}{2} + Q \leq F_{t,Rd}$

where W is the external tension on the joint.

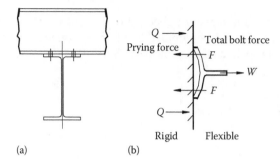

Figure 10.5 (a) Bolts in tension and (b) prying force.

In this method, it is necessary to calculate the prying force. The magnitude of the prying force depends on the stiffnesses of the bolt and the flanges. Theoretical analyses based on elastic and plastic theory are available to determine the values of these prying forces. Readers should consult further specialised references for this purpose. If the flanges are relatively thick, the bolt spacing not excessive and the edge distance sufficiently large, the prying forces are small and may be neglected. Where possible, it is recommended that the simple method is used.

10.2.4 Eccentric connections

There are two principal types of eccentrically loaded connections:

1. Bolt group in direct shear and torsion
2. Bolt group in direct shear and tension

These connections are shown in Figure 10.6.

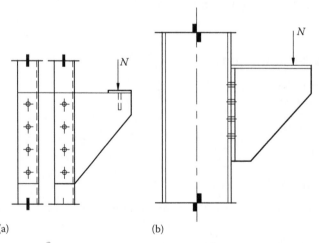

Figure 10.6 Eccentrically loaded connections: (a) bolts in direct shear and torsion; (b) bolts in direct shear and tension.

10.2.5 Bolts in direct shear and torsion

In the connection shown in Figure 10.6a, the moment is applied in the plane of the connection, and the bolt group rotates about its centre of gravity. A linear variation of loading due to moment is assumed, with the bolt furthest from the centre of gravity of the group carrying the greatest load. The direct shear is divided equally between the bolts and the side plates are assumed to be rigid.

Consider the group of bolts shown in Figure 10.7a, where the load P is applied at an eccentricity e. The bolts A, B, etc. are at distances r_1, r_2, etc. from the centroid of the group. The coordinates of each bolt are (y_1, z_1), (y_2, z_2), etc. Let the force due to the moment on bolt A be F_t. This is the force on the bolt farthest from the centre of rotation. Then the force on a bolt r_2 from the centre of rotation is F_{tr2}/r_1 and so on for all the other bolts in the group. The moment of resistance of the bolt group is shown in Figure 10.7.

The load F_t due to moment on the maximum loaded bolt A is given by

$$F_t = \frac{N \cdot e \cdot r_1}{\sum x^2 + \sum y^2}$$

The load F_S due to direct shear is given by

$$F_s = \frac{N}{\text{No. of bolts}}$$

The resultant load F_R on bolt A can be found graphically, as shown in Figure 10.7b. The algebraic formula can be derived by referring to Figure 10.7c.

Resolve the load Ft vertically and horizontally to give

vertical load on bolt $A = F_s + F_t \cos \phi$
horizontal load on bolt $A = F_t \sin \phi$

Resultant load on bolt A

$$F_R = \sqrt{\left(F_t \sin\phi\right)^2 + \left(F_s + F_t \cos\phi\right)^2}$$

$$= \sqrt{F_s^2 + F_t^2 + \left(2 F_s F_t \cos\phi\right)}$$

The size of bolt required can then be determined from the maximum load on the bolt.

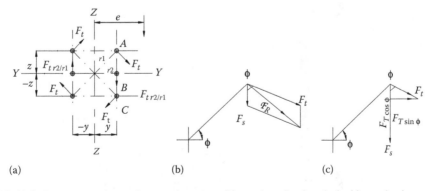

(a) (b) (c)

Figure 10.7 (a) Bolts group in direct shear and torsion; (b) resultant load on bolt; (c) resolved parts of loads on bolt.

10.2.6 Bolts in direct shear and tension

In the bracket-type connection shown in Figure 10.6b, the bolts are in combined shear and tension. BS EN1993-1-8 gives the design procedure for these bolts in Table 3.4. This is described in the following:

The factored applied shear $F_{v,Ed}$ must not exceed the shear capacity $F_{v,Rd}$. The bearing capacity checks must also be satisfactory. The factored applied tension $F_{t,Ed}$ must not exceed the tension capacity $F_{t,Rd}$.

In addition to the aforementioned, the following relationship must be satisfied:

$$\frac{F_{v,Ed}}{F_{v,Rd}} + \frac{F_{t,Ed}}{1.4F_{t,Rd}} \le 1.0$$

An approximate method of analysis that gives conservative results is described first. A bracket subjected to a factored load P at an eccentricity e is shown in Figure 10.8a. The centre of rotation is assumed to be at the bottom bolt in the group. The loads vary linearly, as shown in Figure 10.8a, with the maximum load F_t in the top bolt.

The moment of resistance of the bolt group is

$$M_R = 2\left[F_t \cdot z_1 + F_t \cdot \frac{z_2^2}{z_1} + \cdots \right]$$

$$= 2\frac{F_t}{z_1} \cdot \left[z_1^2 + z_2^2 + \cdots \right]$$

$$= \frac{2F_t}{z_1} \sum z^2$$

$$= N \times e$$

The maximum bolt tension is

$$F_t = \frac{N \cdot e \cdot z_1}{2 \sum z^2}$$

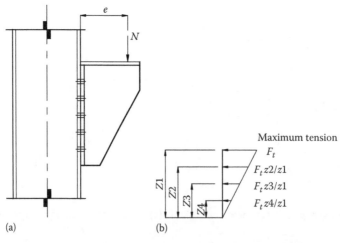

Figure 10.8 Bolts in direct shear and tension: approximate analysis. (a) Joint and (b) bolt loads.

The vertical shear per bolt is

$$F_s = N/\text{no. of bolts}$$

A bolt size is assumed and checked for combined shear and tension as described earlier.

In a more accurate method of analysis, the applied moment is assumed to be resisted by the bolts in tension with uniformly varying loads and an area at the bottom of the bracket in compression, as shown in Figure 10.9b.

For equilibrium, the total tension T must equal the total compression C. Consider the case where the top bolt is at maximum capacity $F_{t,Rd}$ and the bearing stress is at its maximum value f_y.

Referring to Figure 10.11c, the total tension is given by

$$T = F_{t,Rd} + \frac{(D-z-p)}{(D-z)} \cdot F_{t,Rd} + \frac{(D-z-2p)}{(D-z)} \cdot F_{t,Rd} + \cdots$$

$$F_{t,Rd} = \frac{k_2 f_{ub} A_s}{\gamma_{M2}} \quad \text{(see Section 10.2.3 [1])}$$

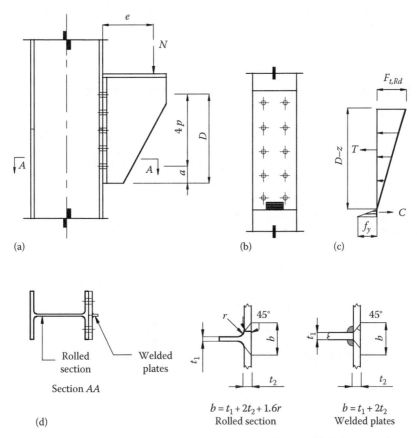

Figure 10.9 Bolts in direct shear and tension: accurate analysis: (a) joint; (b) bolt in tension compression area; (c) internal forces and stresses; (d) stiff bearing width.

The total compression is given by

$$C = 0.5 \times f_y \times b \times z$$

where b, the stiff bearing width, is obtained by spreading the load at 45°, as shown in Figure 10.9d. In the case of the rolled section, the 45° line is tangent to the fillet radius. In the case of the welded plates, the contribution of the fillet weld is neglected.

The expressions for T and C can be equated to give a quadratic equation that can be solved to give z, the location of the neutral axis. The moment of resistance is then obtained by taking moments of T and C about the neutral axis. This gives

$$M_R = F_{t,Rd}\left(D - z\right) + F_{t,Rd}\frac{(D - z - p)^2}{(D - z)} + \cdots + \frac{2}{3} \cdot C \cdot z$$

The actual maximum bolt tension F_t is then found by proportion, as follows:

Applied moment $M = Ne$

Actual bolt tension $F_t = MF_{t,Rd}/M_R$

The direct shear per bolt $F_S = N/$no. of bolts and the bolts are checked for combined tension and shear.

Note that to use the method, a bolt size must be selected first and the joint set out and analysed to obtain the forces on the maximum loaded bolt. The bolt can then be checked.

10.2.7 Examples of non-preloaded bolted connections

Example 10.1

The joint shown in Figure 10.10 is subjected to a tensile dead load of 75 kN and a tensile imposed load of 85 kN. All data regarding the member and joint are shown in Figure 10.10. The steel is Grade S275 and the bolt Class 4.6. Check that the joint is satisfactory.

Using load factors from BS EN1990,
Factored load = $(1.35 \times 75) + (1.5 \times 85) = 229$ kN
Strength of bolts from the blue book for 20 mm diameter bolts

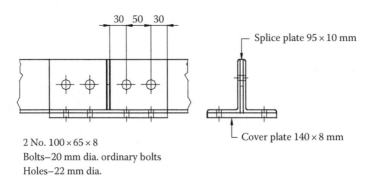

2 No. $100 \times 65 \times 8$
Bolts–20 mm dia. ordinary bolts
Holes–22 mm dia.

Figure 10.10 Double-angle splice.

Single-shear capacity on threads = 47.0 kN
Bearing capacity of bolts on 10 mm ply = 84.2 kN
Bearing capacity on 10 mm splice with 30 mm end distance

$$F_{b,Rd} = \frac{k_1 \alpha_b f_u dt}{\gamma_{M2}} = \frac{2.5 \times 0.45 \times 400 \times 20 \times 10}{1.25} = 72\,kN$$

Bolt capacity – two bolts are in double shear and four in single.
Shear = (4 × 47.0) + (2 × 47.0) + 72 = 354 kN
Note that the capacity of the end-bolt bearing on the 10 mm splice plate is controlled by the end distance (Table 3.4 of BS EN1993-1-8).
Capacity of the angles: Gross area = 12.7 cm² per angle
The angles are connected through both legs. Clause 3.10.3 of BS EN1993–1–8 states that the net area defined in Clause 3.10.3(2) is to be used in design. The standard clearance holes are 22 mm in diameter:
Net area = 2(1270 – 2 × 22 × 8) = 1836 mm²
Ultimate strength f_u = 430 N/mm²

$$N_{u,Rd} = \frac{0.4 \times 1836 \times 430}{1.25} = 252.6\,kN$$

Splice plate and cover plate
Effective area = [(95 – 22)10 + (140 – 44)8] = 1498 mm²
$$< gross\ area = 2070\ mm^2$$
$$N_{u,Rd} = \frac{0.9 \times 1498 \times 430}{1.25} = 463.8\,kN$$

The splice is adequate to resist the applied load.

Example 10.2

Check that the joint shown in Figure 10.11 is adequate. All data required are given in the figure.
Factored load = (1.35 × 60) + (1.5 × 80) = 201 kN
Moment M = (212 × 525)/10³ = 111.3 kN-m
Bolt group $\Sigma y^2 = 12 \times 250^2 = 750 \times 10^3$

$$\Sigma y^2 = 4(35^2 + 105^2 + 175^2) = 171.5 \times 10^3$$

$$\Sigma y^2 + \Sigma z^2 = 921.5 \times 10^3$$

$$\cos \phi = 250/305.16 = 0.819$$

Bolt A is the bolt with the maximum load:

$$\text{Load due to moment} = \frac{111.3 \times 103 \times 305.16}{921.5103} = 36.85\,kN$$

Load due to shear = 212/12 = 17.67 kN

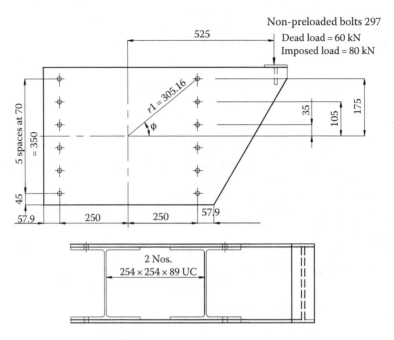

Figure 10.11 Example: bolt group in direct shear and torsion.

Resultant load on bolt

$$= [17.67^2 + 36.85^2 + (2 \times 17.67 \times 36.85 \times 0.819)]^{0.5}$$

$$= 52.31 \text{ kN}.$$

Single-shear value of 24 mm diameter ordinary bolt on the threads
From the blue book,
$F_{v,Rd} = 67.8$ kN
Universal column flange thickness = 17.3 mm
Side-plate thickness = 15 mm
Minimum end distance = 45 mm
Bearing capacity of the bolt = 150 kN
The strength of the joint is controlled by the single-shear value of the bolt.
The joint is satisfactory.

Example 10.3

Determine the diameter of ordinary bolt required for the bracket shown in Figure 10.12.
The joint dimensions and loads are shown in the figure. Use Grade 4.6 bolts.
Try 20 mm diameter ordinary bolts. From the blue book tension capacity, $F_{t,Rd} = 70.6$ kN.
The design strength from Table 3.1 of BS EN1993-1-1

20 mm thick $f_y = 275$ N/mm²

Referring to Figure 10.12c, the depth to the neutral axis is the total tension is

$$T = 70.6\left[1 + \frac{(280 - z)}{(350 - z)} + \frac{(210 - z)}{(350 - z)} + \frac{(140 - z)}{(350 - z)} + \frac{(70 - z)}{(350 - z)}\right]$$

$$= 70.6\left[\frac{(1050 - 5z)}{350 - z}\right] \text{ kN}$$

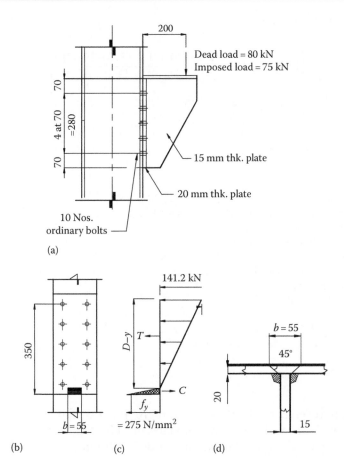

Figure 10.12 Bracket: bolts in direct shear and tension. (a) Joint; (b) tension and compression areas; (c) internal forces; (d) stiff bearing length.

The width of stiff bearing is shown in Figure 10.12d:

$$b = 15 + 2 \times 20 = 55 \text{ mm}$$

The total compression in terms of z is given by

$$C = \frac{1}{2} \times 275 \times \frac{55z}{10^3} = 7.56 \text{ kN}$$

Equate T and C and rearrange terms to give the quadratic equation:

$$7.56z^2 - 2{,}999z + 74{,}130 = 0$$

Solve to give

$$y = 26.49 \text{ mm}$$

The moment of resistance is

$$
M_R = \frac{141.2}{10^3} \left(\begin{array}{l} (350-26.49) + \dfrac{(280-26.49)^2}{(350-26.49)} + \dfrac{(210-26.49)^2}{(350-26.49)} \\[2mm] + \dfrac{(140-26.49)^2}{(350-26.49)} + \dfrac{(70-26.49)^2}{(350-26.49)} \end{array} \right)
$$

$$
+ \frac{7.56 \times 26.49^2 \times 2}{3 \times 10^3} = 98.4\,\text{kN}
$$

Factored load = $(1.35 \times 80) + (1.5 \times 75) = 220.5$ kN
Factored moment = $220.5 \times 200/10^3 = 44.1$ kN-m
Actual tension in top bolts
$F_t = 44.1 \times 70.6/98.4 = 31.64$ kN
Direct shear per bolt = $220.5/10 = 22.1$ kN
Shear capacity on threads $F_{v,Rd} = 47.0$ kN
Tension capacity, $F_{t,Rd} = 70.6$ kN
Combined shear and tension

$$
\frac{F_s}{F_{v,Rd}} + \frac{F_t}{1.4 F_{t,Rd}} = \frac{22.1}{47} + \frac{31.64}{1.4 \times 70.6} = 0.79 \le 1.0
$$

Therefore, 20 mm diameter bolts are satisfactory.

The reader can redesign the bolts using the approximate method. Note that only the bolts have been designed. The welds and bracket plates must be designed and the column checked for the bracket forces. These considerations are dealt with in Section 10.4.5.

10.3 PRELOADED BOLTS

10.3.1 General considerations

Preloaded high-strength friction-grip bolts are made from high-strength steel so they can be tightened to give a high shank tension. The shear in the connected plates is transmitted by friction, as shown in Figure 10.13a, and not by bolt shear, as in ordinary bolts. These bolts are used where strong joints are required, and a major use is in the joints in rigid frames.

The bolts are manufactured in two types to conform to EN 14399:

1. General grade: The strength is similar to Class 8.8 ordinary bolts. This type is generally used.
2. Higher grade: Parallel shank and waisted shank bolts are manufactured. The use of general-grade friction-grip bolts in structural steelwork is specified in BS EN1993-1-8. Parallel and waisted shank bolts are shown in Figure 10.13b. Only general-grade bolts will be discussed here.

The bolts must be used with hardened steel washers to prevent damage to the connected parts. The surfaces in contact must be free of mill scale, rust, paint, grease, etc. that would prevent solid contact between the surfaces and lower the slip factor (see as follows).

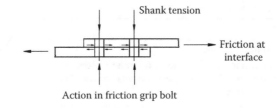

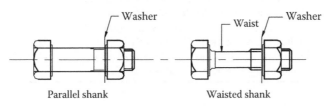

Types of friction grip bolt

Figure 10.13 **Preloaded bolts.**

Care must be taken to ensure that bolts are tightened up to the required tension; otherwise slip will occur at service loads and the joint will act as an ordinary bolted joint. Methods used to achieve the correct shank tension are

1. Part turning – the nut is tightened up and then forced a further half to three quarters of a turn, depending on the bolt length and diameter.
2. Torque control – a power-operated or hand-torque wrench is used to deliver a specified torque to the nut. Power wrenches must be calibrated at regular intervals.
3. Load-indicating washers and bolts – these have projections that squash down as the bolt is tightened. A feeler gauge is used to measure when the gap has reached the required size.

Friction-grip bolts are generally used in clearance holes. The clearances are same as for ordinary bolts given in Section 10.2.2.

10.3.2 Design procedure

Preloaded bolts can be used in shear, tension and combined shear and tension. The design procedure, given in Section 3.9 of BS EN1993-1-8 for general-grade parallel shank bolts, is set out as follows. Long joints are not discussed.

1. Bolts in shear
   The capacity is the lesser of the slip resistance and the bearing capacity. The code states that while the slip resistance is based on a serviceability criterion, the design check is made for convenience, using factored loads. However, the joint can slip and go into bearing at loads greater than the working load, and hence, the bearing capacity must be checked.
   The slip **resistance** of a preloaded Class 8.8 or 10.9 bolt should be taken as

$$F_{s,Rd} = \frac{k_s n \mu}{\gamma_{M3}} F_{p,C}$$

   where
   k_s is given in Table 10.1 and Table 3.6 of BS EN1993-1-8
   n is the number of the friction surfaces
   μ is the slip factor given in Table 10.2 and Table 3.7 of BS EN1993–1–8

Table 10.1 Values of k_s

Description	k_s
Bolts in normal holes	1.00
Bolts in either oversized holes or short slotted holes with the axis of the slot perpendicular to the direction of load transfer	0.85
Bolts in long slotted holes with the axis of the slot perpendicular to the direction of load transfer	0.70
Bolts in short slotted holes with the axis of the slot parallel to the direction of load transfer	0.76
Bolts in long slotted holes with the axis of the slot parallel to the direction of load transfer	0.63

Table 10.2 Slip Factor, μ, for preloaded bolts

Class of friction surfaces to EN1090-2	μ
A	0.5
B	0.4
C	0.3
D	0.2

For Class 8.8 and 10.9 bolts conforming with EN14399 and the tightening procedure conforming with EN1090–2, the preloading force $F_{p,C}$ should be taken as

$$F_{p,C} = 0.7 f_{ub} A_s$$

The slip resistances for Class 8.8 preloaded bolts for various slip factors are given in Table 10.3 The factor k_s is taken as 1.0 for bolts in standard clearance holes.

Table 10.3 Slip factor, μ, for preloaded bolts

	Bolts in tension			Slip resistance							
	Tensile		Min thickness	$\mu = 0.2$		$\mu = 0.3$		$\mu = 0.4$		$\mu = 0.5$	
Diameter of bolt	stress area	Tension resistance	for punching shear	Single shear	Double shear	Single shear	Double shear	Single shear	Double shear	Single shear	Double shear
D	A_s	$F_{t,Rd}$	t_{min}								
mm	mm²	kN	mm	kN	kN	kN	kN	kN	kN	kN	kN
12	84.3	48.6	3.71	7.55	15.1	11.3	22.7	15.1	30.2	18.9	37.8
16	157	90.4	5.59	14.1	28.1	21.1	42.2	28.1	56.3	35.2	70.3
20	245	141	7.36	22.0	43.9	32.9	65.9	43.9	87.8	54.9	110
24	353	203	8.22	31.6	63.3	47.4	94.9	63.3	127	79.1	158
30	561	323	10.7	50.3	101	75.4	151	101	201	126	251

2. Bolts in tension only
   The tension capacity is given by

   $$F_{t,Rd} = \frac{k_2 f_{ub} A_s}{\gamma_{M2}}$$

3. Bolts in combined shear and tension
   If a slip-resistant connection is subjected to an applied tensile force, $F_{t,Ed}$ or $F_{t,Ed,ser}$, in addition to the shear force, $F_{v,Ed}$ or $F_{v,Ed,ser}$, tending to produce slip, the design slip resistance per bolt should be taken as follows:

   For a category B connection,

   $$F_{s,Rd,ser} = \frac{k_s n \mu \left(F_{p,C} - 0.8 F_{t,Ed,ser}\right)}{\gamma_{M3,ser}}$$

   For a category C connection,

   $$F_{s,Rd} = \frac{k_s n \mu \left(F_{p,C} - 0.8 F_{t,Ed}\right)}{\gamma_{M3}}$$

   If, in a moment connection, a contact force on the compression side counterbalances the applied tensile force, then no reduction in slip resistance is required.

10.3.3 Examples of preloaded bolt connections

Example 10.4

Design the bolts for the moment and shear connection between the floor beam and the column in a steel frame building. The following data are given:

Floor beam 610 × 229 UB 140
Column 254 × 254 UC 132
Moment dead load = 180 kN-m
Imposed load = 100 kN-m
Shear dead load = 300 kN
Imposed load = 150 kN

Set out the joint as shown in Figure 10.14.

   1. Moment connection
      The moment is taken by the flange bolts in tension:
         Factored moment = (1.35 × 180) + (1.5 × 100) = 393 kN-m
         Flange force = 393/0.595

              = 660.5 kN

      Provide 24 mm diameter Class 8.8 friction-grip bolts:
         Minimum shank tension = 203 kN
         Tension capacity of four bolts = 4 × 203 = 812 kN

      The joint is satisfactory for moment. Four bolts are also provided at the bottom of the joint, but these are not loaded by the moment in the direction shown.

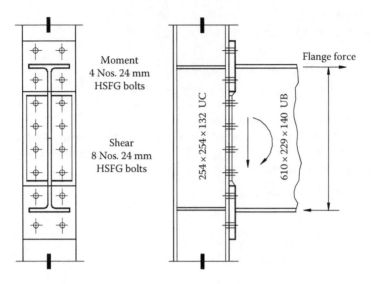

Figure 10.14 Beam-to-column connection.

2. Shear connection
The shear is resisted by the web bolts:
Factored shear = (1.35 × 300) + (1.5 × 150) = 630 kN
Slip resistance of eight no. 24 mm diameter bolts in clearance holes = 8 × 79.1 = 632.8 kN

For the end plate is 12 mm thick, the end distance of the two top bolts is 55 mm and the bearing resistance

= 8 × 180 = 1440 kN.

Therefore, 24 mm diameter bolts are satisfactory.
   Note that only the bolts have been designed. The welds, end plates and stiffeners must be designed and the column flange and web checked. These considerations are dealt with in Section 10.5.

Example 10.5

Determine the bolt size required for the bracket loaded as shown in Figure 10.15a:
   Factored load = (1.35 × 160) + (1.5 × 110) = 381 kN
   Factored moment = 381 × 0.28 = 106.7 kN-m
   Try 20 mm diameter Class 8.8 preloaded bolts:
   Tension capacity from Table 10.3 = 141 kN
   The joint forces are shown in Figure 10.15c. The stiff bearing width can be calculated from the bracket end plate (Figure 10.15d):

$b = 15 + 2 \times 20 = 55$ mm

The total tension in terms of the maximum tension in the top bolts is

$$T = 282\left[1 + \frac{370-z}{470-z} + \frac{270-z}{470-z} + \frac{170-z}{470-z}\right]$$

$$= 280(1280 - 4z)/(470 - z)$$

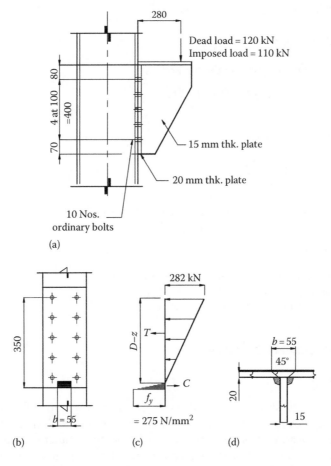

Figure 10.15 Bracket: bolts in shear and tension. (a) Joint; (b) tension and compression areas; (c) internal forces; (d) stiff bearing length.

The design strength from Table 3.1 of EN1993-1-1 in the code for plates 20 mm thick is

$$f_y = 275 \text{ N/mm}^2$$

The total compression is given by

$$C = 0.5 \times 275 \times 55z / 10^3$$

Equate T and C and rearrange to give

$$z^2 - 618.2z + 47407 = 0$$

Solving gives $z = 89.7$ mm.
   The moment of resistance is

$$M_R = \frac{282}{10^3} \left((470 - 89.7) + \frac{(370 - 89.7)^2}{(470 - 89.7)} + \frac{(270 - 89.7)^2}{(470 - 89.7)} + \frac{(170 - 89.7)^2}{(470 - 89.7)} \right)$$

$$+ \frac{7.56 \times 89.7^2 \times 2}{3 \times 10^3}$$

$$= 195.09 + 40.55$$

$$= 235.64 \text{ kN-m}$$

The actual maximum tension in the top bolts is

$$F_{t,Ed} = 106.7 \times 141/235.64 = 63.85 \text{ kN},$$

For combined tension and shear
The shear resistance, $F_{s,Rd}$

$$F_{s,Rd} = \frac{k_s n \mu \left(F_{p,C} - 0.8 F_{t,Ed,ser} \right)}{\gamma_{M3}}$$

$$= \frac{1.0 \times 1 \times 0.5 \left(0.7 \times 800 \times 245 \times 10^{-3} - 0.8 \times 63.85 / 1.4 \right)}{1,25}$$

$$= 40.3 \text{ kN}$$

The bearing resistance is much greater than this value:
Applied shear per bolt $F_s = 381/10 = 38.1 \text{ kN}$
Therefore, 20 mm diameter bolts are satisfactory.

10.4 WELDED CONNECTIONS

10.4.1 Welding

Welding is the process of joining metal parts by fusing them and filling in with molten metal from the electrode. The method is used extensively to join parts and members, attach cleats, stiffeners, end plates, etc. and to fabricate complete elements such as plate girders. Welding produces neat, strong and more efficient joints than are possible with bolting. However, it should be carried out under close supervision, and this is possible in the fabrication shop. Site joints are usually bolted. Though site welding can be done, it is costly, and defects are more likely to occur.

Electric arc welding is the main system used, and the two main processes in structural steel welding are

1. Manual arc welding, using a handheld electrode coated with a flux that melts and protects the molten metal. The weld quality depends very much on the skill of the welder.
2. Automatic arc welding. A continuous wire electrode is fed to the weld pool. The wire may be coated with flux or the flux can be supplied from a hopper. In another process, an inert gas is blown over the weld to give protection.

10.4.2 Types of welds, defects and testing

The two main types of welds, butt and fillet, are shown in Figure 10.16a and b. Butt welds are named after the edge preparation used. Single and double U and V welds are shown in Figure 10.16c. The double U welds require less weld metal than the V types. A 90° fillet weld is shown but other angles are used. The weld size is specified by the leg length. Some other types of welds – partial butt, partial butt and fillet weld and deep penetration fillet weld – are shown in Figure 10.16d. In the deep penetration fillet weld, a higher current is used to fuse the plates beyond the limit of the weld metal.

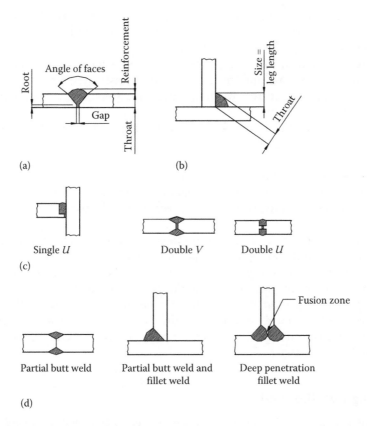

Figure 10.16 Weld types: (a) single *V* butt weld; (b) fillet weld; (c) types of butt weld; (d) other types of welds.

Cracks can occur in welds and adjacent parts of the members being joined. The main types are shown in Figure 10.17a. Contraction on cooling causes cracking in the weld. Hydrogen absorption is the main cause of cracking in the heat-affected zone, while lamellar tearing along a slag inclusion is the main problem in plates.

Faulty welding procedure cart leads to the following defects in the welds, all of which reduce the strength (see Figure 10.17b):

1. Over-reinforcement and undercutting
2. Incomplete penetration and lack of side-wall fusion
3. Slag inclusions and porosity

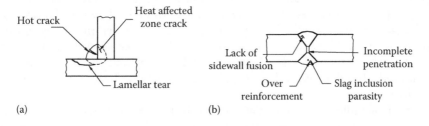

Figure 10.17 Cracks and defects in welds: (a) cracking; (b) defects and faulty welding.

When the weld metal cools and solidifies, it contracts and sets up residual stresses in members. It is not economical to relieve these stresses by heat treatment after fabrication, so allowance is made in design for residual stresses.

Welding also causes distortion, and special precautions have to be taken to ensure that fabricated members are square and free from twisting. Distortion effects can be minimized by good detailing and using correct welding procedure. Presetting, pretending and preheating are used to offset distortion.

All welded fabrication must be checked, tested and approved before being accepted. Tests applied to welding are given in Ref. [9]:

1. Visual inspection for uniformity of weld
2. Surface tests for cracks using dyes or magnetic particles
3. X-ray and ultrasonic tests to check for defects inside the weld

Only visual and surface tests can be used on fillet welds. Butt welds can be checked internally, and such tests should be applied to important butt welds in tension.

When different thicknesses of plate are to be joined, the thicker plate should be given a taper of 1 in 5 to meet the thinner one. Small fillet welds should not be made across members such as girder flanges in tension, particularly if the member is subjected to fluctuating loads, because this can lead to failure by fatigue or brittle fracture. With correct edge preparation if required, fit- up, electrode selection and a properly controlled welding process, welds are perfectly reliable.

10.4.3 Design of fillet welds

Important provisions regarding fillet welds are set out in Clause 4.3.2 of BS EN1993-1-8. Some of these are listed as follows (the complete clause in the code should be consulted):

1. Fillet welds should be returned around corners for twice the leg length.
2. In lap joints, the lap length should not be less than four times the thickness of the thinner plate.
3. In end connections, the length of weld should not be less than the transverse spacing between the welds.
4. Intermittent welds should not be used under fatigue conditions. The spacing between intermittent welds should not exceed 200 mm or $12t$ for parts in compression and 200 mm or $16t$ for parts in tension, where t is the thickness of the thinner plate. These provisions are shown in Figure 10.18.

It is the recognition that the fillet weld is stronger in the transverse direction compared to its longitudinal direction, and this has led to the so-called directional method shown in Figure 10.19. Welds subject to transverse shear force is shown in Figure 10.19a. In a general case, the weld forces have to be analysed to determine the resultant transverse force on the weld as shown in Figure 10.19b to take advantage of this method. However, for simplicity and ease of use, the code still allows the use of the traditional 'simple method' that does not take into the account of the direction of the force acting on the weld.

The strength of a fillet weld is calculated using the throat thickness. For the 90° fillet weld shown in Figure 10.19a, the throat thickness is taken as 0.7 times the size or leg length.

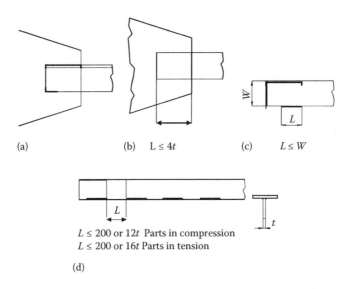

Figure 10.18 Design details for fillet welds: (a) return ends; (b) lap length; (c) end fillet; (d) intermittent fillet welds.

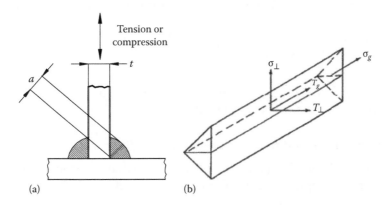

Figure 10.19 Directional method: (a) throat thickness of fillet weld; (b) stresses on a throat section of a fillet weld.

10.4.3.1 Directional method for design resistance of fillet weld

In this method, the force transmitted by a unit length of weld is resolved into components parallel and transverse to the longitudinal axis of the weld and normal and transverse to the plane of its throat. The design throat area A_w should be taken as $A_w = \Sigma a l_{eff}$.

A uniform distribution of stress is assumed on the throat section of the weld, leading to the normal stresses and shear stresses shown in Figure 10.19b and as follows:

$\sigma_\perp$ is the normal stress perpendicular to the throat.
$\sigma_{//}$ is the normal stress parallel to the axis of the weld.
$\tau_\perp$ is the shear stress perpendicular to the axis of the weld.
$\tau_{//}$ is the shear stress perpendicular to the axis of the weld.

The normal stress $\sigma_{//}$ parallel to axis of the weld is not considered when verifying the design resistance of the weld.

Table 10.4 Correlation factor, β_w, for fillet welds

Standard and steel grade			
EN 10025	EN 10210	EN 10219	Correlation factor, β_w
S 235	S 235 H	S 235 H	0.8
S 235 W			
S 275	S 275 H	S 275 H	0.85
S 275 N/NL	S 275 NH/NLH	S 275 NH/NLH	
S 275 M/ML		S 275 MH/MLH	
S 355	S 355 H	S 355 H	0.9
S 355 N/NL	S 355 NH/NLH	S 355 NH/NLH	
S 355 M/ML		S 355 MH/MLH	
S 355 W			
S 420 N/NL		S 420 MH/MLH	1.0
S 420 M/ML			
S 460 N/NL	S 460 NH/NLH	S 460 NH/NLH	1.0
S 460 M/ML		S 460 MH/MLH	
S 460 Q/QL/QL1			

The design resistance of the fillet weld will be sufficient if both equations are satisfied:

$$\left[\sigma_\perp^2 + 3\left(\tau_\perp^2 + \tau_{//}^2\right)\right]^{0.5} \le \frac{f_u}{\beta_w \gamma_{M2}} \quad \text{and} \quad \sigma_\perp \le \frac{0.9f_u}{\gamma_{M2}}$$

where
   f_u is the ultimate tensile strength of the weaker part joined
   β_w is the correction factor taken from Table 10.4

10.4.3.2 Simplified method for design resistance of fillet weld

In this method, the design resistance of a fillet weld may be assumed to be adequate if, at every point along its length, the design resistance per unit length, $F_{w,Rd}$, should be determined from

$$F_{w,Rd} = f_{vw.d}a$$

where $f_{vw.d}$ is the design shear strength of the weld.
   The design shear strength $f_{vw.d}$ of the weld is determined from

$$f_{vw.d} = \frac{f_u/\sqrt{3}}{\beta_w \gamma_{M2}}$$

where
   f_u is the ultimate tensile strength of the weaker part joined
   β_w is the correction factor taken from Table 10.4

10.4.4 Design of butt welds

The design of butt welds is covered in Clause 4.7 of BS EN1993-1-8. The design resistance of the full penetration butt weld should be taken as equal to the design resistance of the weaker of the part joined, and the strength of the weld metal is not less than that of the parent metal.

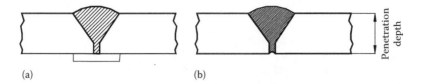

Figure 10.20 Design details for butt welds: (a) single *V* weld with backing plate; (b) single *V* weld made from one side.

Full penetration depth is ensured if the weld is made from both sides or if a backing run is made on a butt weld made from one side (see Figure 10.16a and c). Full penetration is also achieved by using a backing plate (see Figure 10.20a):

1. V weld – depth of preparation minus 3 mm (see Figure 10.20b)
2. J or U welds – depth of preparation unless it can be shown that greater penetration can be achieved (see Figure 10.16c)

10.4.5 Eccentric connections

The two types of eccentrically loaded connections are shown in Figure 10.21. These are

1. The torsion joint with the load in the plane of the weld
2. The bracket connection

In both cases, the fillet welds are in shear due to direct load and moment.

10.4.6 Torsion joint with load in plane of weld

The weld is in direct shear and torsion. The eccentric load causes rotation about the centre of gravity of the weld group. The force in the weld due to torsion is taken to be directly

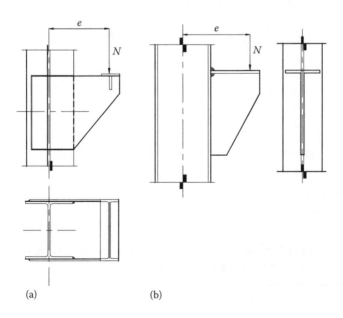

Figure 10.21 Eccentrically loaded connections: (a) load in plane of weld; (b) bracket connection.

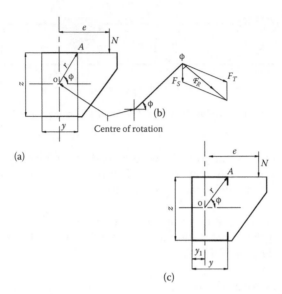

Figure 10.22 Torsion joints load in plane of weld: (a) joint welded all round; (b) resultant force on weld at A; (c) joint welded all round on three side.

proportional to the distance from the centre of gravity and is found by a torsion formula. The direct shear is assumed to be uniform throughout the weld. The resultant shear is found by combining the shear due to moment and the direct shear, and the procedure is set out as follows. The side plate is assumed to be rigid.

A rectangular weld group is shown in Figure 10.22a, where the eccentric load N is taken on one plate. The weld is of unit leg length throughout:

$$\text{Direct shear, } Fs = \frac{N}{\text{Length of weld}}$$

$$= \frac{N}{[2(x+y)]}$$

Shear due to torsion

$$F_T = \frac{N \cdot e \cdot r}{I_p}$$

where

I_p is the polar moment of inertia of the weld group $= I_y + I_z$

$I_y = (z^3/6) + (yz^2/2)$

$I_z = (y^3/6) + (y^2z/2)$

$r = 0.5\sqrt{(y^2 + z^2)}$

The heaviest loaded length of weld is that at A, furthest from the centre of rotation O. The resultant shear on a unit length of weld at A is given by

$$F_R = \sqrt{\left(F_s^2 + F_T^2 + 2F_sF_T \cos\phi\right)}$$

The resultant shear is shown in Figure 10.22b.

If the weld is made on three sides only, as shown in Figure 10.22c, the centre of gravity of the group is found first by taking moments about side BC:

$$y_1 = y^2/(2y + z)$$
$$I_y = z^3/12 + xy^2/2$$
$$I_z = y^3/6 + 2y(y/2 - y_1)^2 + yx_1^2$$

The aforementioned procedure can then be applied.

10.4.7 Bracket connection

Various assumptions are made for the analysis of forces in bracket connections. Consider the bracket shown in Figure 10.23a, which is cut from a universal beam with a flange added to the web. The bracket is connected by fillet welds to the column flange. The flange welds have a throat thickness of unity and the web welds a throat thickness q, a fraction of unity. Assume rotation about the centroidal axis y–y. Then

Weld length	$L = 2b + 2aq$
Moment of inertia	$I_y = bd^2/2 + qa^3/6$
Direct shear	$F_s = Pl\,L$
Load due to moment	$F_T = N\,ed/2I_y$
Resultant load	$F_R = (F_T^2 + F_s^2)^{0.5}$

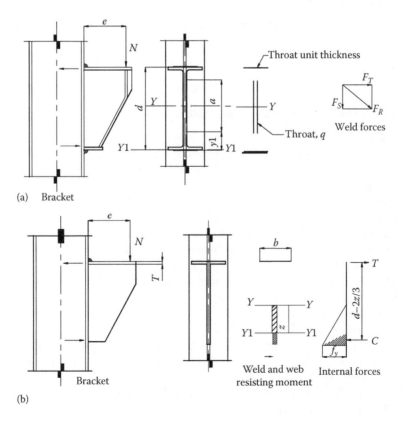

Figure 10.23 Bracket connections: (a) I-section bracket; (b) T-section bracket.

In a second assumption, rotation takes place about the bottom flange Y_1–Y_1. The flange welds resist moment and web welds shear. In this case,

$F_T = P\ e/d\ b$

$F_s = P/2a$

With heavily loaded brackets, full-strength welds are required between the bracket and the column flange.

The fabricated T-section bracket is shown in Figure 10.23b. The moment is assumed to be resisted by the flange weld and a section of the web in compression of depth y, as shown in the figure. Shear is resisted by the web welds.

The bracket and the weld dimensions and internal forces resisting moment are shown in Figure 10.23b. The web area in compression is tz. Equating moments gives

$Ne = T\ (d - 2z/3) = C(d - 2z/3)$

$C = 0.5f_y\ tz = T$

$Ne = 0.\ 5f_ytz(d - 2z/3)$

Solve the quadratic equation for y and calculate C:

The flange weld force $F_T = T/b$
The web weld force $F_s = N/2(d - z)$

The calculations are simplified if the bracket is assumed to rotate about the $y1$–$y1$ axis, when

$F_T = Ne/db.$

10.4.8 Examples on welded connections

1. Direct shear connection
   Design the fillet weld for the direct shear connection for the angle loaded as shown in Figure 10.24a, where the load acts through the centroidal axis of the angle. The steel is Grade S275:
   Factored load = $(1.35 \times 50) + (1.5 \times 60) = 157.5$ kN
   Use 6 mm fillet weld, $F_{w,Rd} = 0.98$ kN/mm
   Length required = $157.5/0.98 = 161$ mm
   Balance the weld on each side as shown in Figure 10.24b:
   Side X, length = $161 \times 43.9/65 = 108.7$ mm
   Add 12 mm, final length = 120.7 mm, say 121 mm
   Side Y, length = $161 - 121 = 40$ mm
   Add 12 mm, final length = 52 mm
2. Torsion connection with load in plane of weld
   One side plate of an eccentrically loaded connection is shown in Figure 10.25a. The plate is welded on three sides only. Find the maximum shear force in the weld and select a suitable fillet weld.
   Find the position of the centre of gravity of the weld group by taking moments about side AB (see Figure 10.25b):
   Length $L = 700$ mm

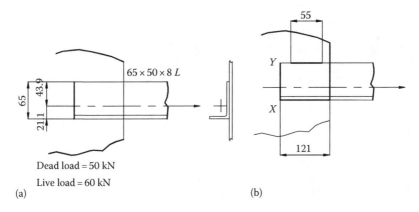

Figure 10.24 Direct shear connection: (a) angle and loads; (b) welds on sides only.

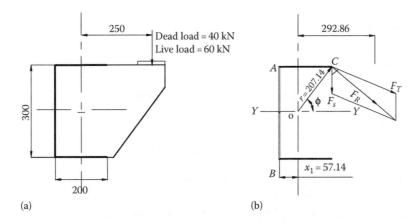

Figure 10.25 Torsion connection loaded in plane of weld: (a) connection; (b) weld group.

Distance to centroid $y_1 = 2 \times 200 \times 100/700 = 57.14$ mm
Eccentricity of load $e = 292.86$

Moments of inertia

$$I_y = (2 \times 200 \times 150^2) + 300^3/12 = 11.25 \times 10^6 \text{ mm}^3$$

$$I_z = (300 \times 57.14^2) + (2 \times 200^3/12) + (2 \times 200 \times 42.86^2)$$

$$= 3.047 \times 10^6 \text{ mm}^3$$

$$I_p = (11.25 + 3.047)10^6 = 14.297 \times 10^6 \text{ mm}^2$$

Angle $\cos \phi = 142.86/207.14 = 0.689$
Factored load $= (1.35 \times 40) + (1.5 \times 60) = 144$ kN
Direct shear $F_s = 144/700 = 0.206$ kN/mm

Shear due to torsion on weld at C

$$F_T = \frac{144 \times 292.86 \times 207.14}{14.297 \times 10^6} = 0.611 \text{ kN/mm}$$

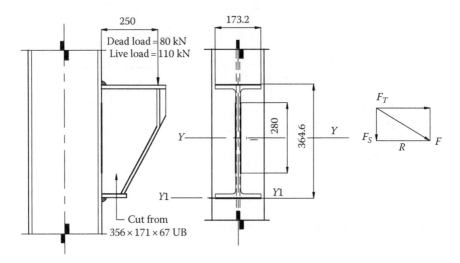

Figure 10.26 Bracket connection.

Resultant shear

$$F_R = [0.206^2 + 0.611^2 + 2 \times 0.206 \times 0.611 \times 0.689]^{0.5}$$

$$= 0.77 \text{ kN/mm}$$

In a 6 mm fillet weld, strength 0.98 kN/mm is required.

3. Bracket connection

   Determine the size of fillet weld required for the bracket connection shown in Figure 10.26. The web welds are to be taken as one-half the leg length of the flange welds. All dimensions and loads are shown in Figure 10.26.

   Design assuming rotation about Y–Y axis:
   Factored load = $(1.35 \times 80) + (1.5 \times 110) = 273$ kN
   Length $L = (2 \times 173.2) + 280 = 626.4$ mm
   Inertia $I_x = (2 \times 173.2 \times 182^2) + 280^3/12 = 13.3 \times 10^6$ mm^3
   Direct shear $F_s = 273/626.4 = 0.44$ kN/mm

$$\text{Shear from moment } F_T = \frac{273 \times 250 \times 182}{13.3 \times 10^6} = 0.985 \text{ kN/mm}$$

   Resultant shear $F_R = [0.44^2 + 0.985^2]^{0.5} = 1.08$ kN/mm
   Provide 8 mm fillet welds for the flanges, strength 1.31 kN/mm. For the web welds, provide 6 mm fillets (the minimum size recommended).
   Design assuming rotation about Y_1 – Y_1 axis.
   The flange weld resists the moment = 273×250

   Provide 8 mm fillet welds, strength 1.31 kN/mm. The web welds resist the shear:

$$F_s = 273/(2 \times 280) = 0.49 \text{ kN/mm}$$

   Provide 6 mm fillet welds. The methods give the same results.

10.5 FURTHER CONSIDERATIONS IN DESIGN OF MOMENT CONNECTIONS

10.5.1 Load paths and forces

The design of bolts and welds has been considered in the previous sections. Other checks that depend on the way the joint is fabricated are necessary to ensure that it is satisfactory. Consistent load paths through the joint must be adopted.

Consider the brackets shown in Figure 10.27. The design checks required are

1. The bolt group (see Section 10.3.3).
2. The welds between the three plates (see Section 10.4.5).
3. The bracket plates – these are in tension, bearing, buckling and local bending.
4. The column in axial load, shear and moment – local checks on the flange in bending and web in tension at the top and buckling and bearing at the bottom are also required.

10.5.2 Other design checks

Some points regarding the design checks are set out as follows.

1. A direct force path is provided by the flange on the bracket in Figure 10.27a. The flange can be designed to resist force when e is the eccentricity of the load and d the depth of bracket:

$$R = (d^2 + e^2)^{0.5} \, N/d$$

2. More rigorous method for moment connections using the concept known as 'component method' is given in Section 6 in the BS EN1993-1-8.

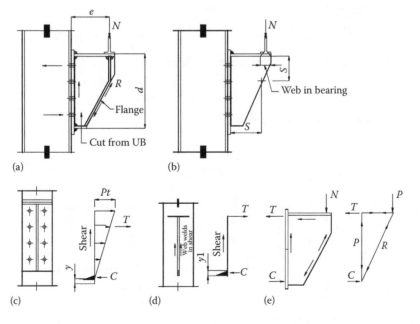

(a)   (b)

(c)   (d)   (e)

Figure 10.27 Brackets: load paths and forces. (a) Bracket with flange; (b) bracket with web only; (c) bolt forces; (d) weld forces; (e) web plate forces.

PROBLEMS

10.1 A single-shear bolted lap joint (Figure 10.28) is subjected to an ultimate tensile load of 200 kN. Determine a suitable bolt diameter using Grade 4.6 bolts.

10.2 A double-channel member carrying an ultimate tension load of 820 kN is to be spliced, as shown in Figure 10.29.
1. Determine the number of 20 mm diameter Grade 4.6 bolts required to make the splice.
2. Check the double-channel member in tension.
3. Check the splice plates in tension.

10.3 A bolted eccentric connection (illustrated in Figure 10.30a and b) is subjected to a vertical ultimate load of 120 kN. Determine the size of Class 4.6 bolts required if the load is placed at an eccentricity of 300 mm.

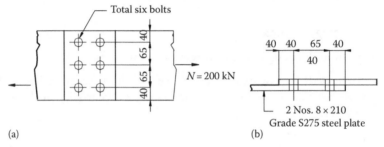

Figure 10.28 (a) Plan and (b) elevation.

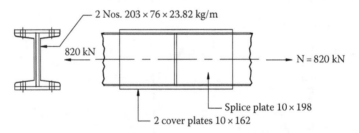

Figure 10.29 Double-channel with cover plates.

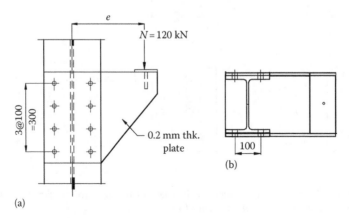

Figure 10.30 Bolted eccentric connection: (a) elevation; (b) plan.

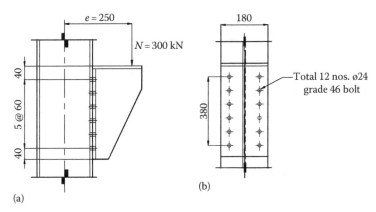

Figure 10.31 Bolted bracket connection: (a) side elevation; (b) front elevation.

10.4 The bolted bracket connection shown in Figure 10.31 carries a vertical ultimate load of 300 KN placed at an eccentricity of 250 mm. Check that twelve no. 24 mm diameter Class 4.6 bolts are adequate. Use both approximate and accurate methods of analysis discussed in Section 10.2.6. Assume all plates to be 20 mm thick.

10.5 Design a beam–splice connection for a 533 × 210 UB 82. The ultimate moment and shear at the splice are 300 kN-m and 175 kN, respectively. A sketch of the suggested arrangement is shown in Figure 10.32. Prepare the final connection detail drawing from your design results.

10.6 The arrangement for a preloaded bolt grip connection provided for a tie carrying an ultimate force of 300 kN is shown in Figure 10.33. Check the adequacy if all the bolts provided are 20 mm diameter.

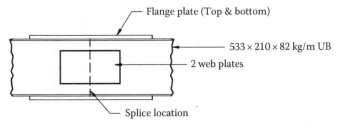

Figure 10.32 Beam-splice connection.

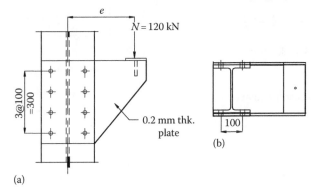

Figure 10.33 Preloaded bolt grip connection: (a) elevation; (b) plan.

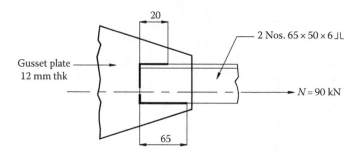

Figure 10.34 Welded angle with gusset plate.

10.7 Redesign the bracket connection in Problem 10.4 using high-strength preloaded bolts. What is the minimum bolt diameter required? Discuss the relative merits of using Class 4.6, Grade 8.8 and preloaded Class 8.8 HSFG bolt for connections.

10.8 The welded connection for a tension member in a roof truss is shown in Figure 10.34. Using an E43 electrode on Grade S275 plate, determine the minimum leg size of the welds if the ultimate tension in the member is 90 kN.

10.9 Determine the leg length of fillet weld required for the eccentric joint shown in Figure 10.35. The ultimate vertical load is 500 kN placed at 300 mm from the centre line. Use an E43 electrode on a Grade S275 plate.

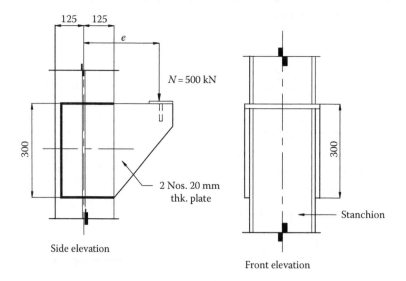

Figure 10.35 Welded eccentric joint.

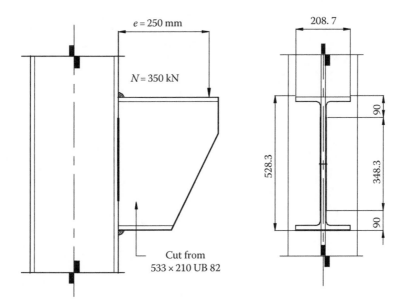

Figure 10.36 Welded bracket connection.

10.10 A bracket cut from a 533 × 210 UB 82 of Grade S275 steel is welded to a column, as shown in Figure 10.36. The ultimate vertical load on the bracket is 350 kN applied at an eccentricity of 250 mm. Design the welds between the bracket and the column.

Chapter 11

Workshop steelwork design example

11.1 INTRODUCTION

An example giving the design of the steel frame for a workshop is presented here and illustrates the following steps in the design process:

1. Preliminary considerations and estimation of loads for the various load cases
2. Computer analysis for the structural frame
3. Design of the truss and crane column
4. Sketches of the steelwork details

The framing plans for the workshop with overhead crane are shown in Figure 11.1. The frames are spaced at 6.0 m centres, and the overall length of building is 48.0 m. The crane span is 19.1 m and the capacity is 50 kN.

Design the structure using grade S275 steel. Structural steel angle sections are used for the roof truss and universal beams for the columns.

Computer analysis is used because plane frame programmes are now generally available for use on microcomputers. (The reader can consult Refs. [17] and [23] for particulars of the matrix stiffness method analysis. The plane frame manual for the particular software package used should also be consulted.)

A manual method of analysis could also be used, and the procedure is as follows:

1. The roof truss is taken to be simply supported for analysis.
2. The columns are analysed for crane and wind loads assuming portal action with no change of slope at the column top. The portal action introduces forces into the chords of the truss, which should be added to the forces in (1) for design.

The reader should consult Refs. [16] and [22] for further particulars of the manual method of analysis.

11.2 BASIC DESIGN LOADS

Details of sheeting and purlins used are given in the following:
   Sheeting: Cellactite 11/3 corrugated sheeting, type 800 thickness 0.8 mm
   Dead load = 0.1 kN/m^2

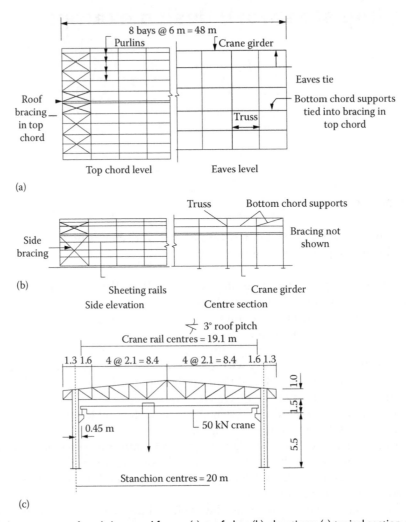

Figure 11.1 Arrangement of workshop steel frame: (a) roof plan; (b) elevations; (c) typical section of frame.

The loads and estimated self-weight on plan are as follows:

Roof dead load	kN/m²
Sheeting	0.11
Insulation and lighting	0.14
Purlin self-weight	0.03
Truss and bracing	0.10
Total load on plan	0.38
Imposed load on plan	0.75

Purlins: Purlins are spaced at 2.1 m centres and span 6.0 m between trusses, purlin loads = 0.11 + 0.14 + 0.75 = 1.0 kN/m², and provide Ward Building Components purlin A200/180, safe load = 1.07 kN/m².

Walls: Cladding, insulation, sheeting rails and bracing = 0.3 kN/m². Stanchion: Universal beam section, say, 457 × 191 UB 67 for self-weight estimation.

Crane data: Hoist capacity = 50 kN, bridge span = 19.1 m, weight of bridge = 35 kN, weight of hoist = 5 kN, end clearance = 220 mm, end-carriage wheel centres = 2.2 m and minimum hook approach = 1.0 m.

Wind load: The readers should refer to the latest code of practice for wind loads, BS EN1991-1-4.

The structure is located in a site in country or up to 2 km into town (terrain category II). The basic wind velocity $v_{b,0}$ is 22 m/s.

11.3 COMPUTER ANALYSIS DATA

11.3.1 Structural geometry and properties

The computer model of the steel frame is shown in Figure 11.2 with numbering for the joints and members. The joint coordinates are shown in Table 11.1. The column bases are taken as fixed at the floor level, and the truss joints and connections of the truss to the columns are taken as pinned. The structure is analysed as a plane frame. The steel frame resists horizontal load from wind and crane surge by cantilever action from the fixed based columns and portal action from the truss and columns. The other data and member properties are shown in Table 11.2.

Elastic modulus, E = 210,000 N/mm^2
Columns: Try 457 × 191 UB 67

$A = 85.5$ cm^2, $E \times A = 1795$ MN

$I_y = 29,380$ cm^4 $E \times I_y = 61.7$ MN/m^2

Top-chord angles: Try 100 × 100 × 12 mm angles

$A = 22.8$ cm^2, $E \times A = 479$ MN

Bottom-chord angles:

Try 90 × 90 × 8 mm angles

$A = 13.9$ cm^2, $E \times A = 292$ MN

All web members:
Try 80 × 80 × 6 mm angles

$A = 9.36$ cm^2, $E \times A = 196$ MN

Note: All $E \times I$ values of the steel angles used in the truss are set to nearly zero.

11.3.2 Roof truss: Dead and imposed loads

For the steel truss, the applied loads are considered as concentrated at the purlin node points. With the purlin spacing of 2.1 m, the applied joint loads at the top chord are

Dead loads per panel point = 0.38 × 6 × 2.1 = 4.8 kN
Imposed loads per panel point = 0.75 × 6 × 2.1 = 9.45 kN

Figure 11.3 shows the dead and imposed loads applied on the steel truss at the top-chord node points. The dead load of crane girder, side rail, cladding and vertical bracing is assumed to load directly onto the column without affecting the roof truss. The estimated load is 18 kN per leg.

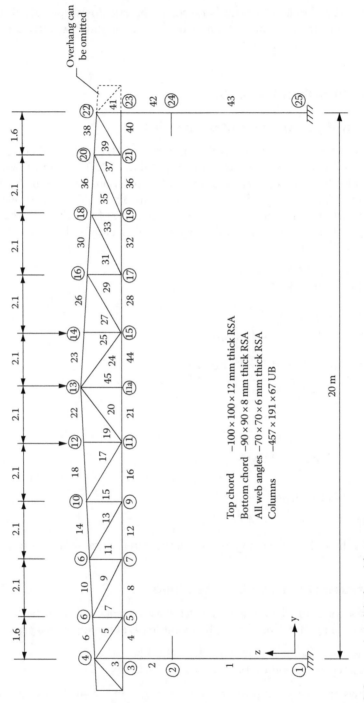

Figure 11.2 Structural model of steel frame.

Table 11.1 Joint coordinates of structure

Joint	Y-distance	Z-distance	Joint	Y-distance	Z-distance
1	0.00	0.000	14	12.1	8.415
2	0.00	5.500	15	12.1	7.000
3	0.00	7.000	16	14.2	8.305
4	0.00	8.000	17	14.2	7.000
5	1.60	7.000	18	16.3	8.195
6	1.60	8.085	19	16.3	7.000
7	3.70	7.000	20	18.4	8.085
8	3.70	8.195	21	18.4	7.000
9	5.80	7.000	22	20.0	8.000
10	5.80	8.305	23	20.0	7.000
11	7.90	7.000	24	20.0	5.500
12	7.90	8.415	25	20.0	0.000
13	10.0	8.525	11a	10.0	7.000

Table 11.2 Structural control data

Number of joints	26
Number of members	45
Number of joint loaded	13
Number of member loaded	4

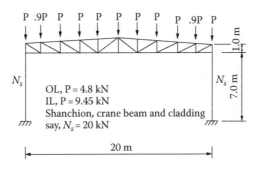

Figure 11.3 Applied dead and imposed loads on truss.

11.3.3 Crane loads

The maximum static wheel load from the manufacturer's table is

Maximum static load per wheel = 35 kN

Add 25% for impact 35 × 1.25 = 43.8 kN

The location of wheels to obtain the maximum reaction on the column leg is shown in Figure 11.4, with one of the wheels directly over the support point:

Maximum reaction on the column through the crane bracket

= 43.8 + 43.8 × (6 − 2.2)/6

= 43.8 × 1.63 = 71.5

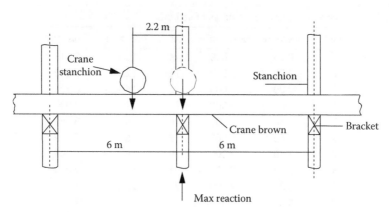

Figure 11.4 Wheels' location for maximum reaction.

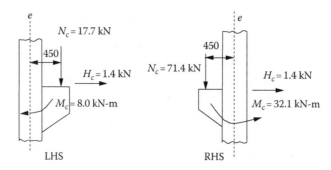

Figure 11.5 Crane loading on steel frame.

Corresponding reaction on the opposite column from the crane

$$= 8.7 \times 1.25 \times 1.63 = 17.7 \text{ kN}$$

Transverse surge per wheel is 10% of hoist weight plus hook load

$$= 0.1 \times (50 + 5)/4 = 1.4 \text{ kN}$$

The reaction on the column is

$$= 1.4 \times 71.4/43.8 = 2.28 \text{ kN}$$

Crane load eccentricity from centre line of column assuming 457 × 191 UB 67

$$e = 220 + 457/2 = 448.5 \text{ say } 450 \text{ mm}$$

Figure 11.5 shows the crane loads acting on the frame. The two applied moments are due to the vertical loads multiplied by the eccentricities.

11.3.4 Wind loads onto the structure

The basic wind speed, $v_b = v_{b,0} \times C_{dir} \times C_{season} \times C_{alt}$, where C_{dir} = direction factor (1.0 for all directions) and $Cseason$ = seasonal factor, should be taken as 1.0. C_{alt} = altitude factor should be determined from NA 2.5 of BS EN1991-1-4.

From NA 3, BS EN1991-1-4, for height of 10 m and closest distance to sea upwind of 10 km, the factor $c_r(z)$ is 1.04.

The mean wind velocity $v_m(z) = v_b \times c_r(z) \times c_0(z) = 22 \times 1.04 \times 1.0 = 23$ m/s

Basic velocity pressure $q_b = 0.5 \times \rho \times v_b^2 = 0.5 \times 1.226 \times 22^2 = 296$ N/m²

Peak velocity pressure $q_p(z) = c_e(z) \times q_b = 2.32 \times 296 = 686$ N/m²

The wind-pressure coefficient for $\theta = 0°$ is shown in Figure 11.6 and Table 11.3. The wind forces applied on the frame and used for the computer analysis are given in Figure 11.7a and b for the internal pressure and suction cases, respectively.

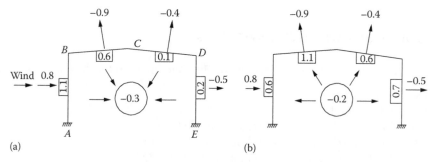

Figure 11.6 Wind-pressure coefficients $\theta = 0°$: (a) internal suction; (b) internal pressure.

Table 11.3 Wind-pressure coefficient on building

External pressure coefficient, $C_{pe,10}$				Internal coefficient	
Roof surfaces		Wall surfaces		Suction	Pressure
GH	IJ	D	E		
−0.9	−0.4	+0.8	−0.5	−0.3	+0.2

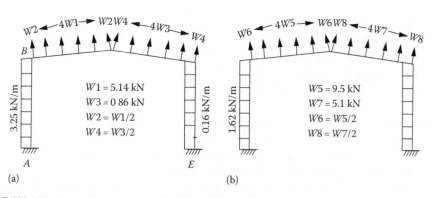

Figure 11.7 Wind loads on structure: $\theta = 0°$: (a) internal suction; (b) internal pressure.

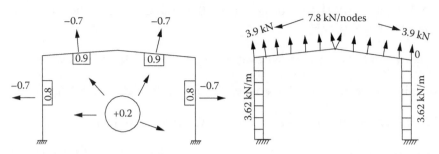

Figure 11.8 Wind loads on structure: θ = 90°.

For θ = 90°, the wind-pressure coefficients and calculated wind loads are shown in Figure 11.8.

11.4 RESULTS OF COMPUTER ANALYSIS

A total of seven computer runs were carried out with one run each for the following load cases:

Case 1: Dead loads (DL)
Case 2: Imposed loads (IL)
Case 3: Wind loads at 0° angle, internal suction (WL, IS)
Case 4: Wind loads at 0° angle, internal pressure (WL, IP)
Case 5: Wind loads at 90° angle, internal pressure (WL, IM)
Case 6: Crane loads, when maximum wheel loads occur (CRWL)
Case 7: Crane surge loads (CRSL)

Table 11.4 gives the summary of the truss member axial forces extracted from the computer output. There is no bending moment in the truss members. It was found that the crane loads do not produce any axial force in truss members listed in Table 11.4, except for those truss members connected directly to the column legs, which have some forces due to crane loads. They are members 3–4, 3–5, 4–5, 4–6, 20–22, 21–22 and 22–23.

Table 11.5 shows the member axial forces and moments for the column legs and the truss members connected directly onto it. There is no bending moment for members 3–4, 3–5, 4–5, 4–6, 20–22, 21–22 and 22–23.

The following five critical load combinations are computed:

1. 1.35 DL + 1.5 LL
2. 1.35 DL + 1.5 LL + 1.5 CRWL
3. 1.0 DL + 1.5 WL (wind at 0°, IS)
4. 1.0 DL + 1.5 WL (wind at 0°, IP)
5. 1.0 DL + 1.5 WL (wind at 90°, IP)

The maximum values from the earlier load combinations are tabulated in Tables 11.4 and 11.6. These will be used later in the design of members.

Note that design conditions arising from notional horizontal loads are not as severe as those in cases 2–5 in Table 11.4. The displacements at every joint are computed in the analyses, but only the critical values are of interest. They are summarized in Table 11.7 and compared with the maximum allowable values.

Table 11.4 Summary of member force (kN) for truss

Member	DL	IL	Wind at 0° (IP)	Wind at 90° (IP)	1.35 DL + 1.5 IL	DL + 1.5 × wind at 0°	DL + 1.5 × wind at 90°
4–6	−1.30	−2.56	17.50	2.70	−5.60	24.95	2.75
6–8	−30.70	−60.45	63.00	50.04	−132.12	63.80	44.36
8–10	−47.50	−93.53	85.60	76.86	−204.42	80.90	67.79
10–12	−54.50	−107.31	90.70	88.11	−234.54	81.55	77.67
12–13	−54.50	−107.31	90.70	−88.11	−234.54	81.55	−186.67
13–14	−54.50	−107.31	90.70	88.11	−234.54	81.55	77.67
14–16	−54.50	−107.31	90.70	88.11	−234.54	81.55	77.67
16–18	−47.50	−93.53	85.60	76.86	−204.42	80.90	67.79
18–20	−30.70	−60.45	63.00	50.04	−132.12	63.80	44.36
20–22	−1.30	−2.56	17.50	2.70	−5.60	24.95	2.75
3–5	40.50	79.74	−50.30	−56.70	174.29	−34.95	−44.55
5–7	−6.10	12.01	−6.00	−1.44	9.78	−15.10	−8.26
7–9	23.30	45.88	−51.40	−45.81	100.28	−53.80	−45.42
9–11	40.00	78.76	−74.00	−72.54	172.14	−71.00	−68.81
11–11a	46.40	91.36	−70.40	−82.71	199.68	−59.20	−77.67
11a–15	46.40	91.36	−70.40	−82.71	199.68	−59.20	−77.67
15–17	40.00	78.76	−47.30	−72.54	172.14	−30.95	−68.81
17–19	23.30	45.88	−18.90	−45.81	100.28	−5.05	−45.42
19–21	−6.10	12.01	−25.10	−1.44	9.78	−43.75	−8.26
21–23	40.50	79.74	−73.40	−56.70	174.29	−69.60	−44.55
4–5	40.60	79.94	−66.40	−65.07	174.72	−59.00	−57.01
5–6	−21.50	−42.33	35.20	34.56	−92.52	31.30	30.34
4–5	40.60	79.94	−66.30	−65.07	174.72	−58.85	−57.01
5–6	−21.50	−42.33	35.20	34.56	−92.52	31.30	30.34
6–7	33.10	65.17	−51.00	−53.10	142.44	−43.40	−46.55
7–8	−15.20	−29.93	23.50	24.39	−65.42	20.05	21.39
8–9	19.30	38.00	−25.90	−30.87	83.06	−19.55	−27.01
9–10	−9.60	−18.90	12.80	15.21	−41.31	9.60	13.22
10–11	8.30	16.34	6.00	−13.23	35.72	17.30	−11.55
11–12	−4.80	−9.45	9.40	7.74	−20.66	9.30	6.81
11–13	0.80	9.45	−10.70	−1.26	15.26	−15.25	−1.09
13–11a	0.00	0.00	0.00	0.00	0.00	0.00	0.00
13–15	0.80	1.58	8.50	−1.26	3.45	13.55	−1.09
14–15	−4.80	−9.45	5.10	7.74	−20.66	2.85	6.81
15–16	8.30	16.34	−19.20	−13.23	35.72	−20.50	−11.55
16–17	−9.60	−18.90	16.10	15.21	−41.31	14.55	13.22
17–18	19.30	38.00	−32.60	−30.87	83.06	−29.60	−27.01
18–19	−15.20	−29.93	22.80	24.39	−65.42	19.00	21.39
19–20	33.10	65.17	−49.60	−53.10	142.44	−41.30	−46.55
20–21	−21.50	−42.33	30.20	34.56	−92.52	23.80	30.34
21–22	40.60	79.94	−57.00	−65.07	174.72	−44.90	−57.01

DL = dead load; IL = imposed load; IP = internal pressure; (−) = compression.

Table 11.5 Summary of member forces (kN) and moments (kN-m) for columns and truss member connected to them

Member		Dead load (DL)	Imposed load (IL)	Crane load (CRWL)	Crane load (CRSL)	Wind at 0° (IS)	Wind at 0° (IP)	Wind at 90° (IP)
1–2(M)		−26.4	−51.9	−16.9	0.4	9.4	40.9	38.4
	Top	−22.1	−43.5	−7.1	−4.0	6.5	−24.1	−42.9
	Bottom	18.6	36.6	11.3	8.6	−35.1	8.3	48.7
2–3(M)		−26.4	−51.9	−0.7	0.4	8.8	38.5	36.6
	Top	−33.1	−64.9	4.2	4.0	−3.5	−38.9	−52.6
	Bottom	−22.1	−43.5	0.9	−4.0	6.5	−24.1	−42.9
3–4(M)		−26.4	−51.9	−0.7	0.4	8.8	38.5	36.6
	Top	0.0	0.0	0.0	0.0	0.0	0.0	3.6
	Bottom	−33.1	−64.9	4.2	4.0	3.5	−38.9	−52.6
3–5		−40.5	−79.7	7.5	4.0	13.2	50.3	56.7
4–5		40.6	79.9	−1.7	−1.0	−14.5	−66.4	−65.1
4–6		−1.3	−2.6	−2.7	−3.1	−8.8	17.5	2.7
20–22		−1.3	−2.6	9.1	3.1	−7.2	17.5	2.7
21–22		40.6	79.9	2.4	1.0	−10.1	−57.0	−65.1
21–23		−40.5	−79.7	−7.8	−4.0	25.5	−73.4	−56.7
22–23(M)		−26.4	−51.9	−0.7	−0.4	5.0	30.8	36.6
	Top	0.0	0.0	0.0	0.0	0.0	0.0	0.0
	Bottom	−33.1	−64.9	1.1	4.0	−17.9	−61.9	−52.6
23–24(M)		−26.4	−51.9	−0.7	−0.4	5.0	30.8	36.6
	Top	−33.1	−64.9	1.1	4.0	−17.9	−61.9	−52.6
	Bottom	−22.1	−43.5	16.0	0.4	6.9	−42.7	−42.5
24–25(M)		−26.4	−51.9	−72.2	−0.4	5.2	32.1	38.4
	Top	−22.1	−43.5	−15.9	−4.0	6.9	−42.7	−42.9
	Bottom	18.6	36.6	2.4	8.6	−17.4	63.6	48.8

11.5 STRUCTURAL DESIGN OF MEMBERS

11.5.1 Design of the truss members

Using grade S275 steel with a design strength of 275 N/mm², the truss members are designed using structural steel angle sections.

1. *Top-chord members 10–12, 12–13, 13–14 and 14–16*
   The maximum compression from load combination (2) is −234.5 kN. The maximum tension from load combination (5) is 81.5 kN.

   Try 100 × 100 × 12.0 mm rolled steel angle (RSA)

   $i_v = 1.95$ cm, $A = 22.8$ cm²

   The section is class 3. Lateral restraint is provided by the purlins and web members at the node points:

   $L_{cr} = 2100$ mm

   Slenderness, $\lambda = 2100/19.5 = 108$

Table 11.6 Load combination for Table 11.5

Member		1.35DL + 1.5IL	1.35DL + 1.5IL + 1.5CRWL	1.0DL + 1.5WL at 0 (IS)	1.0DL + 1.5WL at 0 (IP)	1.0DL + 1.5WL at 90 (IP)
1–2(M)		−113.5	−138.8	−12.3	35.0	31.2
	Top	−95.1	−105.7	−12.4	−58.3	−86.5
	Bottom	80.0	97.0	−34.1	31.1	91.6
2–3(M)		−113.5	−114.5	−13.2	31.4	28.5
	Top	−142.0	−135.7	−38.4	−91.5	−111.9
	Bottom	−95.1	−93.7	−12.4	−58.3	−86.5
3–4(M)		−113.5	−114.5	−13.2	31.4	28.5
	Top	0.0	0.0	0.0	0.0	5.4
	Bottom	−142.0	−135.7	−27.9	−91.5	−111.9
3–5		−174.2	−163.0	−20.7	35.0	44.6
4–5		174.7	172.1	18.9	−59.0	−57.0
4–6		−5.7	−9.7	−14.5	25.0	2.8
20–22		−5.7	8.0	−12.1	25.0	2.8
21–22		174.7	178.3	25.5	−44.9	−57.0
21–23		−174.2	−185.9	−2.3	−150.6	−125.6
22–23(M)		−113.5	−114.5	−18.9	19.8	28.5
	Top	0.0	0.0	0.0	0.0	0.0
	Bottom	−142.0	−140.4	−60.0	−126.0	−111.9
23–24(M)		−113.5	−114.5	−18.9	19.8	28.5
	Top	−142.0	−140.4	−60.0	−126.0	−111.9
	Bottom	−95.1	−71.1	−11.8	−86.2	−85.8
24–25(M)		−113.5	−221.8	−18.6	21.8	31.2
	Top	−95.1	−118.9	−11.8	−86.2	−86.5
	Bottom	80.0	83.6	−7.5	114.0	91.8

Table 11.7 Critical joint displacements

Load case: 1.0DL + 1.0LL

Max. vertical deflection at joint 13	= 32.1 mm
Max. horizontal deflection at joints 2 and 24	= 3.7 mm
Horizontal deflection at joint 22	= 1.8 mm

Load case: 1.0CRWL + 1.0CRSL

Max. horizontal deflection at joint 22	= 2.62 mm

Load case: 1.0CRWL = 1.0wind at 0° IS

Max. horizontal deflection at joint 22	= 7.86 mm

Allowable vertical deflection = $L/200 = 20 \times 1000/200$
$= 100$ mm > 32.1 (satisfactory)

Allowable horizontal deflection = $L/300 - 8 \times 1000/300$
$= 27$ mm

Max. horizontal deflection = $1.8 + 7.86$
$= 9.66$ mm < 16 (satisfactory)

From Table 6.2 of EN1993-1-1, for an S275 rolled RSA section, curve b, $\alpha = 0.34$

$$\lambda_1 = 93.9\varepsilon = 93.9 \times 0.92 = 86.4$$

$$\bar{\lambda} = \frac{\lambda}{\lambda_1} = \frac{108}{86.4} = 1.26$$

$$\Phi = 0.5\,[1 + 0.34(1.26 - 0.2) + 1.26^2] = 1.47$$

$$\chi = \frac{1}{1.47 + \sqrt{1.47^2 - 1.26^2}} = 0.45$$

Buckling resistance, $N_{b,Rd} = 0.45 \times 22.8 \times 275/10 = 282.1$ kN

$$\frac{N_{b,Ed}}{N_{b,Rd}} = \frac{234.5}{282.1} = 0.83 < 1.0 \text{ satisfactory}$$

The section will also be satisfactory in tension.

2. *Bottom-chord members 9–11, 11–11 a, 11 a–15 and 15–17*
The maximum tension from load combination (1) is 199.7 kN. The maximum compression from load combination (5) is –77.7 kN.

Try 90 × 90 × 8.0 mm RSA angle

$$r_v = 1.77 \text{ cm}, \quad A = 13.9 \text{ cm}^2$$

Tension capacity = $275 \times 13.9/10 = 382$ kN > 199.7 kN

Lateral support for the bottom chord is shown in Figure 11.1. The slenderness values are

$$\frac{L_{cr,v}}{i_v} = \frac{2100}{17.6} = 119$$

$$\frac{L_{cr,x}}{i_y} = \frac{4200}{27.5} = 153$$

From Table 6.2 of EN1993-1-1, for an S275 rolled RSA section, curve *b*, $\alpha = 0.34$

$$\lambda_1 = 93.9\varepsilon = 93.9 \times 0.92 = 86.4$$

$$\bar{\lambda} = \frac{\lambda}{\lambda_1} = \frac{153}{86.4} = 1.78$$

$$\Phi = 0.5\,[1 + 0.34(1.78 - 0.2) + 1.78^2] = 2.35$$

$$\chi = \frac{1}{2.35 + \sqrt{2.35^2 - 1.78^2}} = 0.26$$

Buckling resistance, $N_{b,Rd} = 0.26 \times 13.9 \times 275/10 = 99.4$ kN

$$\frac{N_{b,Ed}}{N_{b,Rd}} = \frac{77.7}{99.4} = 0.78 < 1.0 \text{ satisfactory}$$

All web members

Maximum tension = 174.7 kN (members 4–5)
Maximum compression = –92.5 kN (members 5–6)

Try 70 × 70 × 6.0 mm RSA angles

$A = 8.13$ cm^2

$i_v = 1.37$ cm

$i_y = 2.13$ cm

$L_{cr,v}/i_v = 1085/13.7 = 79$

$L_{cr,y}/i_y = 1085/21.3 = 51$

From Table 6.2 of EN1993-1-1, for an S275 rolled RSA section, curve b, $\alpha = 0.34$

$\lambda_1 = 93.9\varepsilon = 93.9 \times 0.92 = 86.4$

$\bar{\lambda} = \dfrac{\lambda}{\lambda_1} = \dfrac{79}{86.4} = 0.92$

$\Phi = 0.5 \, [1 + 0.34(0.92 - 0.2) + 0.92^2] = 1.05$

$\chi = \dfrac{1}{1.05 + \sqrt{1.05^2 - 0.92^2}} = 0.64$

Buckling resistance, $N_{b,Rd} = 0.64 \times 8.13 \times 275/10 = 143.1$ kN

$$\frac{N_{b,Ed}}{N_{b,Rd}} = \frac{92.5}{143.1} = 0.65 < 1.0 \text{ satisfactory}$$

The angle is connected through one leg to a gusset by welding.
  Tension capacity = 275 × 8.13/10 = 223.6 kN > 174.7 kN (satisfactory)

11.5.2 Column design

The worst loading condition for the column is due to ultimate loads from dead load, imposed load and maximum crane load on members 24–25. The design loads are extracted from Table 11.6 as follows:

The maximum column compressive load is –222 kN.
The maximum moment at top of column is –119 kN-m.

The corresponding moment at the bottom is 83.6 kN-m.

$$\frac{N_{Ed}}{\chi_y N_{Rk}/\gamma_{M1}} + k_{yy}\frac{M_{y,Ed}}{\chi_{LT}M_{y,Rk}/\gamma_{M1}} + k_{yz}\frac{M_{z,Ed}}{M_{z,Rk}/\gamma_{M1}} \leq 1$$

$$\frac{N_{Ed}}{\chi_z N_{Rk}/\gamma_{M1}} + k_{zy}\frac{M_{y,Ed}}{\chi_{LT}M_{y,Rk}/\gamma_{M1}} + k_{zz}\frac{M_{z,Ed}}{M_{z,Rk}/\gamma_{M1}} \leq 1$$

Try 457 × 191 UB67, class 1, the properties of which are

$A = 85.4$ cm²
$i_y = 18.5$ cm
$i_z = 4.12$ cm
$W_{pl,y} = 1470$ cm³
$W_{el,y} = 1300$ cm³
Yield strength $f_y = 275$ N/mm²
Buckling lengths $L_{cr,y} = 1.5 \times 7000 = 10{,}500$ mm
Buckling lengths $L_{cr,z} = 0.85 \times 5500 = 4675$ mm

Maximum slenderness ratio

$$\lambda_y = \frac{L_{cr,y}}{i_y} = \frac{10{,}500}{185} = 57$$

$$\lambda_z = \frac{L_{cr,z}}{i_z} = \frac{4675}{41.2} = 114$$

From Table 6.2 of EN1993-1-1, for an S275 rolled H section, t_f less than 40 mm, $h/b > 1.2$, buckling about the major $y–y$ axis, curve a, imperfection factor, $\alpha = 0.21$, buckling about the minor $z–z$ axis, curve b, imperfection factor, $\alpha = 0.34$

$$\lambda_1 = 93.9\varepsilon = 93.9 \times 0.92 = 86.4$$

$$\overline{\lambda_y} = \frac{\lambda_y}{\lambda_1} = \frac{57}{86.4} = 0.66$$

$$\Phi = 0.5\,[1 + 0.21(0.66 - 0.2) + 0.66^2] = 0.77$$

$$\chi_y = \frac{1}{0.77 + \sqrt{0.77^2 - 0.66^2}} = 0.86$$

$$\overline{\lambda_z} = \frac{\lambda_z}{\lambda_1} = \frac{114}{86.4} = 1.33$$

$$\Phi = 0.5\,[1 + 0.34(1.33 - 0.2) + 1.33^2] = 1.58$$

$$\chi_z = \frac{1}{1.58 + \sqrt{1.58^2 - 1.33^2}} = 0.41$$

Buckling lengths $L_{cr} = 0.85 \times 5500 = 4675$ mm

$$M_{cr} = C_1 \frac{\pi^2 EI_z}{L_{cr}^2} \left(\frac{I_w}{I_z} + \frac{L_{cr}^2 GI_T}{\pi^2 EI_z} \right)^{0.5}$$

For ratio of end moment, $\psi = 83.6/{-}119 = -0.7$; $C_1 = 2.910$

$$M_{cr} = 2.91 \times \frac{\pi^2 \times 210,000 \times 14,500,000}{4,675^2} \left(\frac{705 \times 10^9}{14,500,000} + \frac{4,675^2 \times 81,000 \times 371,000}{\pi^2 \times 210,000 \times 14,500,000} \right)^{0.5}$$

$M_{cr} = 1062$ kN-m

$$\overline{\lambda_{LT}} = \sqrt{\frac{W_y f_y}{M_{cr}}} = \sqrt{\frac{1470 \times 10^3 \times 275}{1062 \times 10^6}} = 0.62$$

$$\chi_{LT} = \frac{1}{\Phi_{LT} + \sqrt{\Phi_{LT}^2 - \overline{\lambda}_{LT}^2}}$$

$$\Phi_{LT} = 0.5 \left[1 + \alpha_{LT} \left(\overline{\lambda}_{LT} - 0.2 \right) + \overline{\lambda}_{LT}^2 \right]$$

For $h/b > 2$, curve b, $\alpha_{LT} = 0.34$

$$\Phi_{LT} = 0.5 \times [1 + 0.34(0.62 - 0.2) + 0.62^2] = 0.83$$

$$\chi_{LT} = \frac{1}{0.83 + \sqrt{0.83^2 - 0.62^2}} = 0.72$$

Determination of interaction factors k_{ij} (Annex B in BS EN1993-1-1)
Table B.3. Equivalent uniform moment factors, C_m

For $\psi = -0.7$, C_{my}, $C_{mz} = 0.6 + 0.4(-0.7) = 0.32 \geq 0.4$

$$k_{yy} = C_{my} \left(1 + \left(\overline{\lambda}_y - 0.2 \right) \frac{N_{Ed}}{\chi_y N_{Rk}/\gamma_{M1}} \right) \leq C_{my} \left(1 + 0.8 \frac{N_{Ed}}{\chi_y N_{Rk}/\gamma_{M1}} \right)$$

$$k_{yy} = 0.4 \left(1 + (0.66 - 0.2) \frac{222}{0.86 \times 8540 \times 10^{-3} \times 275/1.0} \right) = 0.42$$

$k_{zy} = 0.6 \quad k_{yy} = 0.25$

$$k_{zz} = C_{mz} \left(1 + \left(\overline{\lambda}_z - 0.2 \right) \frac{N_{Ed}}{\chi_z N_{Rk}/\gamma_{M1}} \right) \leq C_{mz} \left(1 + 0.8 \frac{N_{Ed}}{\chi_z N_{Rk}/\gamma_{M1}} \right)$$

$$k_{zz} = 0.4 \left(1 + (1.33 - 0.2) \frac{222}{0.41 \times 8540 \times 10^{-3} \times 275/1.0} \right) = 0.50$$

$$\le 0.4\left(1+0.8\frac{222}{0.41\times8540\times10^{-3}\times275/1.0}\right)=0.47$$

$$k_{zz}=0.47;\quad k_{yz}=0.6;\quad k_{zz}=0.28$$

$$\frac{N_{Ed}}{\chi_y N_{Rk}/\gamma_{M1}}+k_{yy}\frac{M_{y,Ed}}{\chi_{LT}M_{y,Rk}/\gamma_{M1}}+k_{yz}\frac{M_{z,Ed}}{M_{z,Rk}/\gamma_{M1}}\le1$$

$$\frac{222}{0.86\times2348.5/1.0}+0.42\frac{119}{0.72\times404.3/1.0}=0.28$$

$$\frac{N_{Ed}}{\chi_z N_{Rk}/\gamma_{M1}}+k_{zy}\frac{M_{y,Ed}}{\chi_{LT}M_{y,Rk}/\gamma_{M1}}+k_{zz}\frac{M_{z,Ed}}{M_{z,Rk}/\gamma_{M1}}\le1$$

$$\frac{222}{0.41\times2348.5/1.0}+0.25\frac{119}{0.72\times404.3/1.0}=0.33$$

The 457 × 191 UB 67 provided is satisfactory. The reader may try with a 457 × 152 UB 60 to increase the stress ratio and achieve greater economy. For the design of the crane girder, see Chapter 4.

11.6 STEELWORK DETAILING

The details for the main frame and connections are presented in Figure 11.9.

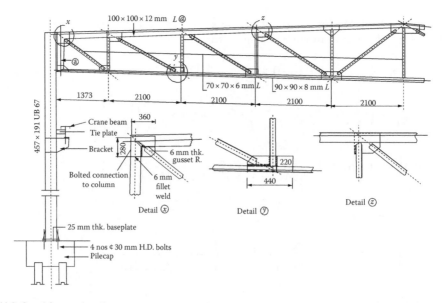

Figure 11.9 Steel frame details.

Chapter 12

Steelwork detailing

12.1 DRAWINGS

Drawings are the means by which the requirements of architects and engineers are communicated to the fabricators and erectors and must be presented in an acceptable way. Detailing is given for selected structural elements.

Drawings are needed to show the layout and to describe and specify the requirements of a building. They show the location, general arrangement and details for fabrication and erection. They are also used for estimating quantities and cost and for making material lists for ordering materials.

Sufficient information must be given on the designer's sketches for the draughtsman to make up the arrangement and detail drawings. A classification of drawings is set out as follows:

Site or location plans: These show the location of the building in relation to other buildings, site boundaries, streets, roads, etc.

General arrangement: This consists of plans, elevations and sections to set out the function of the building. These show locations and leading dimensions for offices, rooms, work areas, machinery, cranes, doors, services, etc. Materials and finishes are specified.

Marking plans: These are the framing plans for the steel-frame building showing the location and mark numbers for all steel members in the roof, floors and various elevations.

Foundation plans: These show the setting out for the column bases and holding-down bolts and should be read in conjunction with detail drawings of the foundations.

Sheeting plans: These show the arrangement of sheeting and cladding on building.

Key plan: If the work is set out on various drawings, a key plan may be provided to show the portion of work covered by the particular drawing.

Detail drawings: These show the details of structural members and give all information regarding materials, sizes, welding, drilling, etc. for fabrication. The mark number of the detail refers to the number on the marking plan.

Detail drawings and marking plans will be dealt with here.

12.2 GENERAL RECOMMENDATIONS

12.2.1 Scales, drawing sizes and title blocks

The following scales are recommended:

    Site, location and key plans 1:500/1:200
    General arrangement 1:200/1:100/1:50
    Marking plans 1:200/1:100
    Detail drawings 1:25/1:10/1:5
    Enlarged details 1:5/1:2/1:1

The following drawing sizes are used:

    A4 210 × 297 – sketches
    A3 297 × 420 – details
    A2 420 × 59 – general arrangement, details
    A1 594 × 841 – general arrangement, details
    A0 841 × 1189 – general arrangement, details

Title blocks on drawings vary to suit the requirements of individual firms and authorities. Typical title blocks are shown in Figure 12.1.

Materials lists can be shown either on the drawing or on separate A4 size sheets. These generally give the following information: item or mark number, description, material, number off and weight.

12.2.2 Lines, sections, dimensions and lettering

Recommendations regarding lines, sections and dimensions are shown in Figure 12.2.

12.3 STEEL SECTIONS

Rolled and formed steel sections are represented on steelwork drawings as set out in Figure 12.3. The first two numbers in the example of Figure 12.3 indicate the size of

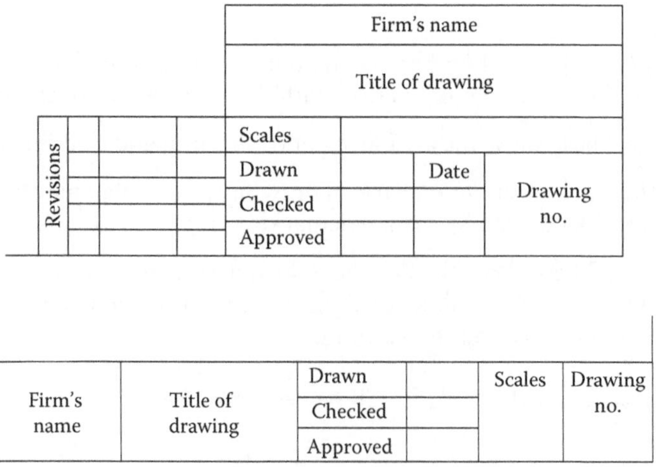

Figure 12.1 Typical title blocks.

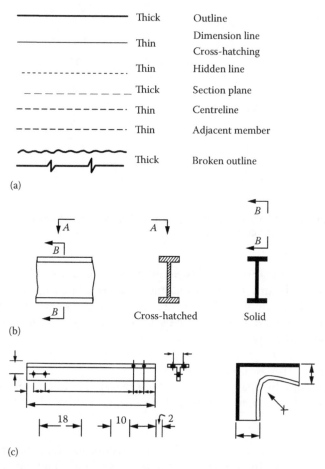

Figure 12.2 Recommendations for (a) lines, (b) sections and (c) dimensions.

Section		Reference	Example
Universal beam	I	UB	$610 \times 178 \times 91$ kg/m UB
Universal column	I	UC	$203 \times 203 \times 89$ kg/m UC
Joist	I	RSJ / Joist	$203 \times 102 \times 25.33$ kg/m Joist
Channel	[	channel	$254 \times 76 \times 28.29$ kg/m [
Angle	L	Angle	$150 \times 75 \times 10$ L
Tee	T	Tee	$178 \times 203 \times 37$ kg/m struct. Tee
Rectangular hollow section	▢	RHS	$150.4 \times 100 \times 6.3$ RHS
Circular hollow section	◯	CHS	76.1 O.D. $\times 5$ CHS

Figure 12.3 Representation of rolled and formed steel sections.

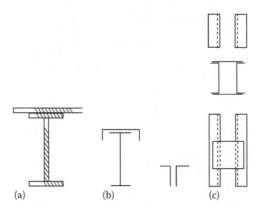

Figure 12.4 Representation of built-up sections: (a) plated UB section; (b) channel welded to UB section; (c) doubled angle and doubled channel.

section (e.g. depth and breadth). The last number in the example of Figure 12.3 indicates the weight in kg/m for beams, columns, channels and tees. For angles and hollow sections, the last number in the example of Figure 12.3 gives the thickness of steel. With channels or angles, the name may be written or the section symbols used as shown. Built-up sections can be shown either by

1. True section for large scale views or
2. Diagrammatically by heavy lines with the separate plates and sections separated for clarity for small-scale views. Here, only the depth and breadth of the section may be true to scale.

These two cases are shown in Figure 12.4. The section is often shown in the middle of a member inside the break lines in the length, as shown in (c). This saves having to draw a separate section.

Beams may be represented by lines, and columns by small-scale sections in heavy lines, as shown in Figure 12.5. The mark numbers and sizes are written on the respective members. This system is used for marking plans.

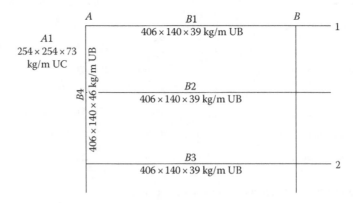

Figure 12.5 Representation of beams and columns.

12.4 GRIDS AND MARKING PLANS

Marking plans for single-storey buildings present no difficulty. Members are marked in sequence as follows:

Columns A1, A2,... (see grid referencing as follows) trusses T1, T2,...
Crane girders CG1, CG2,...
Purlins P1, P2,...
Sheeting rails SR1, SR2,...
Bracing Bl, B2,...
Gable columns GS1, GS2,....

Various numbering systems are used to locate beams and columns in multistorey buildings. Two schemes are outlined as follows:

1. In plan, the column grid is marked A, B, C,... in one direction and 1, 2, 3,... in the direction at right angles. Columns are located at A1, C2,....

   Floors are numbered A, B, C,... for ground, first, second,..., respectively.

   Floor beams (e.g. on the second floor) are numbered B1, B2,.... Column lengths are identified: for example, A4-B is the column on grid intersection A4, the length between the second and third floors.
2. A grid line is required for each beam.

The columns are numbered by grid intersections as aforementioned.

The beams are numbered on the grid lines with a prefix letter to give the floor if required. For example,

Second floor: grid line 1 – C-1a and C-1b
Second floor: grid line B – C-1b and C-b2

The systems are shown in Figure 12.6. The section size may be written on the marking plan.

12.5 BOLTS

12.5.1 Specification

The types of bolts used in steel construction are

Grade 4.6 or black bolts
Grade 8.8 bolts
Class 10.9 bolts

Preloaded HSFG friction-grip bolts
  The European Standards covering these bolts are

BS EN 1993-1-8: Eurocode (EC) 3, design of steel structures and design of joints
BS EN ISO 4016: hexagon head bolts and product grade C
BS EN 14399: high-strength structural bolting assemblies for preloading

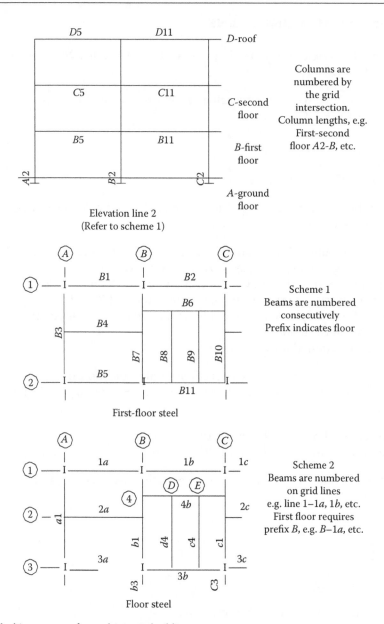

Figure 12.6 Marking systems for multistorey buildings.

The strength grade designation should be specified. BS EN1993-1-8 gives the strengths for ordinary bolts for Grades 4.6 8.8 and 10.9. Minimum shank tensions for friction-grip bolts are given in BS4604. The nominal diameter is given in millimetres. Bolts are designated as M12, M16, M20, M229, M24, M27, M30, etc., where 12, 16, etc. is the diameter in millimetres. The length under the head in millimetres should also be given.

Examples in specifying bolts are as follows:

4 No. 16 mm dia. (or M16) black hex. hd. (hexagon head) bolts, strength grade
4.6 × 40 mm length
20 No. 24 mm dia. friction-grip bolts × 75 mm length

The friction-grip bolts may be abbreviated HSFG. The majority of bolts may be covered by a blanket note. For example,

All bolts M20 black hex. hd
All bolts 24 mm dia. HSFG unless otherwise noted

12.5.2 Drilling

The following tolerances for drilling are used:

For ordinary bolts, holes have a maximum of 2 mm clearance for bolt diameters up to 24 and 3 mm for bolts of 24 mm diameter and over.
For friction-grip bolts, holes are drilled the same as set out earlier for ordinary bolts. Maximum and minimum spacing are given in Table 3.3 of BS EN1993-1-8.

Drilling may be specified on the drawing by notes as follows:

All holes drilled 22 mm dia. for 20 mm dia. ordinary bolts.
All holes drilled 26 mm dia., unless otherwise noted.

12.5.3 Designating and dimensioning

The representation for bolts and holes in plan and elevation on steelwork drawings is shown in Figure 12.7a. Some firms adopt different symbols for showing different types of bolts and to differentiate between shop and field bolts. If this system is used, a key to the symbols must be given on the drawing.

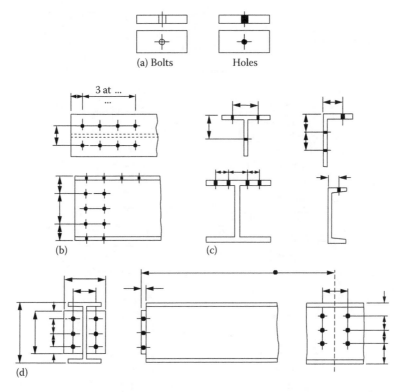

Figure 12.7 Representation of bolts and holes on steelwork drawings: (a) bolts; (b) holes dimensions on plan and elevation; (c) holes dimensions on sections; (d) holes dimensions on end plates.

Gauge lines for drilling for rolled sections are given in the *Structural Steelwork Handbook*. Dimensions are given for various sizes of section, as shown in Figure 12.7c. Minimum edge and end distances were discussed in Section 4.2.2 earlier.

Details must show all dimensions for drilling, as shown in Figure 12.7b and d. The holes must be dimensioned off a finished edge of a plate or the back or end of a member. Holes are placed equally about centrelines. Sufficient end views and sections as well as plans and elevations of the member or joint must be given to show the location of all holes, gussets and plates.

12.6 WELDS

As set out in Chapter 4, the two types of weld are butt and fillet.

12.6.1 Butt welds

The types of butt weld are shown in Figure 12.8 with the plate edge preparation and the fit-up for making the weld. The following terms are defined:

T is the thickness of plate
g is the gap between the plates
R is the root face
α is the minimum angle

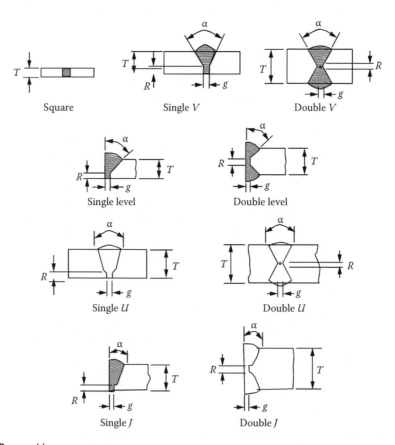

Figure 12.8 Butt welds.

Fillet	△
Square butt weld	⊤⊤
Single *V* butt weld	▽
Double *V* butt weld	⨂
Single *U* butt weld	∪
Double *U* butt weld	⨝
Single level weld	⌐
Double level weld	⌐⌐
Single *J* weld	⇑
Double *J* weld	⊓⊔

Figure 12.9 Symbols for welds.

Values of the gap and root face vary with the plate thickness but are of the order of 1–4 mm. The minimum angle between prepared faces is generally 50°–60° for V preparation and 30°–40° for *U* preparation. For thicker plates, the *U* preparation gives a considerable saving in the amount of weld metal required.

Welds may be indicated on drawings by symbols as shown in Figure 12.9. Using these symbols, butt welds are indicated on a drawing as shown in Figure 12.10a.

The weld name may be abbreviated: for example, DVBW for double *V* butt weld. An example of this is shown in Figure 12.8b. Finally, the weld may be listed by its full description and an enlarged detail given to show the edge preparation and fit-up for the plates. This method is shown in Figure 12.8c.

Enlarged details should be given in cases where complicated welding is required. Here, detailed instructions from a welding engineer may be required and these should be noted on the drawing.

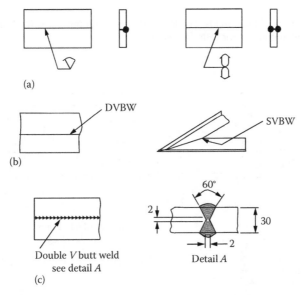

Figure 12.10 Representation of butt welds: (a) symbols for showing single *V* butt weld and double *U* butt weld; (b) other method of showing double *V* butt weld and single *U* butt weld; (c) enlarged detail showing double *V* butt weld.

12.6.2 Fillet welds

These welds are triangular in shape. As set out in Chapter 10, the size of the weld is speci-fied by the leg length. Welds may be indicated symbolically, as shown in Figure 12.11a. BS EN 1993-1-8 should be consulted for further examples. The weld size and type may be written out in full or the words 'fillet weld' abbreviated (e.g. mm FW). If the weld is of limited length, its exact location should be shown and dimensioned. Intermittent weld can be shown by writing the weld size, then two figures that indicate length and space between welds. These methods are shown in Figure 12.11b. The welds can also be shown and speci-fied by notes, as on the plan view in Figure 12.11b. Finally, a common method of showing fillet welds is given in Figure 12.11c, where thickened lines are used to show the weld.

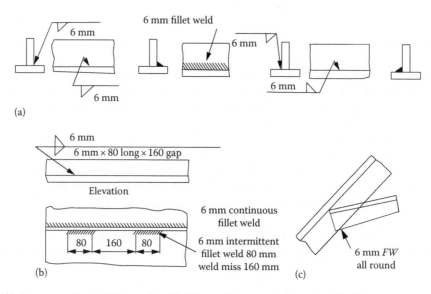

Figure 12.11 Representation of fillet welds: (a) fillet weld near and far sides; (b) fillet welds on plan view; (c) common method of showing fillet welds.

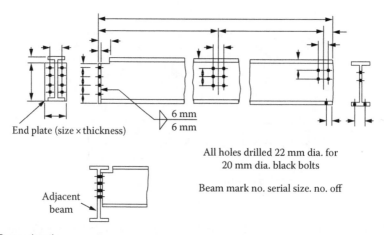

Figure 12.12 Beam details.

12.7 BEAMS

Detailing of beams, purlins and sheeting rails is largely concerned with showing the length, end joints, welding and drilling required. A typical example is shown in Figure 12.12. Sometimes it is necessary to show the connecting member and this may be shown by chain dash lines, as shown in Figure 12.12b.

12.8 PLATE GIRDERS

The detailed drawing of a plate girder shows the girder dimensions, flange and web plate sizes, sizes of stiffeners and end plates, their location and details for drilling and welding. Any special instructions regarding fabrication should be given on the drawing. For example, preheating may be required when welding thick plates, or 2 mm may require machining off flame-cut edges to reduce the likelihood of failure by brittle fracture or fatigue on high-yield-strength steels.

Generally, all information may be shown on an elevation of the girder together with sufficient sections to show all types of end plates, intermediate and load-bearing stiffeners. The elevation would show the location of stiffeners, brackets and location of holes, and the sections complete this information. Plan views on the top and bottom flange are used if there is a lot of drilling or other features best shown on such a view. The draughtsman decides whether such views are necessary.

Part-sectional plans are often used to show stiffeners in plan view. Enlarged details are frequently made to give plate weld edge preparation for flange and web plates and splices and for load-bearing stiffeners that require full-strength welds.

Notes are added to cover drilling, welding and special fabrication procedures, as stated earlier. Finally, the drawing may contain a material list giving all plate sizes required in the girder. Typical details for a plate girder are shown in Figure 12.13.

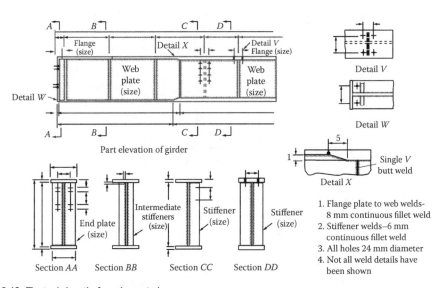

Figure 12.13 Typical details for plate girder.

12.9 COLUMNS AND BASES

A typical detail of a column for a multistorey building is shown in Figure 12.14. In the bottom column length with base slab, drilling for floor beams and splice details is shown.

A compound crane column for a single-storey industrial building is shown in Figure 12.15. The crane column is a built-up section and the roof portion is a universal beam. Details at the column cap, crane girder level and base are shown.

12.10 TRUSSES AND LATTICE GIRDERS

The rolled sections used in trusses are small in relation to the length of the members. Several methods are adopted to show the details at the joints. These are

1. If the truss can be drawn to a scale of 1 in 10, then all major details can be shown on the drawing of the truss.
2. The truss is drawn to a small scale, 1 in 25, and then separate enlarged details are drawn for the joints to a scale of 1 in 10 or 1 in 5.

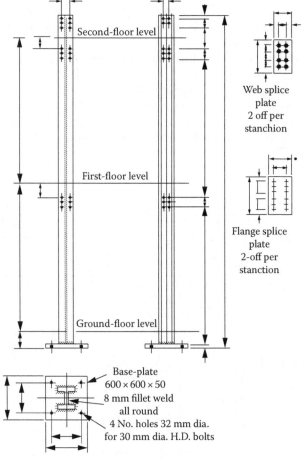

Second-floor level

Web splice plate 2 off per stanchion

First-floor level

Flange splice plate 2-off per stanction

Ground-floor level

Base-plate 600 × 600 × 50
8 mm fillet weld all round
4 No. holes 32 mm dia. for 30 mm dia. H.D. bolts

All holes 22 mm dia. for 20 mm dia. black bolts except as noted

Figure 12.14 Stanchion in a multistorey building.

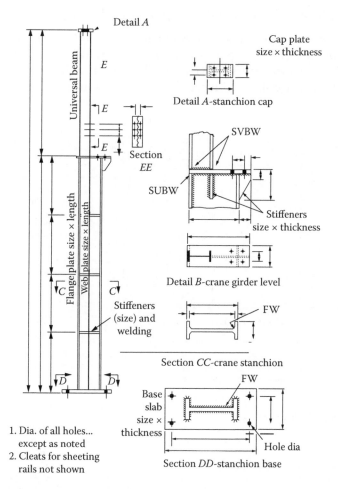

Figure 12.15 Compound crane stanchion for a single-storey industrial building.

Members should be designated by size and length; for example, where all dimensions are in millimetres:

100 × 75 × 10 angle × 2312 long
100 × 50 × 4 RHS × 1310 long

On sloping members, it is of assistance in fabrication to show the slope of the member from the vertical and horizontal by a small triangle adjacent to the member.

The centroidal axes of the members are used to set out the frame and the members should be arranged so that these axes are coincident at the nodes of the truss. If this is not the case, the eccentricity causes secondary stresses in the truss.

Full details are required for bolted or welded splices, end plates, column caps, etc. The positions of gauge lines for drilling holes are given in the *Structural Steelwork Handbook*. Dimensions should be given for edge distances, spacing and distance of holes from the adjacent node of the truss. Splice plates may be detailed separately.

Sometimes the individual members of a truss are itemized. In these cases, the separate members, splice plates, cap plates, etc. are given an item number. These numbers are used to identify the member or part on a material list.

The following figures are given to show typical truss detailing:

Figure 12.16 shows a portion of a flat roof truss with all the major details shown on the elevation of the truss. Each part is given an item number for listing.

Figure 12.17 shows a roof truss drawn to small scale where the truss dimensions, member sizes and lengths are shown. Enlarged details are given for some of the joints.

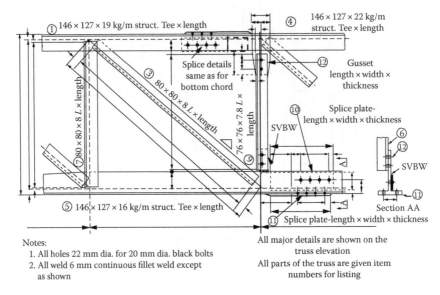

Notes:
1. All holes 22 mm dia. for 20 mm dia. black bolts
2. All weld 6 mm continuous fillet weld except as shown

All major details are shown on the truss elevation

All parts of the truss are given item numbers for listing

Figure 12.16 Portion of a flat roof truss.

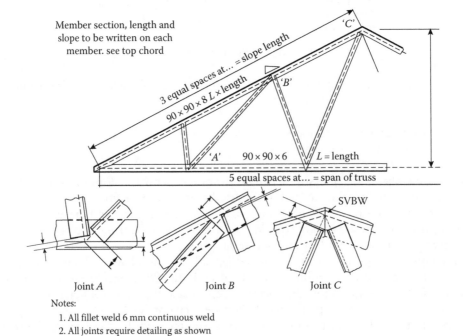

Notes:
1. All fillet weld 6 mm continuous weld
2. All joints require detailing as shown

Figure 12.17 Small-span all-welded truss.

12.11 COMPUTER-AIDED DRAFTING

Computer-aided drafting (CAD) is now being introduced into civil and structural drafting practice. It has now replaced much manual drafting work. CAD can save considerable man-hours in drawing preparation, especially where standard details are used extensively.

In computerized graphical systems, the drawing is built up on the screen, which is divided into a grid. A menu gives commands (e.g. line, circle, arc, text, dimensions). Data are input through the digitizer and keyboard. All drawings are stored in mass storage devices from which they can be retrieved for subsequent additions or alterations. Updating of drawings using the CAD system can be accomplished with little effort. Standard details used frequently can either be drawn from a program library or created and stored in the user's library.

Some comprehensive steel design and detailing software can automate the entire process for standard type of steel structures.

Overlay of drawings in CAD software is a very useful feature, which allows repeated usage of common drawing templates (e.g. grid lines, floor plans) and results in considerable time saving by reducing input time required. Texts and dimensions can be typed from the keyboard using appropriate character sizes.

References

1. Richards, K.G., *Fatigue Strength of Welded Structures*, The Welding Institute, Cambridge, U.K., 1969.
2. Boyd, G.M., *Brittle Fracture in Steel Structures*, Butterworths, London, U.K., 1970.
3. *Building Regulations and Associated Approved Document B*, HM Government, London, U.K., 2000.
4. *Fire Protection of Structural Steel in Building* (3rd edn.), Association for Specialist Fire Protection (ASFP), London, U.K., 2002.
5. Steel building design: Design date. SCI & BCSA, London, U.K., 2009.
6. Case, J. and Chiler, A.H., *Strength of Materials*, Edward Arnold, London, U.K., 1964.
7. Trahair, N.S., Bradford, M.A., Nethercot, D.A., and Gardner, L., *Behaviour and Design of Steel Structures to EC3* (4th edn.), Spon Press, London, U.K., 2007.
8. Nethercot, D.A. and Lawson, R.M., *Lateral Stability of Steel Beams and Columns – Common Cases of Restraint,* Steel Construction Institute, Ascot, Berks, U.K., 1992.
9. Horne, M.R., *Plastic Theory of Structures*, Pergamon Press, Oxford, U.K., 1979.
10. Davison, B. and Owens, G.W., *Steel Designers Manual*, Wiley-Blackwell, U.K., 2012.
11. Ghali, A. and Neville, A.M., *Structural Analysis*, Chapman & Hall, London, U.K., 1989.
12. *Joints in Steel Construction. Simple Joints to Eurocode 3 (P358)*, SCI & BSCA, London, U.K., 2002.
13. Ward Building Components, *Multibeam Products*, Ward Building Components Ltd, Gloucestershire, U.K., 2002.
14. Timoshenko, S.P. and Gere, J.M., *Theory of Elastic Stability*, McGraw-Hill, New York, 1961.
15. Porter, D.M., Rockey, K.C., and Evans, H.R., The collapse behaviour of plate girders loaded in shear, *The Structural Engineer*, 53(8), 313–325, August 1975.
16. Pratt, J.L., *Introduction to the Welding of Structural Steelwork*, Steel Construction Institute, Ascot, Berks, U.K., 1989.
17. Holmes, M. and Martin, L.H., *Analysis and Design of Structural Connections – Reinforced Concrete and Steel*, Ellis Horwood, London, U.K., 1983.
18. Salmon, C.G. et al., Laboratory investigation on unstiffened triangular bracket plates, *Journal of Structural Division, American Society of Civil Engineers*, 90(2), 357–378, March/April, 1964.
19. Horne, M.R. and Morris, L.J., *Plastic Design of Low Rise Frames*, Collins, London, U.K., 1981.
20. Coates, R.C., Coutie, M.G., and Kong, F.K., *Structural Analysis*, Spon E & FN, London, U.K., 1998.
21. Southcombe, C., *Design of Structural Steelwork: Lattice Framed Industrial Buildings*, Steel Construction Institute, Ascot, Berks, U.K., 1993.
22. Lee, C.K. and Chiew, S.P., *An Efficient Modified Flange-Only Method for calculation of Bending Moment Resistance of Plate Girder*. Paper submitted to the Journal of Constructional Steel Research, 2013.

Index